# Systematic Layout Planning

# Systematic Layout Planning

## Second Edition


**Richard Muther**

President, Richard Muther and Associates, Inc.; Senior Partner, Muther International
Author or co-author of the books *Production Line Technique, Practical Plant Layout, Systematic Handling Analysis (SHA)* and *Simplified Systematic Layout Planning*


*Foreword* by
Allan H. Mogensen
Director, Work Simplification Conferences
Consultant and Author


Cahners Books, *division of Cahners Publishing Company, Inc.*
89 Franklin Street, Boston, Massachusetts, U.S.A. 02110

Third Printing, January 1976


Printed in the United States of America.


Library of Congress Catalog Card Number: 72-91983

International Standard Book Number: 0-8436-0814-5

# Dedication

Systematic Layout Planning (SLP) evolved from many years of research, experience, and practice. Three persons contributed significantly to its development. This book is respectfully dedicated to them:

Harold Bright Maynard
Erwin Haskell Schell
René de Vallière

Their belief in sound basic principles, their insistence on meeting practical realities, and their personal encouragement are warmly appreciated.

Since its initial publication in book form, there have been many users of SLP who have contributed to its further development. Special appreciation should go to the professional consultants on the staff of Richard Muther & Associates, Inc., and to Derek Manderson, Knut Haganaes, and Hiroshi Ito – all members of the Muther International group. They have constructively applied SLP on numerous projects and have diligently brought to light several improvements and clarifications included in this Second Edition.

# Contents

# Preface

This book has one objective: to tell and show managers and layout planners how to plan the arrangement of their facilities. It does not dwell on principles nor delve into academic viewpoints. Rather, it presents *a universal approach* and *a specific set of procedures to follow* for use by the man who is facing a layout planning project and who wants to execute it in a sound, effective, and systematic way.

Layouts of working areas must be as old as trades and crafts themselves. As factory systems and modern business developed, more attention had to be given to space utilization. German ore-processing and chemical engineers, meat packers in Chicago, Canadian rail-car producers, automobile makers in Detroit, and British shipbuilders all made advances with unusual layouts. Industrial architects learned to relate their structures to the long-term functional use of the space inside them. But it took industrial engineers like Taylor, the Gilbreths, Barnes, Maynard, and Mogensen to bring us efficiency concepts and process-visualizing techniques that we could use as a basis for attacking industrial layouts.

Even with these significant contributions and with modern industry's recognition of the cost savings that result from sound planning, layout planning never developed into a clear procedure. On many layout planning jobs, I groped for an effective pattern to follow. Like most analysts, I wrestled with the problem of having so many factors, considerations, physical features, and objectives in my layout projects that I often felt bewildered. Therefore, I have striven to develop a universal layout planning procedure. I wanted a logical approach, sound techniques of analysis, a simple set of conventions or sign language, and a straightforward, easy-to-follow pattern of procedures—a pattern that would integrate and put into perspective the many good, though isolated, techniques already in use.

Systematic Layout Planning (SLP) basically attains these objectives. It is not perfect; it is not a substitute for judgment or intelligence; it is still not complete. Indeed, the facts that this Second Edition includes several new developments, lays increased emphasis on certain points, and incorporates both a slightly revised pattern of procedures and the recently approved color code for layout planning, all indicate that SLP is being refined and added to as it continues to be used.

Actually, Systematic Layout Planning (SLP) consists of *a framework of phases* through which each layout project passes; *a pattern of procedures* for step-by-step planning; and *a set of conventions* for identifying, visualizing, and rating the various activities, relationships, and alternatives involved in any layout project.

The *phases* were originally identified and described in my earlier book, *Practical Plant Layout* (New York: McGraw-Hill Book Co., 1955). They seem well recognized today.

The *pattern* of procedures is a newer development. Based originally on experience gained in over 200 layout planning projects, the SLP pattern has now been applied by the author and his associates to more than 1000 projects. Today there is no doubt that, properly applied, the procedures pattern helps to avoid errors, save time, and produce better solutions.

The *conventions* are, whenever possible, adopted directly from standards approved by professional societies in North America or from other long-established codes. However, layout planners do not have to adopt these conventions in order to follow the phases and pattern. In fact, there are times when the SLP conventions may be modified slightly to provide special meaning.

This book is really an instruction manual. Perhaps it is too specific—but we have designed it so. We have included many illustrations, check lists, and examples from a wide variety of projects. The working forms in the last section (including cross-references to their text explanations) in particular are aimed at providing direct, practical assistance. But it is the total SLP package—the complete planning system—that should bring experienced planners their biggest gain.

During the twelve years since the first printing of *Systematic Layout Planning*, it has been translated into seven foreign-language editions; its techniques have been widely adopted by planners throughout commerce and industry. I sincerely hope that this edition will prove of direct and significant benefit to you.

Richard Muther
Kansas City, Missouri

# Foreword

Back in the early 30's, *Factory Magazine* published a special section on Plant Layout in which, I believe, I was one of the first to suggest the use of the flow process chart as the essential first step in making a good layout. Since then, several books on the subject have appeared, but in my opinion, they all leave much to be desired.

At last, it seems that Richard Muther, an internationally recognized authority on plant layout, has combined the many tools used by industrial engineers into a "package" that offers a direct sequence of attacking and planning any layout project. In addition he has developed several new techniques of his own.

A big contribution is the breakdown of the entire subject into a four-phase framework and a detail pattern for Phase II—the general overall layout—and Phase III—the detailed layout plan. The book presents a universal approach to solving a series of reasonably complex problems for a wide variety of industrial conditions.

The emphasis placed on product-quantity analysis has been largely overlooked in other books on this subject. Several of the various flow-analysis techniques are described, and suggestions are made on selecting the proper one as a result of product-quantity analysis.

The technique of the rated and reason-supported relationship chart is a systematic way of getting data into usable form. It serves as a check list to insure that a decision is made for each activity in relation to each other activity.

The book is directly oriented toward the layout planning engineer, but, because of the straight-forward approach and the inclusion of the forms and worksheets contained in the text, operating people and others not trained as engineers should be able to go into the business of layout planning quickly and effectively.

Certainly, one of the greatest means to possible cost reduction lies in properly planning the layout at the outset, in order to apply Work Simplification before the plant or office has been built. This book should go a long way in helping to ensure this.

Allan H. Mogensen, Director
Work Simplification Conferences

# Part One

# Chapter 1
# Fundamentals

This chapter covers the fundamentals of layout planning. Chapter 2 sets forth the actual procedure to follow.

## Why Layout Planning

Many managers ask "Why plan layouts at all?" In some cases it would seem to be about as easy to move the furnishings into an area and then have the fun of arranging them and rearranging them until you are satisfied. For the housewife who likes the freshness of rearranging her house occasionally, this makes sense. But for industry, merely rearranging will, in practically every case, result in lost time, idle equipment, and disruption of personnel. In addition, it may well lead to serious blunders in the use of a company's available land, in costly re-arrangements, in actually tearing down buildings, walls, or major structures which are still usable but which subsequently turn out to be roadblocks to efficiency and low-cost operation.

A little time spent in planning the arrangement before it is installed can prevent such losses. Moreover, it allows the integration of subsequent moves and rearrangements into a logical overall program. Planning makes facilities arrangements an orderly, logical sequence. Layout planning pays off: Obviously, it is much easier to move templates or replicas of facilities and equipment around on a piece of paper than it is to move the actual buildings, machinery, or equipment around. As Professor Schell used to say, "You can make as many mistakes as you want in layout planning, and they'll all pay for themselves if they avoid mistakes in the physical installation."

Actually, from an installation standpoint, it is about as inexpensive to put in a good layout as to put in a poor one—frequently much less expensive. However, once a poor layout is installed, the cost of rearranging, disrupting production, and fighting your way through a new financial appropriation prohibit remaking it into a good layout.

## The Key to Unlocking Layout Problems

There are two basic elements on which every layout problem rests:

1. Product (or material or service)—what is to be made or produced
2. Quantity (or volume)—how much of each item is to be made

Directly or indirectly, these two elements underlie all other features or conditions in layout work. Therefore, facts, estimates, or information about these two elements are essential.

By *Product* (or material or service) we mean the goods produced by the company or area in question, the starting materials (raw materials or purchased parts), the formed or treated parts, the finished goods, and/or service items supplied or processed.

Products may be termed items, varieties, models, styles, part numbers, formulations, product groups, or material classes.

By *Quantity* (or volume) we mean the amount of goods or services produced, supplied, or used.

Quantity may be termed number of pieces, tons, cubic volume, or value of the amount produced or sold.

In terms of unlocking layout problems (see Figure 1-1), these two elements represent the handle of any key we must grasp. For it seems obvious that if we are planning the layout of a plant or department, the layout must accomplish something. That "something" is certain products in certain quantities.

After obtaining the product and quantity information, we must next learn about the routing (or process). The routing refers to how the product or material will be made.

By *Routing* we mean the process, its equipment, its operations, and their sequence.

Routing may be defined by operation-and-equipment lists, process sheets, flow sheets, and the like.

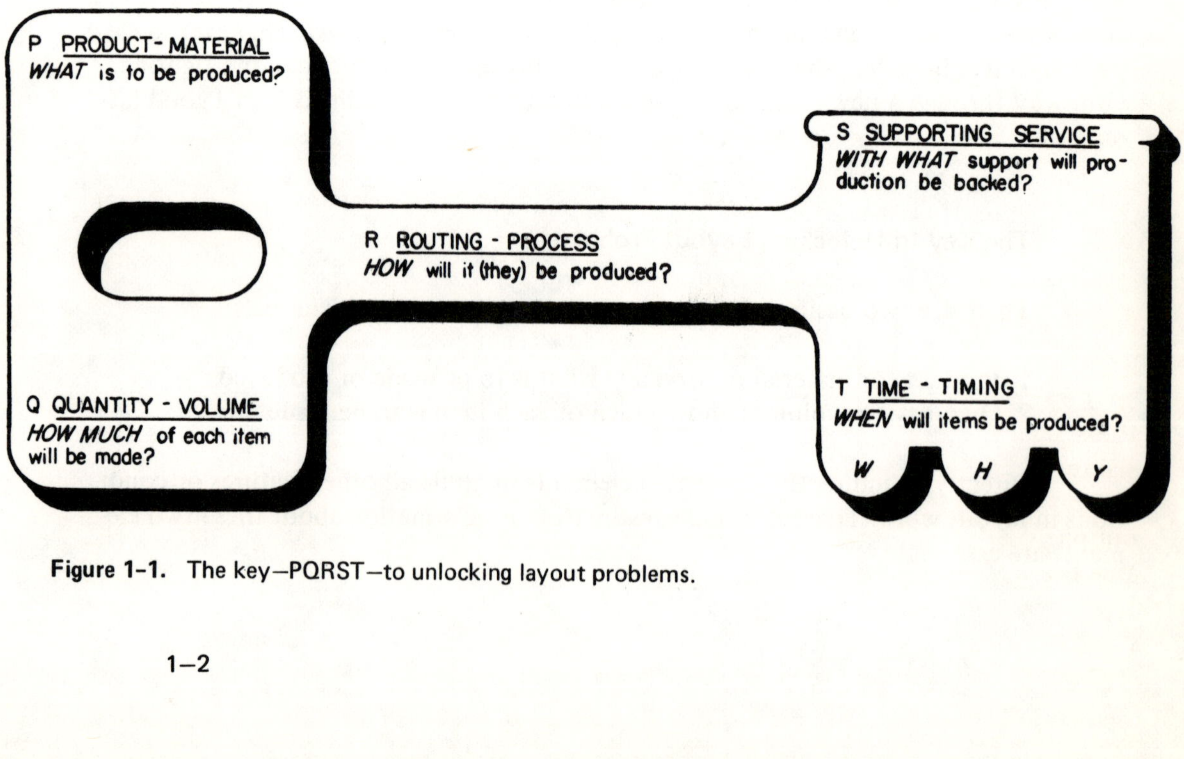


**Figure 1-1.** The key—PQRST—to unlocking layout problems.

The machinery and equipment used will depend on the operations selected to change the form or characteristics of the material. Similarly, the movement of work through the area to be laid out is dependent upon the sequence of the operations. Therefore, the operations involved in the process and their sequence become the body (or stem) of our key.

Backing up the direct forming or assembly operations—the producing activities or areas—are a number of supporting services. In a sense these are the things that give strength to the producing operations, for without adequate support, the producing equipment and workers could not function adequately.

By *Supporting Services* we mean the utilities, auxiliaries, and related activities or functions that must be provided in the area to be laid out, so that it will function effectively.

Supporting services include maintenance, machine repair, tool room, toilets and locker rooms, cafeteria, first aid, and frequently shop offices, rail siding, receiving dock, shipping dock, receiving (or "in area"), and shipping (or "out area"). It is common to include storage areas as a part of the supporting services as well.

Taken all together, the supporting services often occupy more floor area than the producing departments themselves. Therefore, adequate attention must be given to them.

One other basic element of the key to unlocking layout problems is Time (or timing).

By *Time* (or timing) we mean when, how long, how often, and how soon.

Time or timing involves *when* products will be produced or *when* the layout being planned will operate (one shift only, during harvest season, Christmas rush). Operating times for the producing operations determine how many of a given piece of machinery are required, which in turn determines the space required, man-power staffing, and operation balancing. Urgency (of delivery or action) is also a part of timing, as are the frequency of lot or batch "run" and the response of supporting services.

Perhaps the most important of all, time affects us—the layout planners. Every layout project takes a certain amount of time to accomplish, and usually there is a deadline to meet.

Figure 1–1 shows these elements as a key—to have an easy method of remembering them. But note three letters at the business end of the key: W H Y. These are an essential reminder to the layout man to question the basic data—to check with reliable sources or have his top management "say grace over" the basic figures on which his layout planning will depend. Therefore, a few challenging "why's" may be necessary to be sure the starting data is sound.

Five elements–P (Product), Q (Quantity), R (Routing), S (Supporting Services), and T (Timing)–form the basis for layout planning. They are recapped in Figure 1–2.

This sequence of letters is essentially a new alphabet for the layout planner. For him, the alphabet no longer begins with A, B, C. From the start, each one of his projects involves P, Q, R, S, and T.

A play on letters, true– but it is the heart and soul of what eventually becomes millions of dollars of new construction, modernization, and plant rearrangement.

With this alphabet, everyone planning new or rearranged facilities has a place to start–and starting is sometimes the hardest part of a project.

| | |
|---|---|
| P | Product (Material) |
| Q | Quantity (Volume) |
| R | Routing (Process sequence) |
| S | Supporting Services |
| T | Time (Timing) |

**Figure 1–2.** The layout planner's alphabet, which begins with the basic input data (or key input) used in layout planning.

### Product-Quantity Data Sheet

Based on the above, it is easy to recognize that, for the man who is planning a layout, the basic sources of data must be the company's design engineering group or product planning group on the one hand and the sales or marketing department or research people on the other. A plant layout, therefore, does not begin in the "plant." Rather, the planner must first seek out the P and Q information.

For the layout planner who does not have an organized system for securing his product and quantity information, we suggest the Product-Quantity Data Sheet (see Figure 1–3). (Note: For each Figure in the text having a form number in the lower left corner, the reader will find a fresh unused form in the Working Forms section at the back of this book.)

The Product-Quantity Data Sheet merely offers an organized way of gathering P and Q information. The form is divided into several parts. The left side deals with the product data; the right side, with the quantity data. The top two-thirds of the form covers the data for one product; the lower third, for several products. Only the top or the lower portion of the form is used on any one project.

The product data at the top left of the form is for one product where only forming and treating is involved. That is, when no assembly or disassembly takes place. This calls out the identification of the product or material and its physical characteristics, nature, quality features, and degree of standardization in design.

On the right side is space for gathering information about the quantities to be produced.

At the middle left are statements similar to those on the top left, although the middle left is prepared so that it is easy to get information for a product involving assembly operations as well as forming and treating.

The bottom of the form calls for the same information regarding products at the left and quantities at the right for a company producing a line of several products.

Generally speaking, the "plan layout for" product(s) and quantity(ies) should be endorsed by management representatives higher than the layout man himself. He should get the formal endorsement of figures in question from the company officials in charge of these activities. Obviously, to plan a layout which will not produce the desired products in the desired amounts is not achieving management's overall objective.

In addition, trends in products and quantities for the future are incorporated on this form to give the layout planner a picture of the direction in which things are going. Seasonal fluctuations, peak loads, and number of shifts are picked up also so that they will not be overlooked in the planning and so that official approval or endorsement can be given at the outset of any project.

While this form alone is seldom sufficient for a thorough analysis of product and quantity information, it does serve as a universal point of beginning. It has been said that the way to begin is to begin. Therefore, if you are an inexperienced layout planner this form can get you started on your project quickly—although you may soon depart from it to get more specific information on worksheets having more room and more meaning for the particular project in question.

## Layout as a Product

From the standpoint of the person responsible for planning a layout, the output of his labors is the layout plan. Therefore, he is a designer of a product and his plan becomes the blueprint and specification for that product. This product is the physical, installed layout.

# PRODUCT-QUANTITY DATA SHEET

Plant *Chicago* Project *472-C3*
Data gathered by *R.C. Mills* With
Date *8-9* Sheet *1* of *1*

## PRODUCT INFORMATION

Fill-in as applicable.

**For ONE PRODUCT** – Form and/or Treat only

Product Name & Description *Blade Forging*
Finished condition (fluid, delicate, hazardous, etc.) *Nick Free*
Size-shape *2" to 8" long; .10" to .375" thick; ½" to 2½" wide*
Normal unit of measure *Piece (blade)* Weight/unit *¼ to ½ lb.*
Starting material condition *Extrusion*
Size-shape *3½" to 8" long; 1" to 2" head dia.* Weight/unit *¼ to ½ lb.*
Normal container: as received *On skid* as shipped *On skid*

**For ONE PRODUCT** – Assemble and/or Disassemble involved
Product Name & Description
Finished condition
Size-shape Weight/unit

| Major Components: | Material Condition | Size-shape | Weight/unit |
|---|---|---|---|
| a. | | | |
| b. | | | |
| c. | | | |
| d. | | | |
| e. | | | |

See Parts List (s) or Component Breakdown ☐

## PRODUCTION REQUIREMENTS

Quantity produced this year *150,000* Source *Sales*
Quantity anticipated next year *150,000* Approved *J.C. Allen*
Quantity anticipated in 5 yrs. *250,000* Est. by *W.N. Weker*
Length of time present product or model will be produced *Indefinitely*
Seasonal variation *Negligible*
Expansion Plans *Next addition at new site: not here*

Trends in product:
Size *None* Diversification *None*
Weight *None* Simplification *None*
Materials *Higher temperature resistance*
Rec'g. & Shipping amounts and frequencies *Smaller & more frequent*
Refinements *More accurate finish*
Other

Operating hours *8* per shift *16* per day
*80* per week

Plan Layout for (no. of units) *25,000* ⓐ per ~~hour~~, ~~day~~, (week)

**For SEVERAL PRODUCTS**

| Name of Product or Group | Condition | Size-shape | Weight/unit | Quantity: This Yr. | Last Yr. | Next Yr. | 5 Yrs. | Per Order or Lot | % of Production | Plan Layout for |
|---|---|---|---|---|---|---|---|---|---|---|
| A. | | | | | | | | | | |
| B. | | | | | | | | | | |
| C. | | | | | | | | | | |
| D. | | | | | | | | | | |
| E. | | | | | | | | | | |
| F. | | | | | | | | | | |
| G. | | | | | | | | | | |

Trends in Product:
Seasonal Variation
Expansion Plans

NOTES: ⓐ *Excess capacity due to military requirements.*

One difference lies in the fact that a design engineer usually creates specifications and prints for a product to be made in lots or batches. The plant layout engineer designs a product which will be individual—for fundamentally each layout is unlike any other. In this sense, the layout engineer is more like an architect.

If we compare the work of a product design engineer with the work of the layout engineer we find the following sequences of steps:

| *Product Design Engineer* | *Layout Engineer* |
|---|---|
| 1. Research & Development | 1. Location of area to be laid out |
| 2. Engineering & Design | 2. General Overall Layout |
| 3. Production Design | 3. Detailed Layout Plans |
| 4. Processing & Tooling | 4. Installation (Planning & Moves) |

In still another sense, a plant layout is a product. A product design is the consolidation of a number of components, parts, or ingredients into one functional end-product. It is assumed that this product, either because of its shape, its chemical characteristics, or integration of components, has value such that it can be sold and used profitably by the customer. In the same way, the layout is a combination of different departments, areas, or functioning activities which when put together function properly. It is assumed that this layout has value when installed for the benefit of the users, which in this case would be the operating people and the owners of the company.

It should be noted that the layout plan is not an end in itself. There is a great tendency on the part of planners, particularly in fairly large organizations, to think of their layout plan as the "end of the line". "Once the layout is done, that's it," they say—and they feel no further responsibility for it. This is a great pity, for actually no product does any good merely in the design stage. It is not until the designed product is put into production and distributed to users that it really does anyone any good and really becomes a profitable venture for the manufacturing company. In the same sense, a layout plan does little if any good until it is sold to management and operating personnel, installed, and put to work by them.

Therefore, the layout man or planner must realize that he has a responsibility to sell his plan and to have it installed and used as it was conceived and designed to be put to use. Not until this is done does the money spent for him to do his layout planning come home as savings for his company.

**Figure 1-3.** Data gathered for Product and Quantity. At left, this example shows the most significant information about the product in question—a blade forging for aviation gas turbine. At right is basic information about the quantity, together with design trend data.

This form is a starting point for getting P-Q data. It provides space on left for product (or material) information; on right for quantity (or volume) requirements. The appropriate area of the form will be used depending on whether P is a single product, form only or assembly, or includes several products.

## Phases of Layout Planning

The four steps that the layout planner takes may be translated into what is known as the "Four Phases of Layout Planning." These include the following:

*Phase I–Location*

Determine the location of the area to be laid out.

This is not necessarily a new site problem. More often it is one of determining whether the new layout (or re-layout) will be in the same place it is now, in a present storage area which can be made free for the purpose, in a newly acquired building, or some other potentially available space.

*Phase II–General Overall Layout*

Establish the general arrangement of the area to be laid out.

Here the basic flow pattern(s) and the areas allocated are brought together in such a way that the general size, relationships, and configuration of each major area is roughly established.
Phase II is sometimes termed block layout or area allocation or merely rough layout.

*Phase III–Detailed layout Plans*

Locate each specific piece of machinery and equipment.

In detail planning, the actual placement of each specific physical feature of the area to be laid out is established. And this includes utilities and services as well. The detailed layout plan is customarily a sheet or board with replicas of the individual machines or equipment placed or drawn thereon.

*Phase IV–Installation*

Plan the installation, seek the approval of the plan, make the necessary physical moves.

Once the detailed layouts are completed (Phase III), considerable detailing of installation drawings and planning of moves must be worked out. Funds for the installation must be appropriated and the actual moves to install the machinery, equipment, and the services as planned must be made.

These four phases come in sequence, but, for best results, they should overlap each other. This is indicated by Figure 1-4. Every layout project passes through these four phases–even though the layout planning analyst may not be specifically charged

**Figure 1-4.** The four phases of Systematic Layout Planning. Every layout planning project passes through the four phases. As shown against the time scale, these phases should generally come in sequence and for best results the planner should make them overlap.

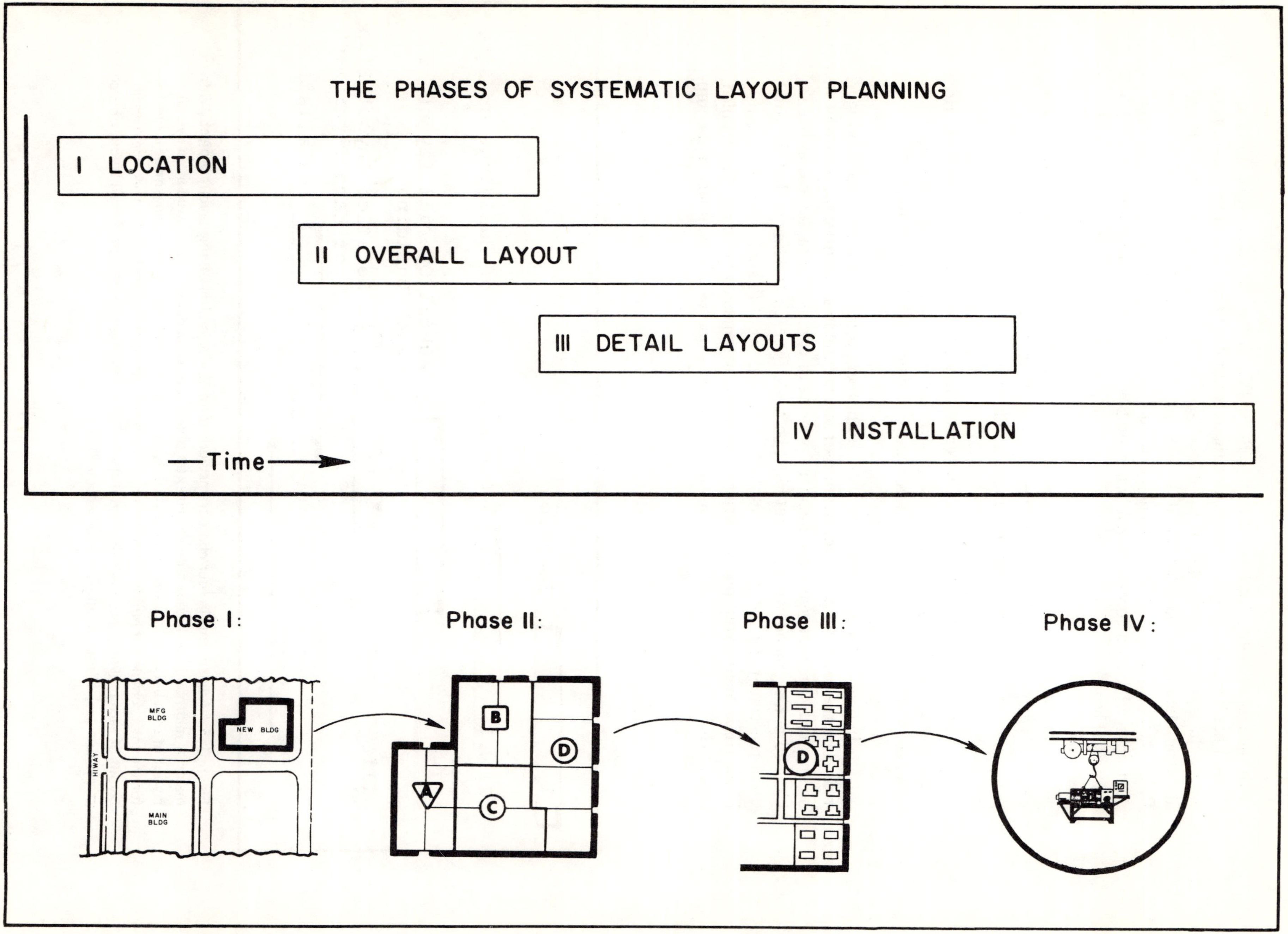
THE PHASES OF SYSTEMATIC LAYOUT PLANNING
I LOCATION
II OVERALL LAYOUT
III DETAIL LAYOUTS
IV INSTALLATION
Time
Phase I:
Phase II:
Phase III:
Phase IV:
MFG BLDG
NEW BLDG
HIWAY
MAIN BLDG
A
B
C
D
D

with the responsibility for Phase I and/or Phase IV. That is, he must make sure that Phase I has been agreed to or that a specific decision has been, or will be, made as to where the layout he is planning is to be located. Obviously, he cannot be very specific about his detailed layout planning if he does not have information about number of floors, ceiling heights, column spacing, and building features. All are generally dependent upon a location—or a reasonably acceptable assumption as to the location—having been established.

In many cases, the Phase I work actually involves a plant location study or a new site analysis. In such cases, the person actually responsible for making the layout plan may or may not be involved directly in Phase I.

Likewise, in Phase IV some other group may do the physical installation. However, in any case the layout planning engineer should be aware of this four-phase sequence and should be prepared to integrate his work with Phases I and IV.

In plans for a major project, there is another phase. This is explained in Figure 1-5. Here the total site or the integration of several buildings at different locations may be involved. Once the major arrangement of areas or complex of buildings has been blocked out, a general overall layout must be made for each building within the total site arrangement. This blocks-within-blocks enlargement of Phase II precedes the usual Phase III detailed layout plan for each department within each building—the detailed layout thus always remaining "the location of specific pieces of machinery and equipment."

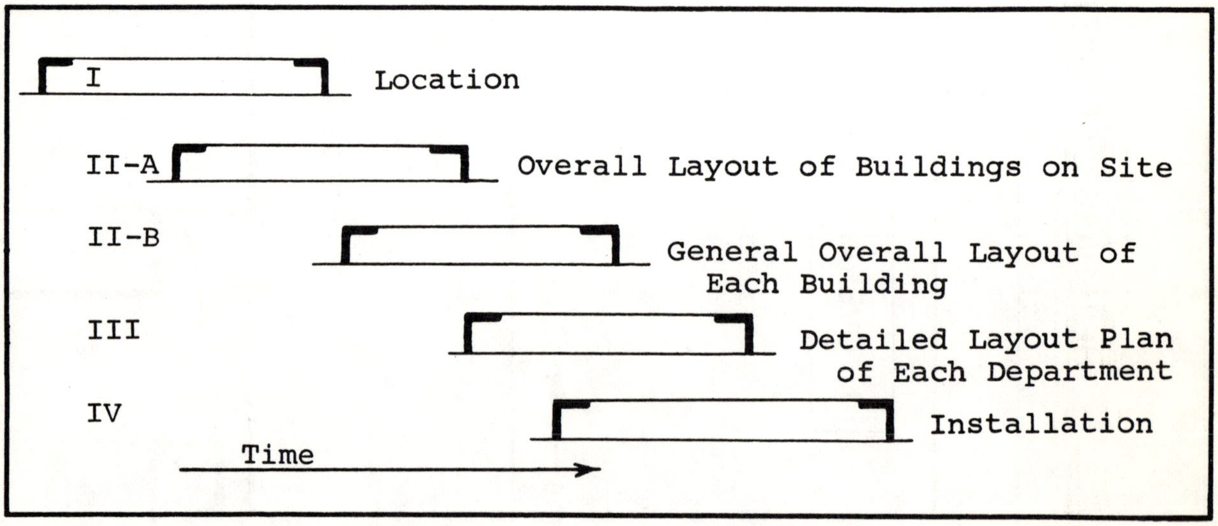


**Figure 1-5.** Adjustments in four-phase framework for major or multi-building layout project. With major or multi-building layout projects, an extra phase enters the picture, for as well as an overall layout of each building into which the layout details can be fitted, an overall arrangement is needed for the total complex of buildings on the site. The overall site plan becomes Phase II-A; the overall layout of each building becomes II-B. And if the buildings are multi-story, the overall layout of each story (or floor) of each building would become II-C.

Similarly, it is important to note that the amount of detailed information required for planning increases as the layout project progresses. In the location phase, rough estimates and general considerations are involved. but in Phase IV installation drawings must be accurate, in some cases to a fraction of an inch. This degree of detailed information as a function of the layout phases is indicated in Figure 1-6.

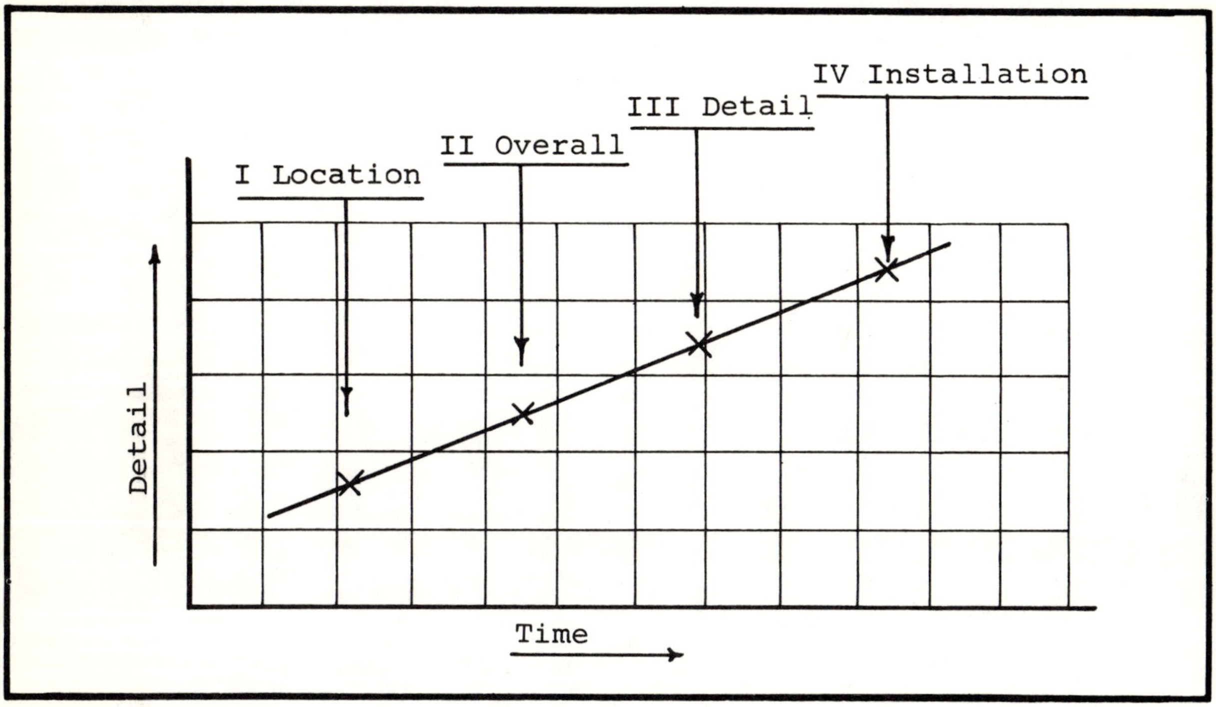


**Figure 1-6.** Degree of detailed information. The amount of detailed information, and the specificity of layout data and of planning techniques, increases as the layout project progresses through the framework of phases toward actual physical reality.

From the standpoint of planning the work to be done in making a new layout plan, the planner can use the four phases as a guide. First of all, it is possible–indeed logical–to schedule the project in terms of the four phases. This means that the layout man should seek approval from his superiors at least at these four places. Likewise, the person responsible for the overall project should check with his associates and subordinates at these same four places.

Chapter 2

# The Systematic Layout Planning Pattern

Systematic Layout Planning is an organized way to conduct layout planning. It consists of a framework of phases, a pattern of procedures, and a set of conventions for identifying, rating, and visualizing the elements and areas involved in planning a layout.

We explained the framework of four phases in Chapter 1. In this chapter, the Systematic Layout Planning *pattern of procedures* is described. The conventions will be introduced at the appropriate places in later chapters.

The strictly "layout planning" phases of any facilities rearrangement involve creating a general overall layout (Phase II) and, subsequently, a detailed layout plan for each portion of the general overall layout (Phase III). In both Phase II and Phase III, the pattern to be followed is essentially the same.

Every layout rests on the three fundamentals:

1. Relationships—the relative degree of closeness desired or required among things
2. Space—the amount, kind, and shape or configuration of the things being laid out
3. Adjustment—the arrangement of things into a realistic best fit

These three are always the heart of any layout planning project, regardless of products, processes, or size of project. It is therefore logical and to be expected that the pattern of layout planning procedures is based directly on these fundamentals. See Figure 2–1.

### The SLP Pattern

In the previous chapter, we indicated the importance of Product (P) and Quantity (Q) to any layout. An analysis of them individually and in their "mix" is a necessary preliminary to any real layout planning. Process routing and equipment, supporting services, and timing information are also basic input data. And in addition, identifying the various activities (or areas) included in the layout is a preliminary planning step.

Box 1 of the pattern—and in process-dominated industries often the most significant aspect of layout planning—is Flow of Materials. By planning the layout around the sequence and intensity of material moves, we attain a progressive flow through the area(s) involved.

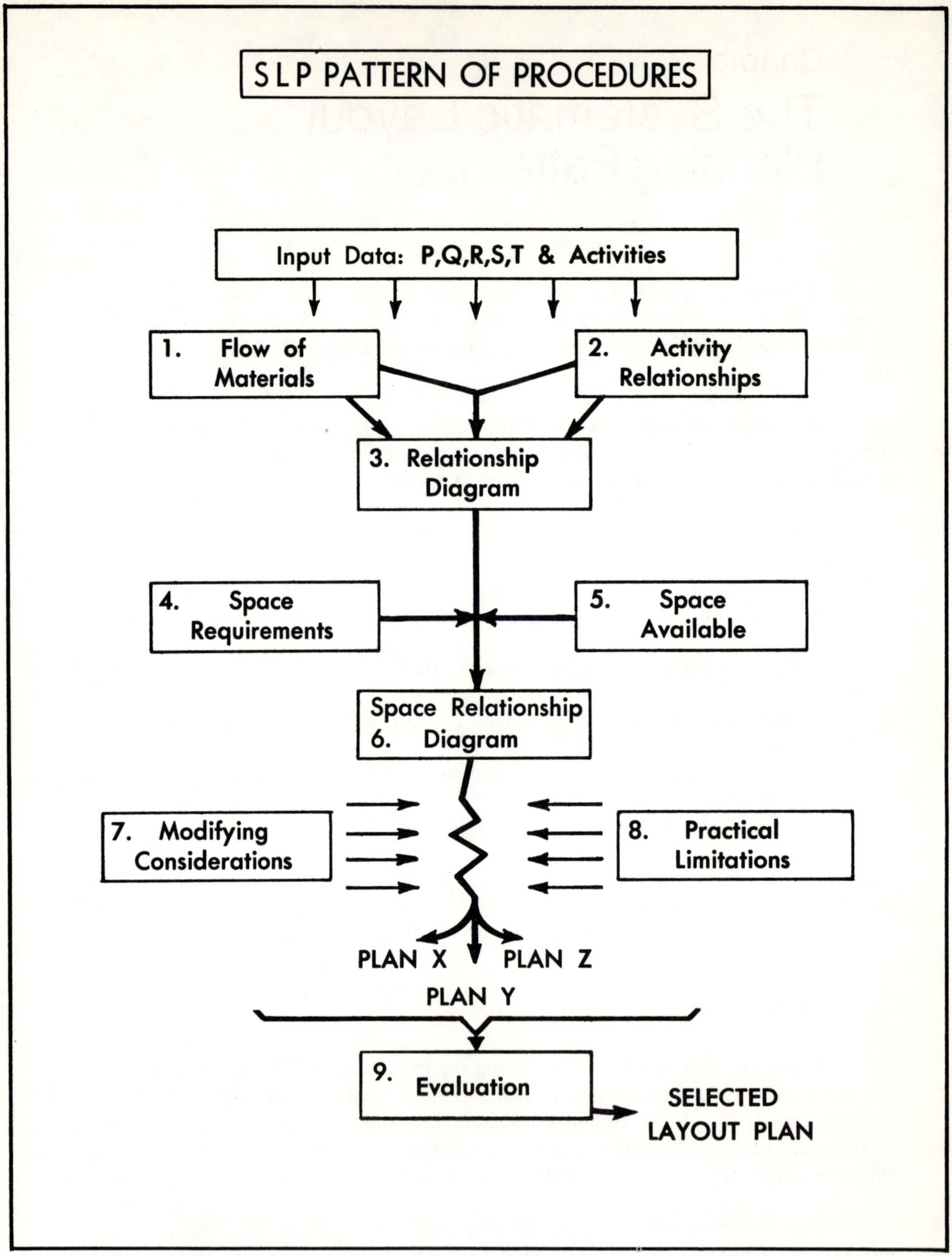


**Figure 2-1.** The Systematic Layout Planning Pattern of Procedures.

In addition to the operating or producing areas, many supporting-service areas must be integrated and planned. As a result, developing or charting the Activity Relationships—that is, the relationships among the service or support activities or functions—is frequently of equal or greater importance than relationships based on flow of materials alone.

These two investigations are then combined into a Flow and/or Activity Relationship Diagram. Here the various activities, departments, or areas are geographically related to each other, without regard to the actual space each requires.

Next, the Space Requirements: These are developed from analysis of the process machinery and equipment necessary and from the service facilities involved. Area requirements must, however, be balanced against the Space Available (Box 5).

Then the area allowed for each activity is "hung" on the activity relationship diagram to form a Space Relationship Diagram (Box 6).

The space relationship diagram is essentially a layout. But, in all likelihood, it is not an effective layout until it is adjusted and manipulated to integrate with its space any Modifying Considerations. These include such basic considerations as the handling method, operating practices, storage scheduling, and the like. As each potentially good consideration or idea concerning these features is thought up, it must be tested against Practical Limitations like cost, safety, and employee preference.

As the integrating and adjusting of the various modifying considerations and their limitations are worked out, one idea after another is probed and examined. The ideas that have practical value are retained and those that do not stand the test are discarded. Finally, after abandoning those plans which do not seem worthy, we end up with two, three, four, or five alternative layout proposals. Each of them will work; each has value. The problem lies in deciding which of these plans should be selected. These alternative plans may be termed Plan X, Plan Y, and Plan Z.

At this point, a cost analysis of some kind should be made for purposes of comparison and justification. In addition, some evaluation of intangible factors should also be made. This is called an evaluation of alternative layouts or an Evaluation of Costs and Intangibles. As a result of this evaulation, one of the alternatives is chosen—although frequently a modification or combination of two or more layouts may actually result from the evaluation process itself.

The alternative layout that is chosen becomes the Selected Layout, the General Overall Layout. With the selection of this general overall layout, Phase II is completed. (In Phase III, basically the same pattern holds, once for each area blocked out in the overall layout, but we shall discuss that later.)

### Tie-in of P, Q, R, S, and T

We have seen how the pattern of Systematic Layout Planning is constructed. Now let us relate it to the basic input data, P, Q, R, S, and T.

P, Q, R, S, and T underlie most of the calculations needed for layout planning. The preparation of the data for the various boxes in the SLP pattern starts with these five base elements. The product designs and sales forecasts must be woven together and integrated with a P-Q analysis—sometimes called volume-variety analysis or study of product mix. The logical splits and combines of various products or product groups or layout groupings are derived from the P-Q analysis. Specifically, this analysis of product mix, along with analyses of Routing (R), Services (S), and Times (T), leads us to an identification or delineation of the individual activities (areas, machine groups, work places) involved, and thus often to the actual type of layout.

P, Q, and R are then woven together to develop the flow of materials (Box 1). P, Q, and S are woven together to develop a service activity relationship (Box 2). From the flow of materials or the activity relationship chart, or a combination of the two, the relationships are then diagramed (Box 3).

It is Routing (R), together with Time (T), which essentially determines the machinery and equipment required. Similarly, the services (S) called for are translated into the various service facilities required. The process machinery and equipment and the service facilities are then translated into space requirements. These space requirements are then worked into the SLP pattern as described above.

### Tie-in of Phase III—Detailed Layout Plans

The general overall layout is developed then in Phase II. The next phase, III, involves making detailed layout plans of each piece of machinery and equipment, aisle, and storage area, for each of the activities, departments, or areas which have been roughed out in the general overall layout.

By reference to Figures 1–4 and 2–2, we see that Phase II overlaps Phase III. This means that before actually finalizing the general overall layout, certain details will have had to be considered. For example, the orientation of a conveyor system may have to be analyzed in detail in order to arrive at a satisfactory solution to the general overall layout. This kind of overlapping investigation takes the layout planning engineer into detailed Phase III work in certain areas before his Phase II is selected. Such overlapping is noted earlier, but it should not be lost sight of merely because we have discussed here the general overall layout, Phase II, as a distinct pattern, separate from Phase III, detailed layout planning.

Note also that a detailed layout plan must be made for *each* of the departmental areas involved. This means not only that adjustments are made within these detailed areas, but that some readjustment of the general overall layout selected may

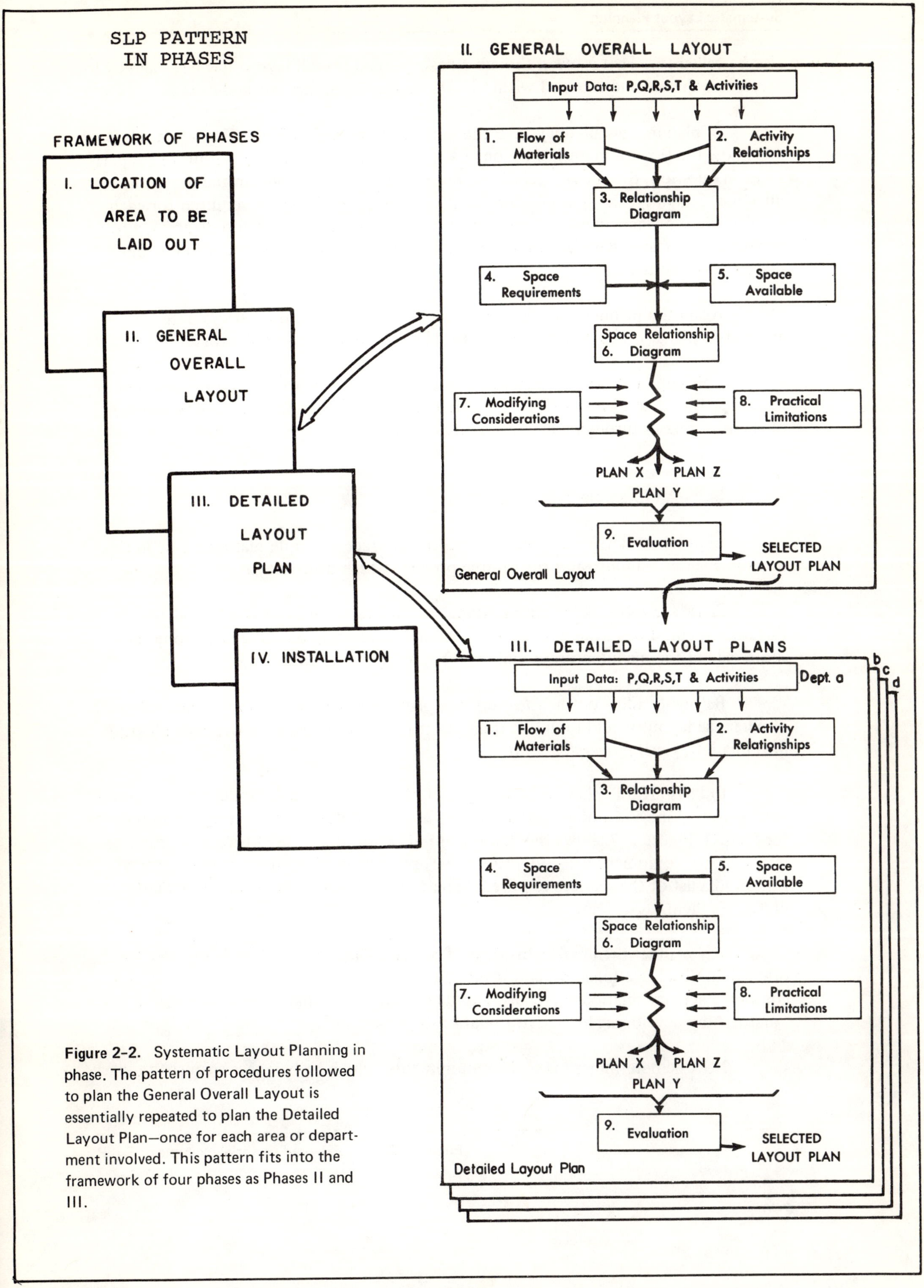


**Figure 2-2.** Systematic Layout Planning in phase. The pattern of procedures followed to plan the General Overall Layout is essentially repeated to plan the Detailed Layout Plan—once for each area or department involved. This pattern fits into the framework of four phases as Phases II and III.

also be called for. That is, even though a basic general overall layout has been agreed to, it can be adjusted and changed within limits, as the details are worked out.

In planning detailed layouts, the same pattern used in Phase II is repeated. However, now the flow of materials becomes the movement of materials within the department in question; the activity relations, those of the activities within the department in question. Similarly, the space requirements now become the space required for each specific piece of machinery and equipment; and the space relationship diagram, a rough arrangement of templates or other replicas of machinery and equipment, men, and materials or products.

Again the planner ends up with several alternatives, and must evaluate the alternative plans (for each department) to select the one most satisfactory.

This same pattern is repeated for each departmental area which is to be laid out in detail. And, as shown in Figure 2–2, these patterns fit into the logical development of the four phases through which each layout project passes.

### SLP—An Example

In order to clarify the concept of SLP, we trace a sample planning case in this section. Figure 2–3 illustrates the sample project, which utilizes SLP.

Phase I involves determining where the area to be laid out will be located. In this case, it could be in the north side of the manufacturing building (X), along the south side of the manufacturing building (Y), or in a new building (Z).

Before deciding on the *location,* the planner should have a fairly reasonable idea of what the layout would look like if he put it there. As a result, he overlaps his study into Phase II, *general overall layout.*

In Phase II, the planner pulls together his basic input data; indeed he has probably already done so (though perhaps not in as much detail) in collecting the input data for Phase I. Figure 2–3 shows the Phase II inputs as products, sales forecast, product mix analysis, equipment lists, operations lists, projection of product changes against time, and a list of the services required. (The example cannot show all inputs, but it illustrates them.)

The planner then makes his flow-of-materials analysis, gathers his service (or other-than-flow) relationships, and makes a combination of the two in the form of a combined activity relationship chart. He then converts his charted *relationships* into a "picture" by developing his activity relationship diagram.

Next he determines his space requirements, which he balances against the space

**Figure 2–3.** A condensed pictorial example of S L P.

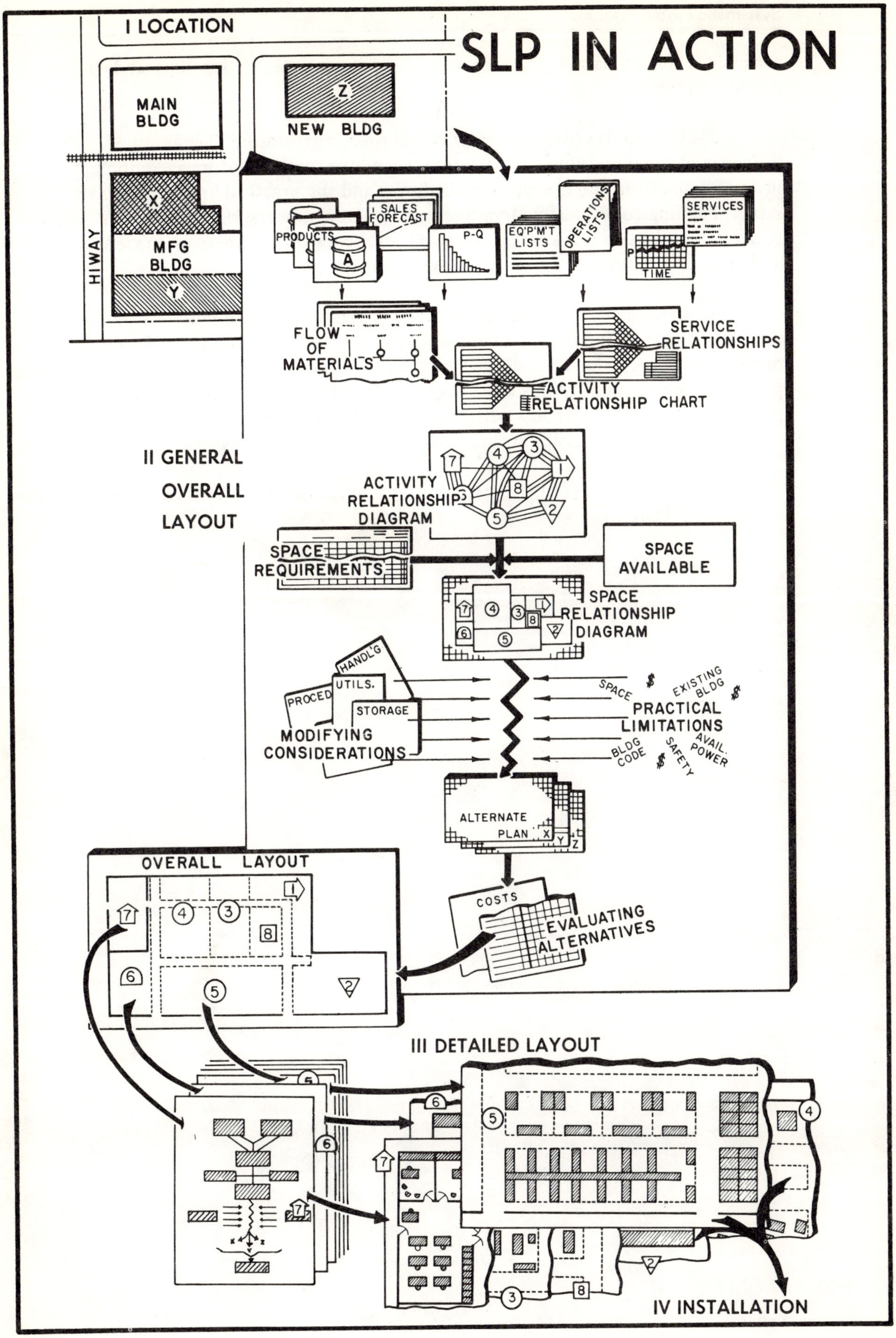
SLP IN ACTION
I LOCATION
MAIN BLDG
NEW BLDG
Z
X
MFG BLDG
Y
HIWAY
PRODUCTS
A
SALES FORECAST
P-Q
EQ'P'M'T LISTS
OPERATIONS LISTS
SERVICES
TIME
FLOW OF MATERIALS
SERVICE RELATIONSHIPS
ACTIVITY RELATIONSHIP CHART
II GENERAL OVERALL LAYOUT
ACTIVITY RELATIONSHIP DIAGRAM
SPACE REQUIREMENTS
SPACE AVAILABLE
SPACE RELATIONSHIP DIAGRAM
HANDL'G
UTILS.
PROCED
STORAGE
MODIFYING CONSIDERATIONS
SPACE
EXISTING BLDG
PRACTICAL LIMITATIONS
BLDG CODE
SAFETY
AVAIL. POWER
ALTERNATE PLAN
X
Y
Z
OVERALL LAYOUT
COSTS
EVALUATING ALTERNATIVES
III DETAILED LAYOUT
IV INSTALLATION

available. Then he marries his *space* data to his diagram of relationships, and so derives a space relationship diagram.

He then makes his *adjustment* of the space relationship diagram under the influence of the modifying considerations (handling methods, storage equipment, utility distribution, and operating procedures) and the practical limitations (cost, safety, building code, existing building, and available power). His adjustments lead him to several alternative block layout plans. He evaluates the alternatives, in view of costs and intangible factors, to arrive at his overall layout plan.

In arriving at the overall layout, the planner has undoubtedly looked ahead at certain critical features or critical areas of Phase III–*detailed layout plans.* Regardless of the planning schedule, he next subjects each area blocked out in Phase II to the same pattern of procedures. By this time, his location will have been selected so that he can proceed with his detail layouts with full knowledge of building features, column locations, position of main aisles, main utility distribution, and the like.

With detailed layouts for each area approved, he then moves his project toward *installation,* Phase IV.

# Part Two

# Introduction to Part Two

*Logically, in this next part of the book, we would discuss Phase I, Location of the Areas to be Laid Out, the first of the phases in the SLP framework, which would in practice be planned before Phases II and III. However, in our effort to concentrate on* layout *and how to plan it, a discussion of* location *of the layout will be more helpful after Phases II and III are understood. Therefore, Location–along with Phase IV, Installation–appear in Part Four of this book.*

*The next eight chapters, which make up Part Two, take us through the SLP pattern. They describe step-by-step the procedure and techniques of analysis to use–generally, one chapter is devoted to each "box" in the pattern. (See the accompanying sketch.)*

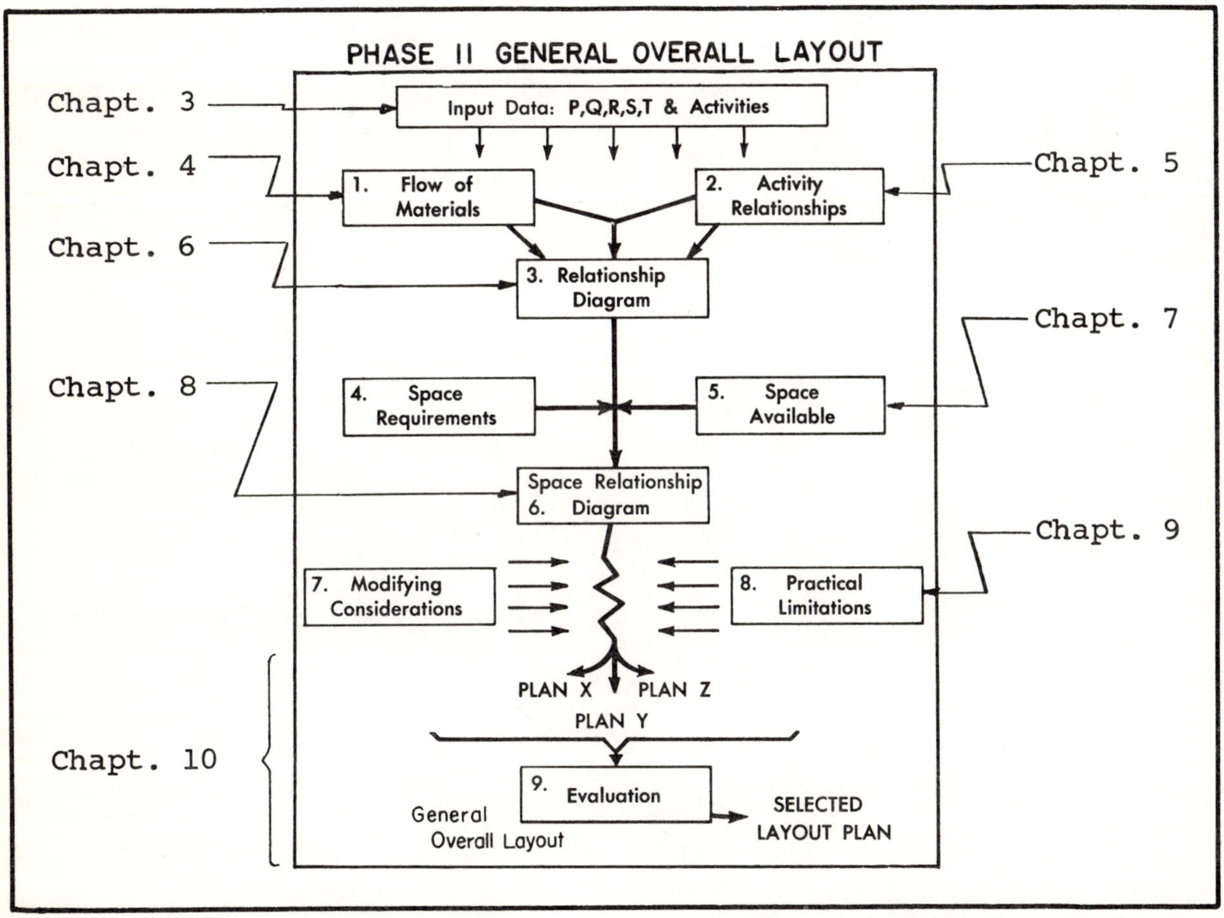


*Note that in these chapters in Part Two procedures and applications for Phase II are discussed. Procedures and applications for Phase III are somewhat different. (They are discussed in Part Three.) In most layout projects, the big savings in both*

*investment and operating costs are made in the overall planning, Phase II—General Overall Layout, rather than in working out the Phase III details. So, logically, we place the greater emphasis on Phase II planning.*

*This emphasis on Phase II should be recognized by both the novice and the trained layout planner. Most popular literature on layout planning leads to overemphasis of detailed techniques—not infrequently drawing even the experienced analyst into using templates or three-dimensional models long before he has developed his reasons for placing them in a meaningful, thought out way.*

# Chapter 3
# Input Data

Chapter 1 set forth the key input data on which every layout problem rests: P, Q, R, S, and T. In this chapter, the interrelationship of P and Q, the division of total areas into distinct activities or sub areas, and the projection of input data into the future are explained.

The first key data involve

1. Product (or material or service)—what is to be produced
2. Quantity (or volume)—how much of each item is to be produced

Figure 1–3 showed a Product-Quantity Data Sheet—one simple way to gather facts about Product and Quantity. This is a beginning point, but veteran layout analysts know they must dig far deeper.

### Volume-Variety Analysis

In almost all aspects of industry, there is a disproportionate balance between P and Q. In a typical situation, 20% of a company's customers account for 80% of the net sales, or 30% of the output is spread over 70% of the products.

This disproportion is well known by economists, statisticians, and market-research analysts. For example, in a mill producing mixed feeds for animals, 10% of the formulae (mixes) account for 65% of the tonnage produced; 20% of the formulae account for 82%; and the last 1% of the tonnage is made up of 31% of the formulae.

In terms of production planning, this relative proportion is the basis of product-mix problems and the so-called ABC concept of inventory control (control according to value). Others term investigation of these relationships "volume-variety analysis." Relatively, the same proportions usually prevail—depending in part on how the items, groups, or lines of product are divided.

For the facilities planner or layout analyst, this volume-variety balance has a significant meaning, because it is the basis for deciding which fundamental type of layout arrangement to use—line production, job-shop, a combination, or a split of two or more basic arrangements.

### The P-Q Chart

Generally, Product-Quantity analysis takes the form of (1) Some division or grouping of the various products, materials, or items involved; and (2) A tally or count of the quantity of each division or grouping, or of each product, item, or variety within each division or grouping.

For ease in visualizing the findings, the count is arranged in sequence and the results plotted on a graph. The result is the P-Q chart, so called for the terms Product and Quantity. (See Figure 3–1.) Note that the curve is plotted in the order of decreasing quantities, items having the largest quantities placed first, and that individual quantities (rather than cumulative figures) are used. The typical P-Q chart approximates a hyperbola—generally being asymptotic to both axes.

In identifying P and in counting Q for any product, item, variety, or product group, the actual count is ordinarily used rather than percentage conversion of the count. The largest item is placed first, then the next largest, and so on. The count is generally best given in number of pieces, weight, or cubic volume, rather than in dollars. The actual unit of measure depends on the nature of the product or items involved and on the unit used in company records available or in forecast projections. Often two or three measures for Q are plotted.

### Significance of the P-Q Chart

The P-Q chart has a fundamental relationship to the layout being planned. At one end of the curve are large quantities of relatively few products or varieties: essentially mass production conditions. Such products tend to favor mass production methods, especially production lines or layouts by product.

At the other end of the curve are a great many different products, each with but small quantities: job-order or job-shop methods—especially layout by process for forming and treating and layout by fixed position (fixed location) for assembly. (See Figure 3–2 for descriptions of the basic types of layout plans.)

The curve demonstrates, then, that some products lend themselves to a conveyorized, automatic type of layout while others require highly flexible handling methods and standardized equipment, arranged for universal operations.

As a result, by dividing the products and producing them in two different types of layouts, efficiency may be obtained for both, whereas to make one layout plan for all products may be a less effective compromise.

A "shallow" P-Q curve indicates, on the other hand, one general layout for all the items with universal handling methods. (See Figure 3–3.) Because most of the production is in the center area of the curve, we plan for efficiency on those quantities, even though items at both ends of the curve may not be produced with complete effectiveness.

Whereas a "deep" curve suggests dividing the products and the production areas into two different layouts and handling systems, and a shallow curve, one general layout,

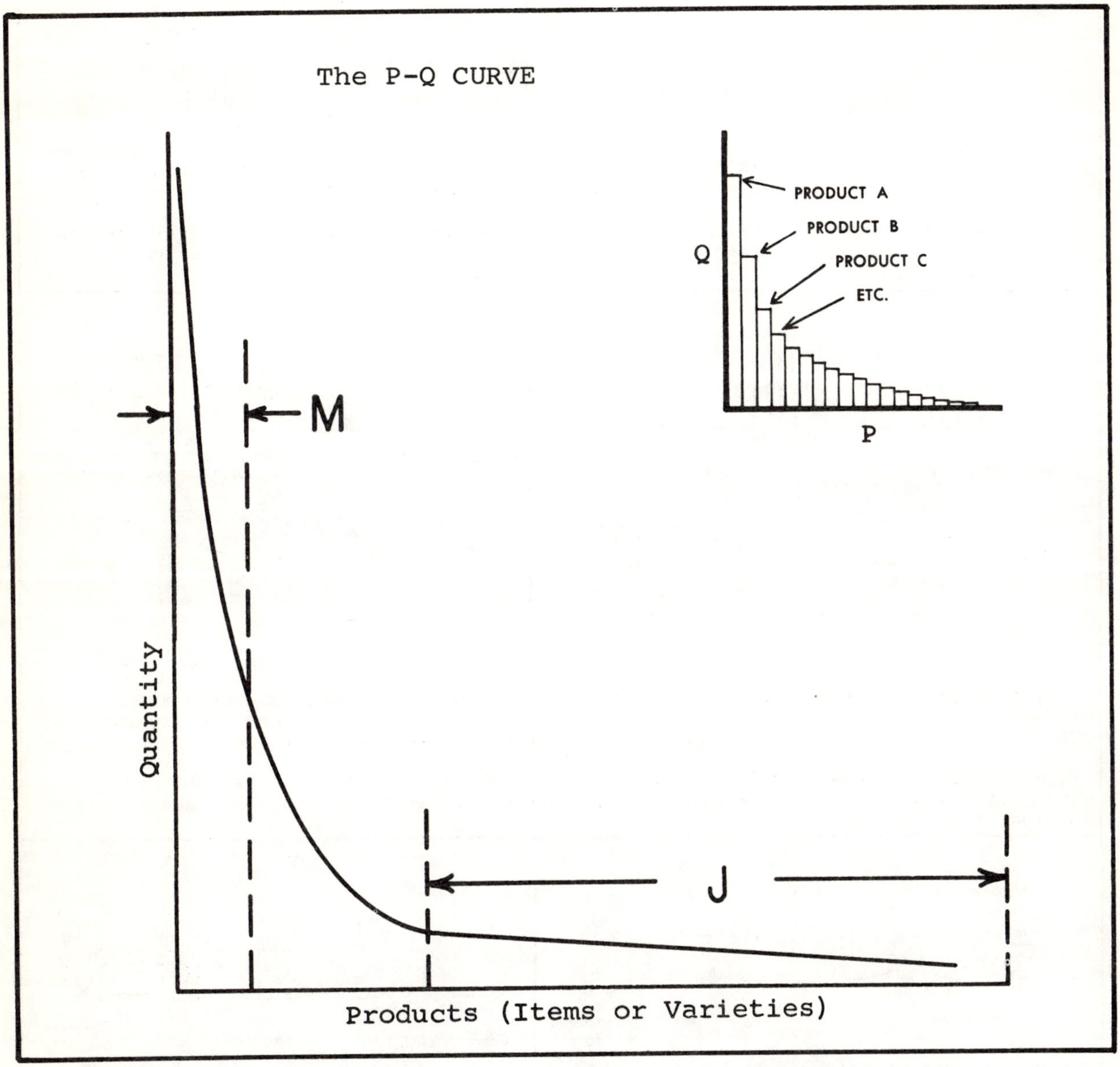


**Figure 3-1.** The Product-Quantity Curve (P-Q Chart) drawn on a noncumulative basis. Typically, this curve reveals the product varieties that are "fast-movers" and those that are "slow-movers". The items in area M frequently lend themselves to Mass Production techniques while those in area J must consistently be produced in a Jobbing or job-lot type of layout. Items falling in the area between the M and J areas and overlapping these areas will generally mean a combination or in-between type of layout such as a modified production line, lined-up process departments, or group production. Insert at top shows the usual method of construction of the curve, with each individual item or variety arrayed in order of decreasing quantity.

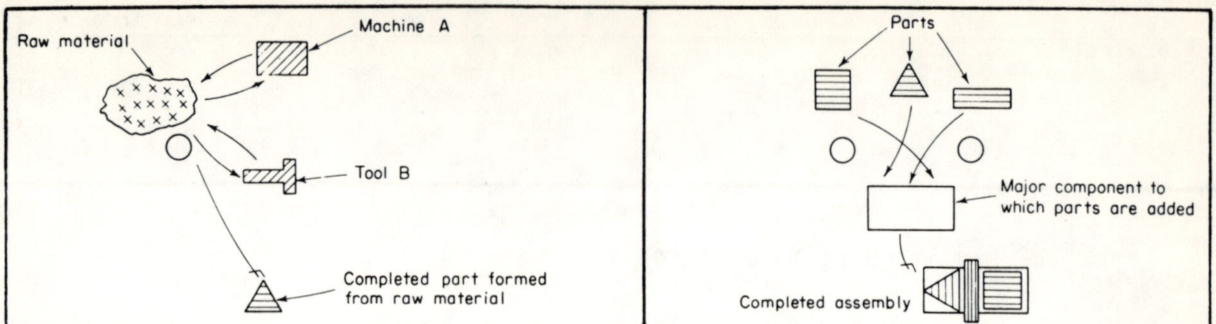


<u>Layout by fixed position</u>. All operations are performed with the material (in the case of forming or treating) or major component (in the case of assembly) remaining in one fixed location. That is, hold the material at a fixed position.

--- Examples ---

Forming and treating--speciality shoemaking; toolmaking; any artisan making a complete unit.

Assembling--hand-embroidery work; building a battleship or constructing a special machine.

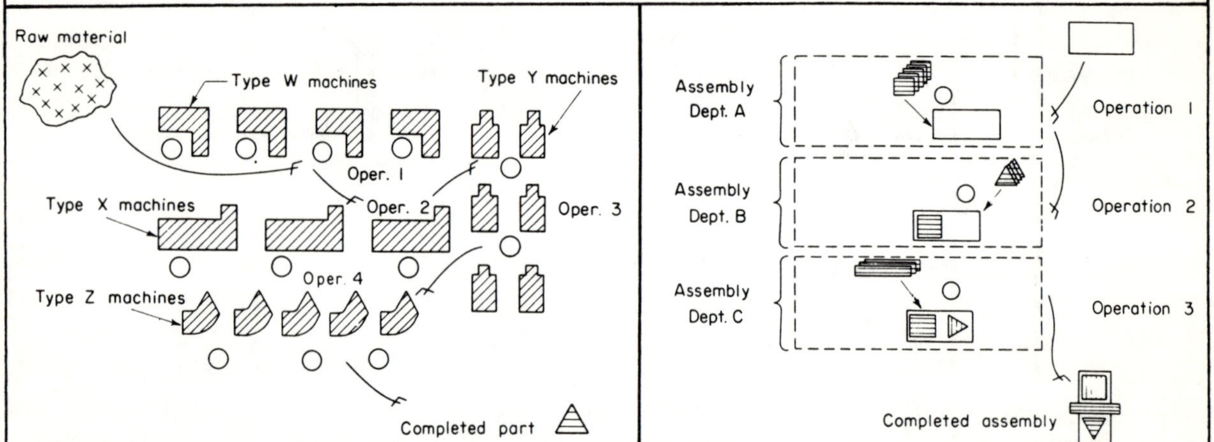


<u>Layout by process (function)</u>. All operations (processes) of the same type are performed in the same area; like machines or similar assembly operations are grouped together. That is, move the material through process departments or areas.

--- Examples ---

Forming and treating--normal machine-shop work; most textile and clothmaking; job printing.

Assembling--sheet-metal assembly by spot welding, riveting, stapling, and soldering.

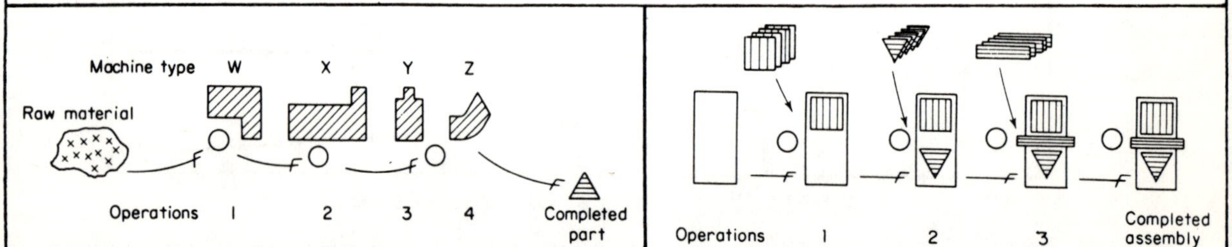


<u>Layout by product (line production)</u>. Machines or assembly work stations arranged in the sequence of operation, successive operations being performed immediately adjacent to each other. That is, move the material from one operation directly to the next.

--- Examples ---

Forming and treating--machining a motor block; quick car-washing line.

Assembling--automobile assembly line; assembling a tray of food in a cafeteria.

The left-hand sketches show the four <u>forming</u> or <u>treating</u> operations necessary to make part △

The right-hand sketches show the three <u>assembly</u> operations necessary to assemble parts △, ▤, and ▭ to major component ▭

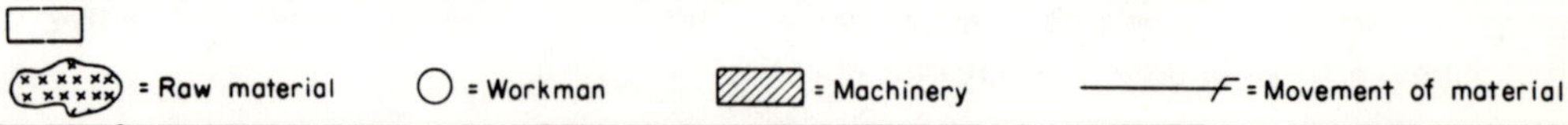


**Figure 3-2.** The classical types of plant layouts. From R. Muther *Practical Plant Layout* (New York: McGraw-Hill Company, 1955).

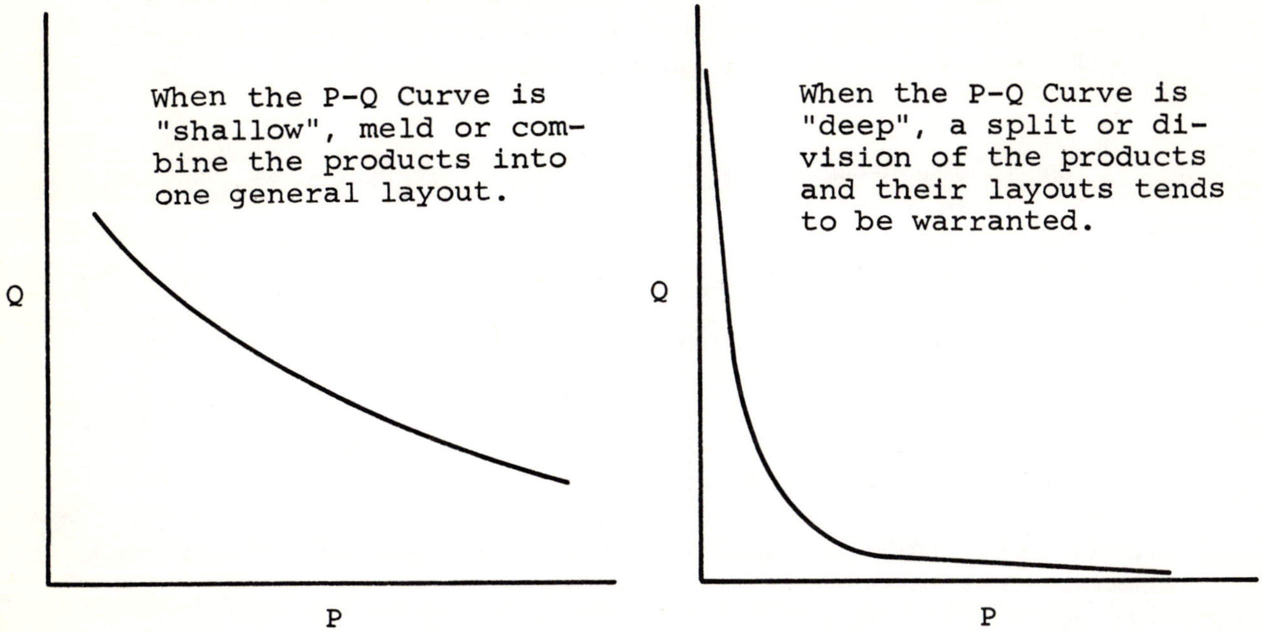


**Figure 3-3.** Significance of the P-Q Chart in layout planning is indicated here.

it is sometimes practical to make a three-division split of the items involved, as in the following example.

### Examples of P-Q Application

One firm making lighting fixtures found its P-Q curve led to considerable change in layout. The plant originally was planned with one assembly department consisting of individual workplaces, that is, a series of fixed-position assembly workplaces with each man assembling the entire fixture.

On analysis, the plant manager found a typical P-Q curve. This encouraged him to rearrange his assembly department accordingly. In one area he set up assembly lines; in another, teams of two to four men each; in a third area, individual assembly benches as before.

The division of his assembly area was approximately as follows:

| | | |
|---|---|---|
| 1. Production line | 19 items | 63% of volume |
| 2. Team assembly | 35 items | 28% of volume |
| 3. Individual bench | 100 items | 9% of volume |

The result of this rearrangement alone led to an increase in output of 17 percent, for no other changes were involved.

Still another example of improved layout based on P-Q analysis is shown in

Figure 3–4. Here is a case where the layout was divided into fast and slow-moving items. This layout allowed efficiency at both ends of the P-Q curve. And, as might be supposed, there were other savings in delivery service and inventory-carrying charges.

### Splits and Combines of Activities

Simply stated, an analysis of product and quantity frequently leads to splitting plant areas or departments into at least two logical areas:

(1) high-volume, low-variety "fast movers," and
(2) low-volume, high-variety "slow movers."

For the first, a high degree of mechanization, special-purpose equipment, and heavy investment in handling equipment are justified. Savings potential is greatest here. On the other hand, for the slow movers, changeovers are frequent, and savings on each piece do not accumulate too much. This condition calls for a high degree of hand work, general-purpose machinery, and low investment in handling equipment. Pricing potential is greatest here.

Within a grouping of more or less *similar* products, splits based on quantity are practical. In such cases, the planner is frequently better off to remove completely the jobbing work from the mass production areas. Besides differences in arrangement, these different types of layouts call for differences in equipment, in production control system, in skill of workers, in pre-production engineering, in supervisory outlook, and in many other factors.

We have emphasized P-Q analysis as a basis of dividing areas because it is so often overlooked. But the products involved in one layout project may be quite dissimilar. In such cases, there are other, more important, reasons for splitting or combining areas. We implied this when we said to divide or group products or items before making the P-Q analysis. Actually, the planner should not attempt to apply the P-Q chart when the products or processes are entirely different. If you must plan for the production of such diverse items as wooden sailboats and electronic components, you already know you would plan for separate plants because of the nature of the products themselves.

Actually, the decision to split or combine plant facilities can be based on one or a combination of several factors:

1. Size, weight, shape, or physical characteristics of the items
2. Basic material of the items
3. Processes, routing, or sequence of operations
4. Equipment involved or the type of building structure to house the equipment
5. Quality or workmanship required
6. Value or risk-of-loss of the item
7. Hazard or danger to personnel or property
8. Type of power, utilities, or auxiliary services

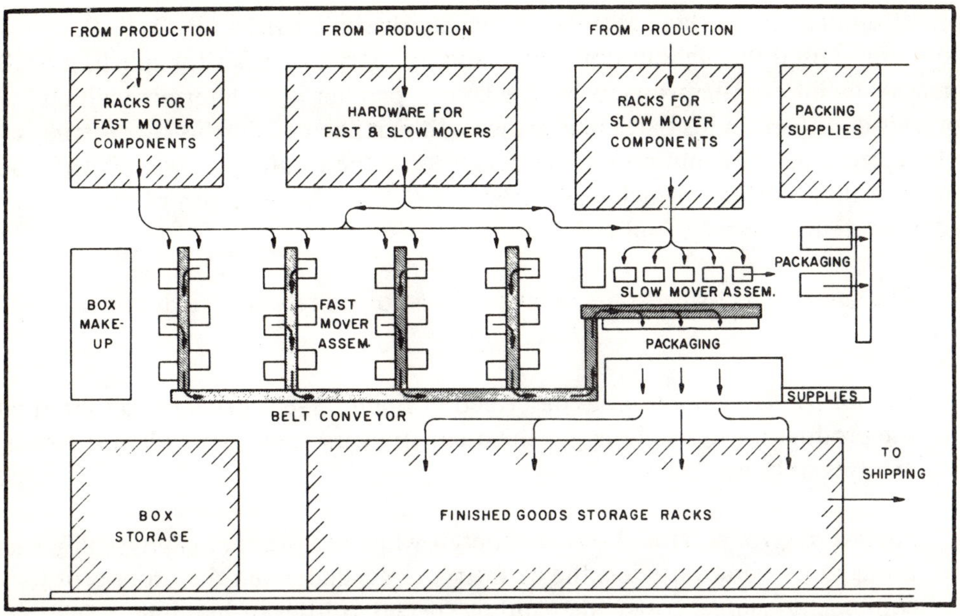


**Figure 3-4.** How P-Q analysis leads to layout effectiveness. The company involved here produces electrical connectors, primarily for the power utilities. Fabricating involves presses, screw machines, milling machines, threaders and tappers, and the like. Subsequently, all items move to storage before assembly. From here they are drawn out in generally the right proportions to make an assembly batch.

Before analysis, the pieces were assembled on individual benches by girls, packaged and/or master packed, stored as finished goods, order filled, and shipped. A P-Q analysis showed that 13% of the total catalog items accounted for 88% of the volume in finished pieces (80% of the sales dollars).

The assembly time is relatively small for all items, so any customer's order can usually be assembled and shipped within one day's time, if the order is not too large and if the parts are ready. However, many parts are used in several different end-products.

Based on these facts, the assembly and finished goods storage was laid out accordingly. The "fast-movers" were to be handled on one basis and the "slow-movers" split off and handled on a different basis. This was done as follows:

> *Fast-movers*—made to schedule by girls working on incentive; taken away by conveyorized methods to packaging; stored in finished-goods inventory on pallet racks waiting for order filling.
>
> *Slow-movers*—made to order by girls with more diverse skill and knowledge not working on incentive; delivered to its own packaging area by hand or push cart; placed on shelves by customer order for immediate order filling.

9. Organizational structure of the company
10. External considerations (property resale, appearance, taxes, and the like)

*Dissimilarities* in these factors—especially dissimilarities in P, Q, R, S, and T—are commonly used throughout commerce and industry as a basis for splitting or dividing areas or facilities. The problem for the layout planner is to determine which factor(s) is dominant in his particular project. In part, he can do this by plotting say "product size" against P and noting the deepness of the resulting curve. If other factors are equal or unimportant—which they seldom are—the relative depth of the curve would indicate the degree of split.

The so-called classical splits in layout most often rely on the dominance of P, Q, or R.

> *Layout by fixed position*, as described in Figure 3–2 is generally adopted when the product is physically large; the quantity, very small; and when the process involves only hand tools.
>
> *Layout by process (function)* is adopted when the product is physically smaller though of several varieties; the quantity, large; and when the process or its equipment is large, costly, demanding special utilities or building.
>
> *Layout by product (line production)* is adopted when the product is special in some way; the quantity, very high; and the process, relatively simple.

Splits determined by several factors in the same project are common. For example, a pharmaceutical plant may divide its production packaging and sample packaging because of the difference in end product (package) and shipping sequence. The same company may divide its sample packing from its packing of salesmen's samples or specially ordered samples because of difference in quantity. Typically, therefore, the most effective analysis lies in dividing areas by dominant factors and then applying P-Q curve. Secondary factors may be similarly identified to reinforce or serve as a further basis for segregation or for subsequent split into sub-areas.

The influence of the product characteristics on layout reemphasizes the back-pressure which layout can put on product design. It is well known that the effectiveness with which diverse products can be designed to be produced with the same materials, equipment, operation sequence, and the like will in large measure control the degree to which line production and automation can be adopted.

### Identification of Activity-Areas

By the term *activity*, or *activity-area*, we mean the various areas or things to be included in the layout. An activity may be a building or the main entrance for the layout of a new site; it may be a department within general offices; it may be a specific testing machine within a small laboratory layout—all depending on the phase or level of planning of the particular project. Activity, thus, is a generally usable term for any layout planning project.

Many layout planners take for granted the division of space into activity-areas. It is convenient to plan a new layout using current department names. But a big opportunity for improvement may be overlooked if the layout planner fails to challenge existing area identifications. Indeed, the possibility of a major cost breakthrough may be lost if new splits or combines of activity-areas are not considered.

Regardless of the depth and extent of activity-area division, or whether the project is for new, renovated, or existing space, the layout planner must specifically identify the activity-areas of his project. They not only define the scope of the total area, which is needed in Phase I, but from this point through all of the SLP pattern of procedures, all calculations and planning are done for the specific activities. For example, to consider Sales Office, when planning the relationships, as including the material-purchase requisitioning function, but to consider that function in Production Planning, when determining space requirements, can only lead to errors in the layout (or their costly correction during its adjustment into alternative plans). Similarly, to consider Receiving and Shipping as one combined activity automatically builds into the layout a U-shaped flow of materials, which may not be intended or at all desirable.

Therefore, clearly and consistently identifying all activities early in the project saves time later on in the planning. Preparing the list of activities is not complex in itself, but it is important, for it represents the result and culmination of the planner's analysis of his input data. Figure 3–5 shows part of a listing for the layout of a new steel mill.

### Projections into Future

Obviously, in planning a layout for the future, the planner must base his work on projected input figures. This means that input data, to which the layout will be designed, must be projected for conditions which will exist beyond—or at least at the time—the layout is installed.

The extent of the projections varies with the nature of the project. On a total, new facility involving several buildings, it is entirely practical to attempt projections extending over 25 or even 50 years, although precise forecasts can seldom be made so far into the future.

When projecting many years ahead, the planner usually can use only policy statements regarding the company's basic direction. That is, you may have to substitute for specific figures an assumption such as:

> The company is dedicated to a steady growth—in the neighborhood of 15 percent per year in dollar sales. Its products will be general industrial chemicals produced through process equipment involving mixing, blending, refining, and the like. Filling or packaging oeprations will continue to be a highly significant

LIST OF ACTIVITIES

North Mill Project 71046
R. T. Jones 5-11

Ident. & Name / Definition

1. Main Entrance(materials)

   Entrance for over-the-road vehicles, including gate house and weigh scale.

2. Entrance for Personnel

   Entrance and exit gate for employees; whether by car or public transportation. (Note: this may be split or divided on actual site, and/or may be combined with Office and Visitor Entrance).

3. Office and Visitor Entrance

   Entrance and exit point for main office and visitors to office including all salesmen and prospective employees. (Service visitors will be given special permits to enter at Main Entrance.)

4. Scrap Yard

   Storage area for all incoming scrap metal and including purchased pig or pellets, croppings, and ends, but excluding purchased billets and in-process metal.

5. Melting

   Furnace area for metal making, including oxygen generating plant.

**Figure 3-5.** Partial listing of activities for a new steel mill.

aspect of the labor content and the area required. Engineering and test laboratories will undoubtedly be the backbone of product development and process control. Marketing and distribution is anticipated to take quite a different form in 20 or 30 years' time. Fresh air and an abundant supply of cooling water are anticipated as being major considerations for the plant during the next 50 years.

Most often, however, the planner is not concerned with such long-range projects. Rather, he is concerned with what will happen in the next two, five, or ten years. And on short-range rearrangements, perhaps a three-to-six months' projection is all that is necessary.

In any event, the layout planner must be able to set forth some statement of what is anticipated at some logical future time, in terms of product designs, marketable quantities, processes and their machinery, plant and personnel service, and timing considerations. This is where most layout projects begin. And this is one reason that the "Inputs" box is set forth as "preliminary" in the SLP pattern—preliminary to actual investigation of relationships and space.

In developing plans for the future, it is frequently practical to see what has been done in the past. The history of the total company or of specific items can be plotted in terms of quantities over past periods. A trend thus derived can help the planner project into the future. Figure 3-6 shows such a plot. (Sound market research is, to be sure, a far better basis for future projections.)

Whether or not the projection is broken down into product groups or remains one total figure, it is best to have both dollars and numerical figures (quantities). The table in Figure 3-7 breaks down the data which give the most aid to layout planners. That is, it contains a quantity break down by product type and gives further break downs, projecting these for several periods in the future.

### Master Plans

It is usually helpful to develop an ultimate layout, or *master plan,* for the future and to work backwards into various increments of expansion, even though the time schedule may not be specific. There are various reasons for this: logical and systematic expansion to save the cost of utilities; utilization of real estate, lining up columns, building facings, and doorways to reduce maintenance and material handling costs and to improve the appearance of the facility; permitting of piecemeal expansion for growth within an overall plan so that capital investments are committed on a more conservative basis. A logical master plan into which expansions can be fitted when and if they come along affords many future savings—not the least of which is the engineering manhours devoted to planning layouts and new facilities.

An overall master plan, such as that shown in Figure 3-8, is a basic ground plan into which subsequent growth can be fitted when it is called for.

### Product Changes

Changes in the inherent characteristics of products and materials must also be anticipated by the planner. Changes in product design can defeat the best layout planning. One layout, for example voted the finest in its industry, became obsolete within 10 years and was completely abandoned, primarily because of a change in size of the product. It is well worthwhile, therefore, to check past trends in nature or characteristics of the products or materials involved in a layout plan. An example of how this can be done simply, and in a meaningful way, is shown in Figure 3-9—though it represents but one way of helping anticipate changes in products or materials.

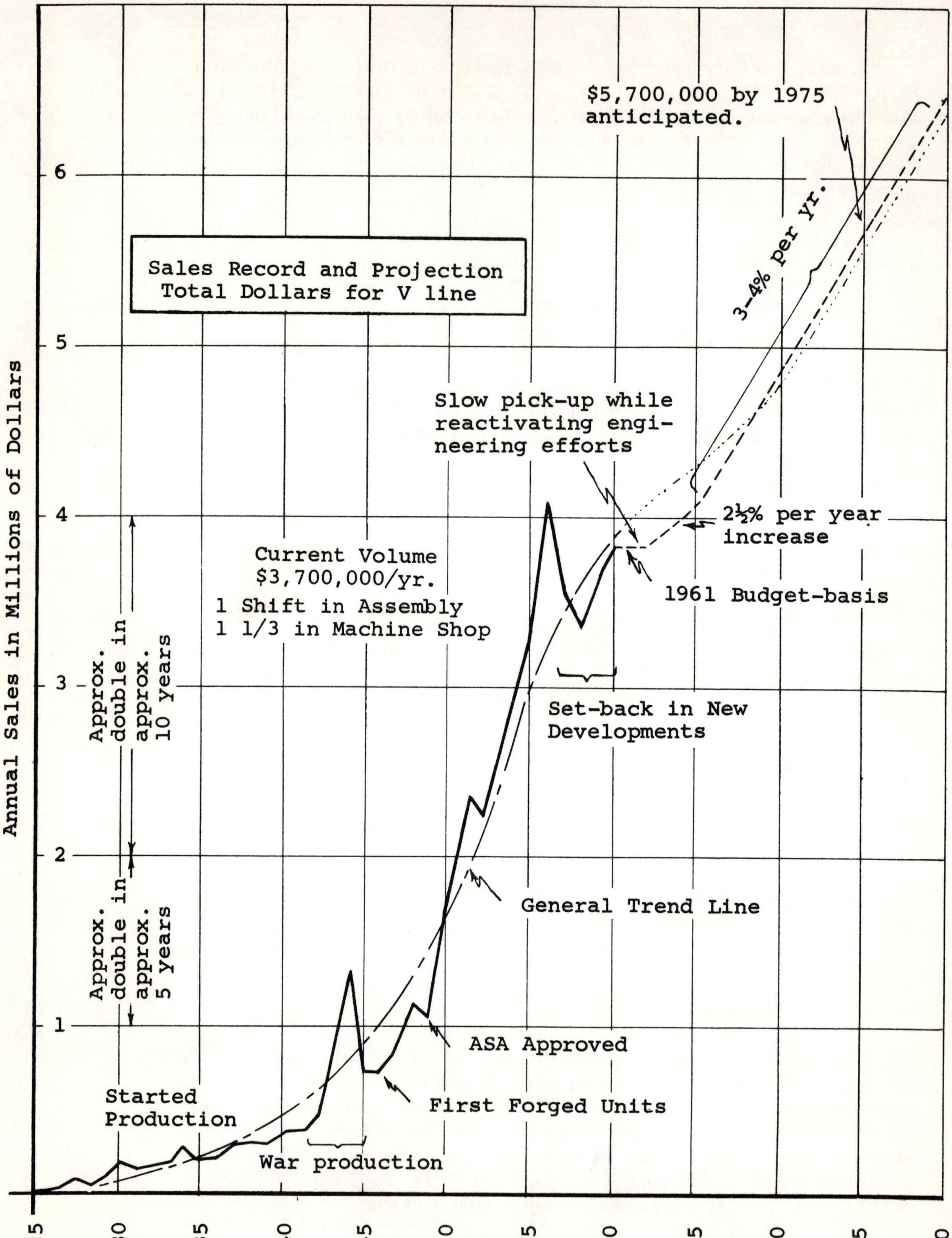

Sales Record and Projection
Total Dollars for V line
Annual Sales in Millions of Dollars
1
2
3
4
5
6
1925
1930
1935
1940
1945
1950
1955
1960
1965
1970
1975
1980
$5,700,000 by 1975
anticipated.
3–4% per yr.
Slow pick-up while
reactivating engi-
neering efforts
2½% per year
increase
1961 Budget-basis
Current Volume
$3,700,000/yr.
1 Shift in Assembly
1 1/3 in Machine Shop
Approx.
double in
approx.
10 years
Set-back in New
Developments
General Trend Line
Approx.
double in
approx.
5 years
ASA Approved
Started
Production
First Forged Units
War production

FIFTEEN YEAR/ESTIMATED PRODUCTION

TEN YEAR/ESTIMATED PRODUCTION

FIVE YEAR/ESTIMATED PRODUCTION

THIS YEAR/ESTIMATED PRODUCTION

SINGLE PHASE TRANSFORMERS — P = Power; all others Distribution

| KVA | PRODUCTION OIL - FILLED | | | | | SERVICE OIL & AIR | PRODUCT'N AIRCOOLED |
|---|---|---|---|---|---|---|---|
| | < 5 KV | 5-21 KV | 22-91 KV | 92-200 KV | >200 KV | ALL KV | ALL KV |
| 0 - 2 | | | | | | | |
| 3 - 10 | 230 | 16,600 | | | | 3300 | 1000 |
| 15 - 25 | 5750 | 39,100 | | | | 2200 | 3500 |
| 35 - 50 | 4375 | 32,250 | | | | 1800 | 600 |
| 75 - 100 | 575 | 5,680 | | | | 350 | 250 |
| 172 | 167 | 3,910 | | | | 135 | |
| 250 - 500 | | 150 | 400P | | | | |
| 600-2000 | | 200P | 325P | | | | |
| 2500-5000 | | 10P | 60P | 40P | | | |
| 6M-10M | | | 15P | 15 P | | | |
| 11M-20M | | | | | | | |
| 25M-50M | | | | | 12P | | |
| 50M & up | | | | | 6P | | |
| THREE PHASE TRANSFORMERS | | | | | | | |
| 6 - 30 | 300 | 300 | | | | | 750 |
| 45 - 112 | 1120 | 1120 | | | | | 1000 |

**Figure 3-7.** Projection of future quantities as a basis for planning a new plant for electrical transformers. Future estimates are set forth for twenty years and the product and quantity data are broken down by phases, by type (P or D), by oil or air, by production or service requirements, by capacity (KVA), and by KV range.

**Figure 3-6.** Graphical record of past sales with projection into the future. This type of picture is very helpful in establishing the plan-for quantities—especially when a new site or major commitment on "location" is involved. Although the long historical and long future figures are needed for master planning or long-range layout, projections for a much shorter period would generally be used for smaller relayout projects. This kind of forecasting, when done for each product line and integrated into total company objectives, can serve as an aid—along with other more sophisticated techniques—to developing long-range requirements for space, equipment, and capital expenditures. This illustration shows such interesting features it has been retained in this edition by your author, even though it is not a current example.

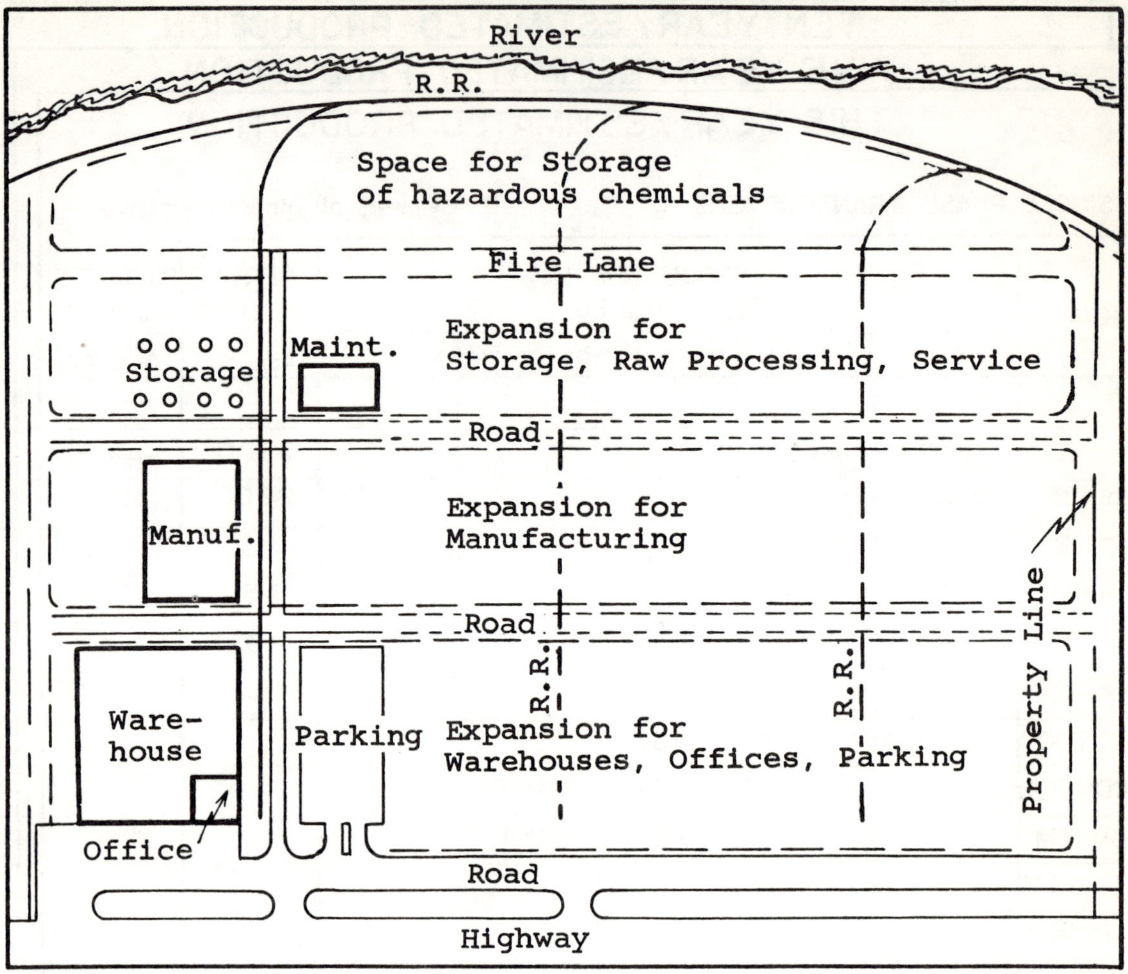


**Figure 3-8.** Master plan for long-range expansion of facilities. The current, "going in" arrangement of facilities—storage tanks, maintenance, manufacturing, warehouse, and office—are shown as solid lines. Areas reserved for future expansion are shown as dotted lines and are set aside on a planned basis by type of activity.

In any event, the product-planning or design-engineering group (speaking for P) and the sales or distribution group (speaking for Q) should provide figures projected over a logical period, approved by top management, broken down sufficiently to serve the layout planner as a specification to which he can plan his facility.

### The Danger of Getting Too Specific

One warning should be made—the tendency to get too specific. Layout planners must have data to plan their layouts; this has been emphasized. However, even though a company's management or product-planning group has set such figures down in writing, they cannot be considered as either unchangeable or precise. In fact, if any projection figures come out exactly as set down, it is a coincidence—except perhaps where contractual customers are involved.

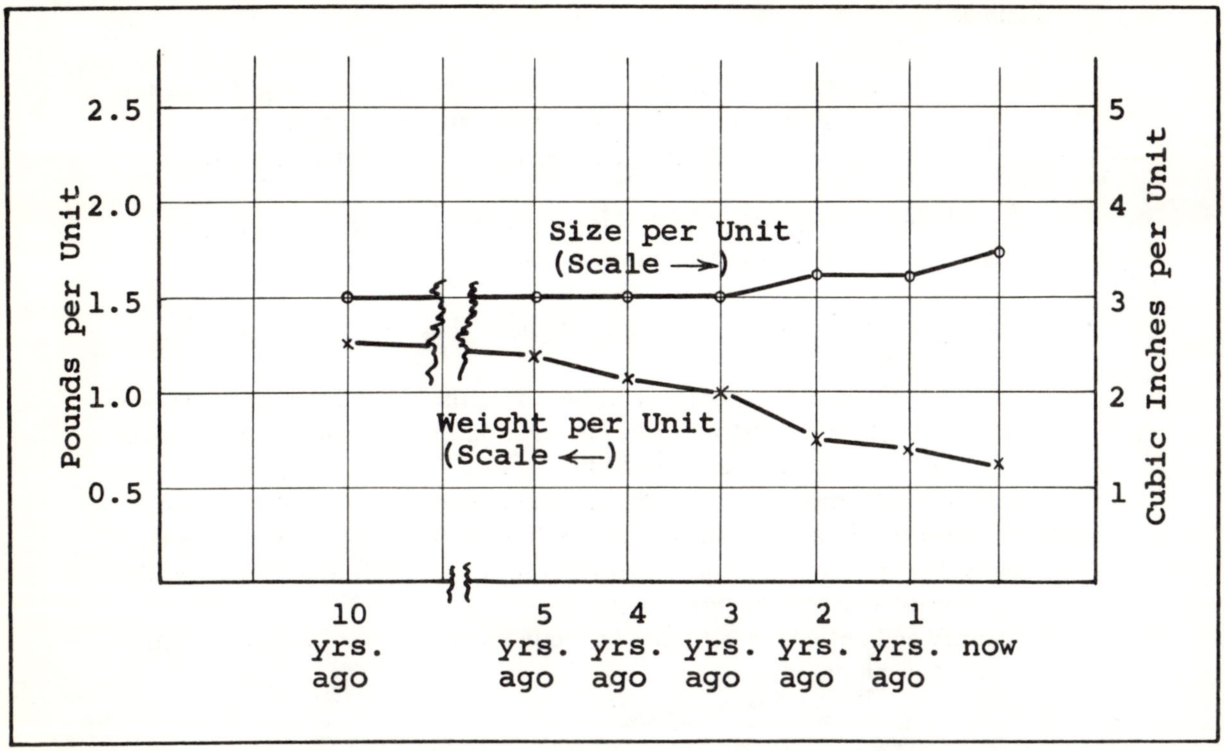


**Figure 3-9.** Trends in products' size and weight. Trends in the nature or the characteristics of the products being produced can seriously affect a layout—especially when sudden or major changes occur. Here a plot of the trends in size and weight of this company's product shows little appreciable increase in the size but a rather significant decrease in the weight per unit. Will these trends stabilize, continue, or change abruptly? And what effect will such changes have on storage facilities, handling container, and clearance or set-down area around machines? These questions should be answered—to the best of the company's ability—before the layout project moves ahead very far.

Trends in process machinery or equipment and in the ratio of production to non-production areas and man-power similarly should be appraised. Trends in product diversification or simplification will normally show up in the P-Q analysis discussed in earlier pages of this chapter.

Usually, the set of plan-for input data must include some kind of cushion, or safety factor, or it must be recognized that overtime or a second or third shift will be used to pick up peak conditions. This should be understood by everyone before specific plan-for figures are agreed upon.

Similarly, once the figures are agreed upon, the layout planner must recognize that these figures are, at best, estimates and that some condition may occur which will change completely the nature of his design criteria—his projected input data. Changes to meet competitor's product design, to take advantage of a new process, to conform to new sanitation or labor-union provisions are bound to occur. The rapidity of such changes is becoming more important each year. If the layout planner relies too literally on the plan-for figures, his layout will be too inflexible or too costly to adapt to the inevitable minor changes.

### Summary

Based on the above discussion, we conclude that the layout planner must take a step preliminary to his planning of relationships, space, and adjustment. That preliminary step involves gathering and analyzing data to which he will plan his layout.

This preliminary data-gathering-and-analysis step is termed Input Data & Activities, and it generally includes the following sequence:

1. Identify specific elements of input data needed as design criteria for the project at hand.
2. Project these data into the future. (Note that this will undoubtedly involve the layout planner's restructuring information that is available from or supplied by others in the organization.)
3. Seek general approval and top-management endorsement of the input data.
4. Examine the data for distinctive dissimilarities, using P–Q chart and variations thereof to arrive at basic-typed layout and/or definitive bases for dividing (splitting or combining) activity-areas.
5. Identify and define the activities (or activity-areas) to be used in the subsequent planning.

# Chapter 4
# Flow of Materials

The third letter of our Key to unlocking layout planning problems (see Figure 1-1) is R (Routing). Routing means *how* an item is made—its process. The process is established essentially by selecting the operations and sequences that will best produce P and Q wanted in the optimum operating T—although many other considerations may be involved in the determination.

The routing yields the basic data for analyzing the flow of materials. But before utilizing the routing handed him, the planner should recall the meaningful little word *why*, the business end of our key. The routing should be examined and proved reasonably right; it should be restudied when the planner feels it can be improved.

The standard work-simplification check originally developed by Allan H. Mogensen—and discussed in all industrial engineering texts or handbooks—is especially applicable. Mogensen's check challenges each step in the process routing with these words:

1. *Eliminate*—Is the operation necessary, or can it be eliminated?
2. *Combine*—Can it be combined with some other operation or action?
3. *Change sequence, place, or person*—Can these be changed or rearranged?
4. *Improve details*—Can the method of performing the operation or action or its equipment be improved?

Once satisfied with the process routing, the planner can begin flow-of-materials analysis. This is Box 1 of the SLP pattern.

### Flow of Materials—Heart of Many Layouts

The analysis of materials flow involves determining the most effective *sequence(s)* of moving materials through the necessary steps of the process(es) involved and the *intensity* or magnitude of these moves. An effective flow means that materials move progressively through the process, always advancing toward completion and without excessive detours or back-tracking (counterflow).

Flow-of-materials analysis is the heart of layout planning wherever movement of materials is a major portion of the process. This is especially true when materials are large, heavy, or many in quantity or when transport or handling costs are high compared with costs of operation, storage, or inspection. In extreme cases of this kind, the desired flow is developed and then diagrammed directly. The space requirements are hung on the flow diagram. Little investigation of supporting services is made, and no activity relationship chart (Box 2 of the SLP pattern) is constructed. The services and other-than-flow relationships are simply picked up as part of the Modifying Considerations (Box 7).

Analyzing materials flow, therefore, is one of the primary steps every layout planner should understand and know how to do.

### Determining Method of Flow Analysis

There are several different methods of analyzing flow of materials. Part of the problem of course is knowing which method to use for a given project. The P-Q chart can be used as a guide, for the method of flow analysis varies with the volume and variety (quantity and number of types) of the items being produced. See Figure 4–1.

1. For one or a few standardized products or items, use *operation process chart* or some similar flow chart.
2. For several products or items, use *multi–product process chart,* if assembly and disassembly are not involved.
3. For many products or items,
   a. *Combine* them into logical groups and analyze as 1 or 2 above; or
   b. *select* or sample products or items and apply 1 or 2 above.
4. For very many diversified products or items, use the *from-to chart.*

Each of these flow-analysis techniques are discussed further in this chapter. The chief point here is that different methods of flow analysis should be used for different product volume and variety conditions and that the curve can show which type of analysis should be made.

### The Operation Process Chart

If the planner can picture the materials flow—if he can *see* it—he can plan its layout. That is why the visual aspects of analysis are continually emphasized in this

**Figure 4–1.** Guide to analysis of flow. Besides being a direct aid in planning a layout, the P-Q Chart is a guide to the type of flow analysis to use.

A plant having only a few high-volume items (A) will analyze its flow by the Operation Process Chart. When several such charts are necessary for a given project, it becomes difficult to integrate these charts. Therefore, for several high-volume items (B), a Multi-Product Process Chart is a better technique for flow analysis.

When many items are involved (C), we follow still another course of analysis—selecting or grouping. Either we group the items—usually by like products or like equipment characteristics—or we select representative, sample, or "worst-condition" items, and then apply one or the other of the two techniques above. Finally, if the project involves a great many diversified items of relatively small volume each (D), we use the From-To Chart—or Cross Chart.

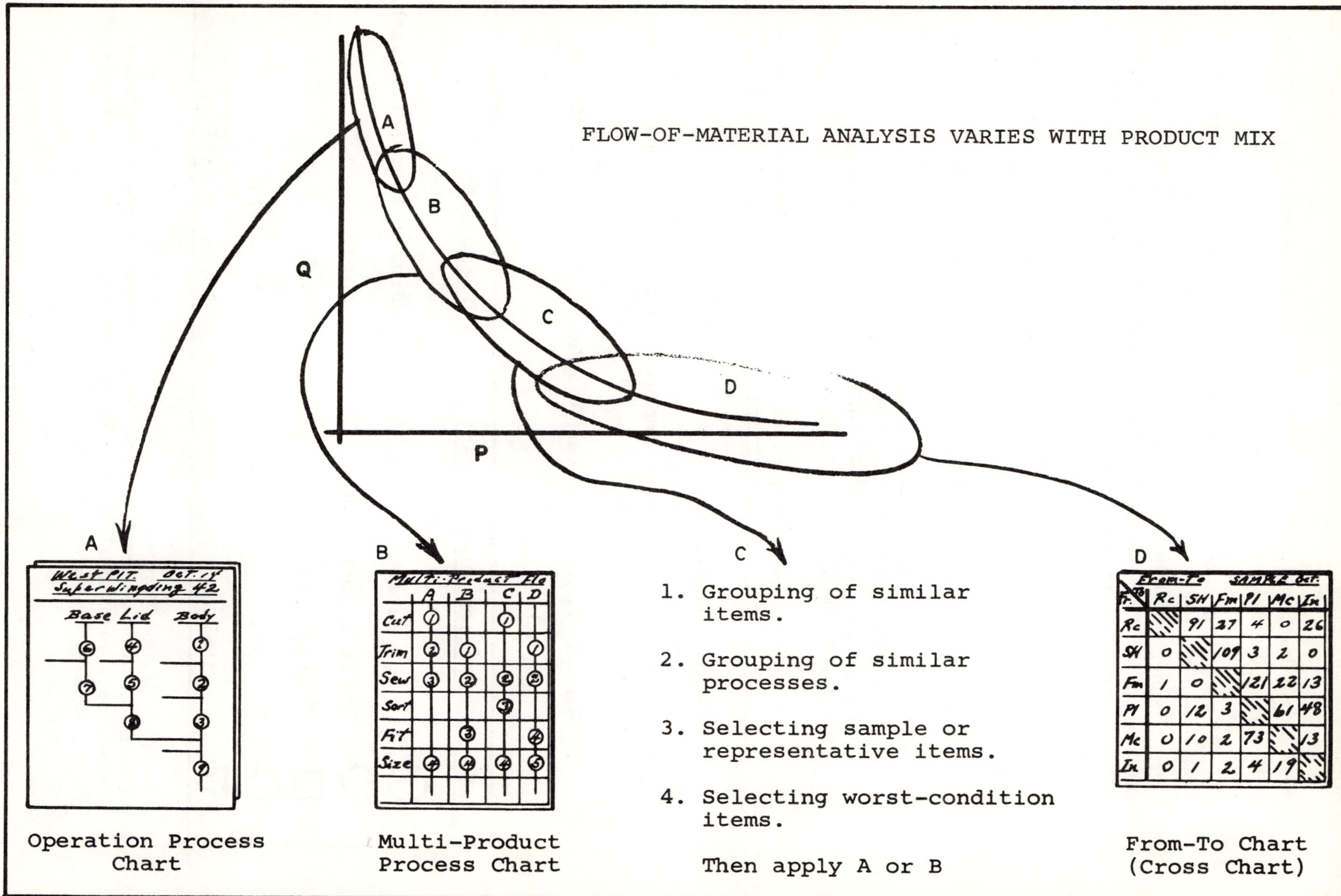
FLOW-OF-MATERIAL ANALYSIS VARIES WITH PRODUCT MIX
Q
P
A
B
C
D
1. Grouping of similar items.
2. Grouping of similar processes.
3. Selecting sample or representative items.
4. Selecting worst-condition items.
Then apply A or B
Operation Process Chart
Multi-Product Process Chart
From-To Chart (Cross Chart)

book. In fact, this necessity to "see the picture" is perhaps the underlying reason that SLP has developed into its present form.

To help the planner see, a system of sign language is used—equivalent to those used by the mathematician, the chemical engineer, or the procedures analyst. The sign language of process charting is well known to trained industrial engineers. It was originally developed by Frank and Lillian Gilbreth. Subsequently, two different committees sponsored by the American Society of Mechanical Engineers modified it to the form given in Figures 4–2 and 4–3.

Essentially, just five things can happen to any material as it moves through its process:

1. It can be formed, treated, or be assembled or disassembled with other items or materials.
2. It can be moved or transported.
3. It can be counted, tested, checked or inspected.
4. It can wait for some other action or for the rest of its batch.
5. It can be stored.

By using a symbol for each of these five actions and connecting the symbols with lines, the movement and process sequence of any product or material can be charted.

The conventions for making an operation-process chart are given in Figure 4–3. A sample chart is illustrated in Figure 4–4. For more details on techniques of process charting, see *Industrial Engineering Handbook*, edited by H. B. Maynard (New York: McGraw-Hill Book Company, Third Edition, 1971).

Some clarification is needed here: We are dealing in Part Two with Phase II

| Symbol | Action Classification | Predominant Result |
|---|---|---|
| ○ | Operation | Produces or Accomplishes |
| ⇨ | Transportation | Moves |
| □ | Inspection | Verifies |
| D | Delay | Interferes |
| ▽ | Storage | Keeps |

**Figure 4–2.** Symbols, action classification, and predominant result of the "sign language" used for process charting. Extracted from ASME Standard No. 101, Operation and Flow Process Charts, with the permission of the publisher, The American Society of Mechanical Engineers, United Engineering Center, 345 East 47th Street, New York, N.Y. 10017.

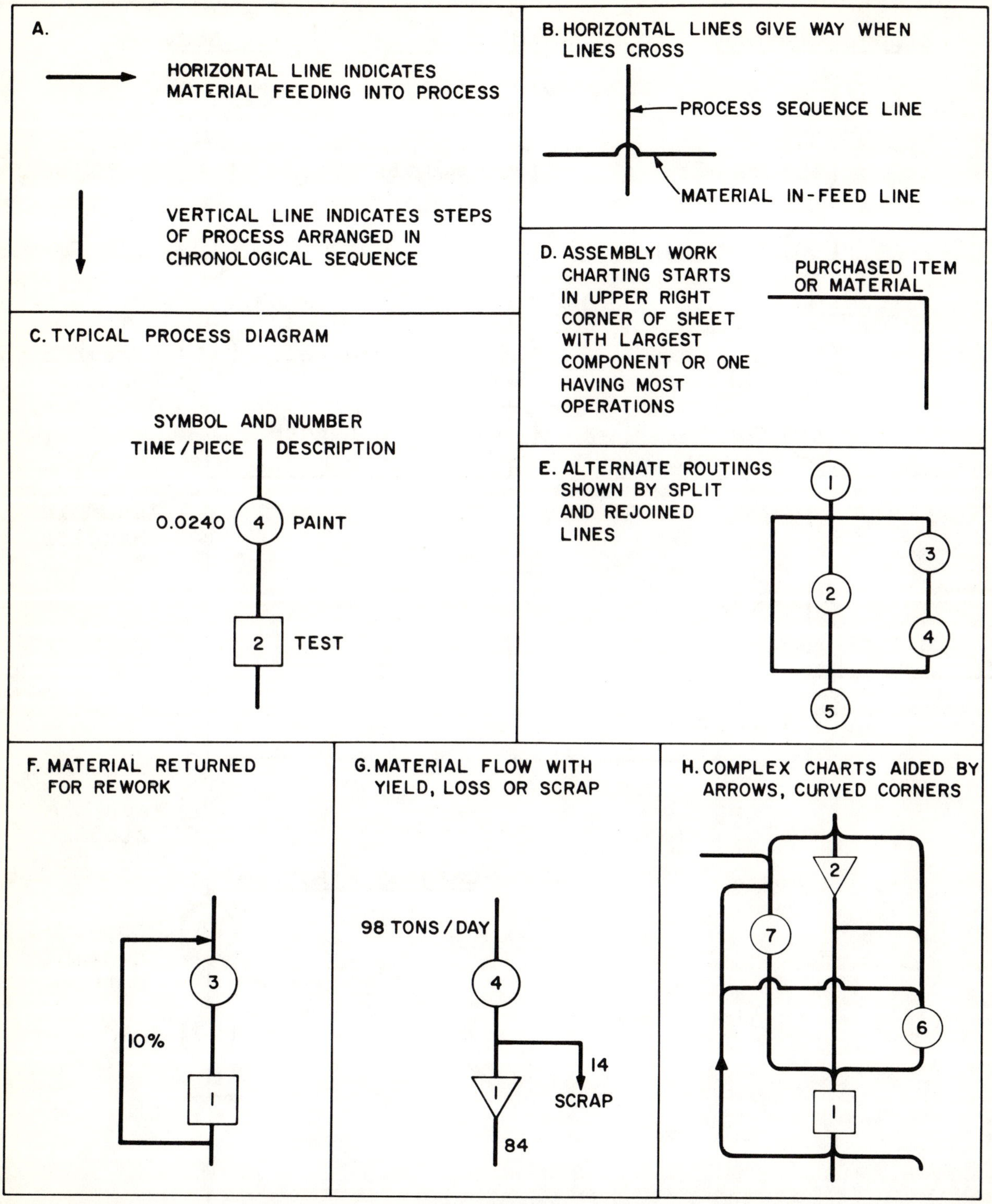


**Figure 4-3.** Conventions and modifications in making process charts. Extracted from ASME Standard No. 101, Operation and Flow Process Charts, with the permission of the publisher, The American Society of Mechanical Engineers, United Engineering Center, 345 East 47th Street, New York, N.Y. 10017.

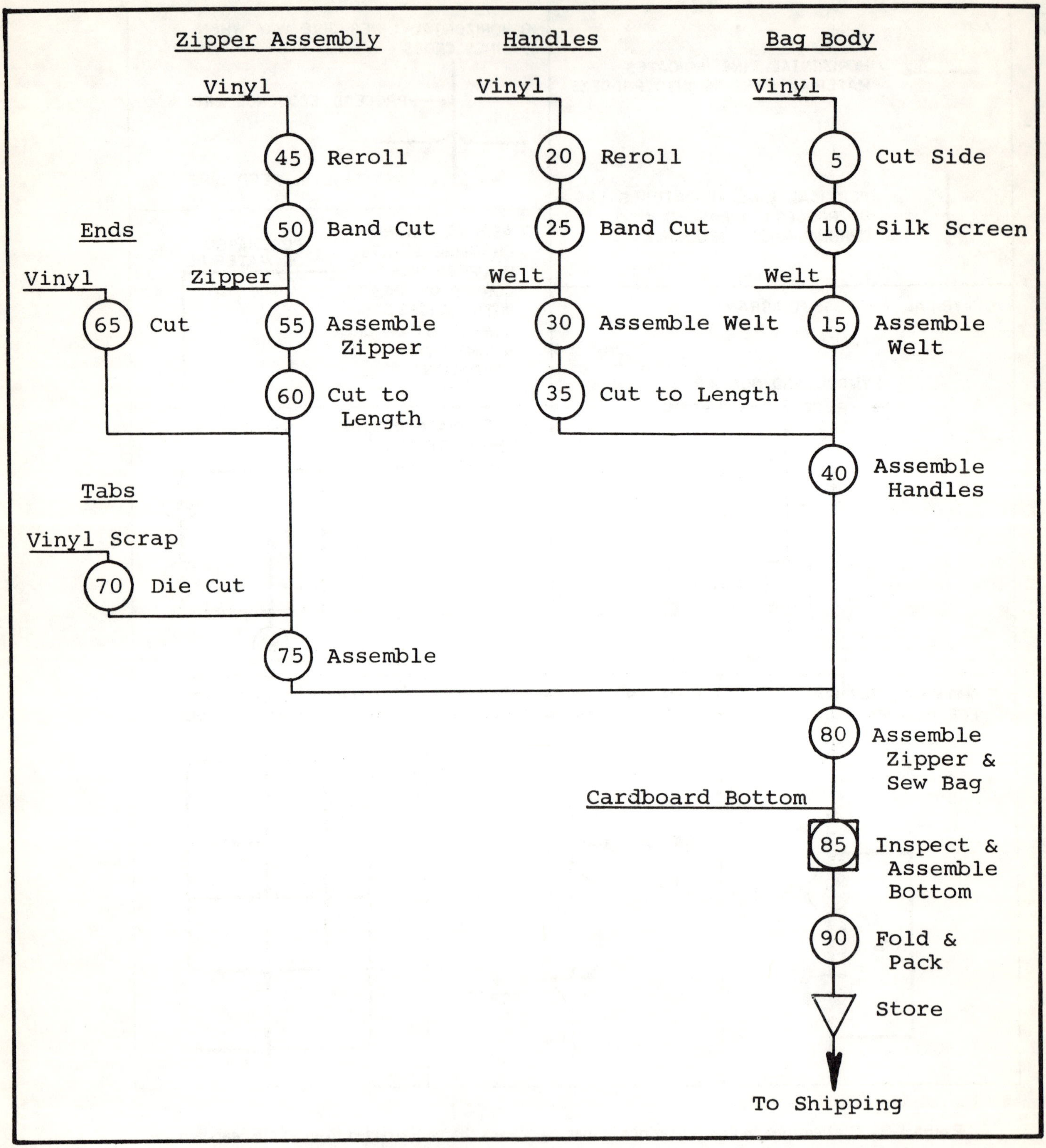


**Figure 4-4.** Operation Process Chart showing the chronological sequence of operations and how and where the components go together.

planning: General Overall Layout, and, therefore, considering only Phase II analysis techniques. Many of these same techniques are also used in Phase III, Detailed Layout Planning, but usually either they are changed to give more detail, or other techniques are used. For this reason, the operations only—or at least primarily—are emphasized in this chapter. We focus on the *operation* process chart. Later on, when integrating the materials handling methods or when planning the detail layout, the more detailed *flow* process chart is used. It charts every action in which the material becomes involved.

To assist in making the operation process chart, especially where complex assembly is involved, it is frequently helpful first to rough out how the materials go together. By taking a completed product and disassembling it, piece by piece, in this way learning how it should be assembled, one can draw a rough chart of the assembly sequence, and then prepare an operation-process chart for flow-of-materials analysis.

Various modifications of the conventions and charting practice can be made. Sometimes there isn't time to include all the information; other times a different form will be more meaningful. Most people agree on sticking to the standard conventions, but they modify or adjust the chart to get the clearest picture possible of the project at hand.

### Intensity of Flow

In this connection, we suggest considering two factors which are too often overlooked: (1) the intensity of materials flow, and (2) the outflow of waste, offal, trimmings, and the like.

Flow-of-materials analysis, by our earlier definition, includes both the sequence and the intensity or magnitude of materials movement. If the flow analysis is made in order to arrange operations or activities in the correct relationship to one another, the magnitude of movement (or intensity of flow) over the various routings or paths is the basic measure of relative importance of each route—and therefore of relative closeness of operations to one another. Figure 4–5 shows this intensity of flow as a number beside each flow line and as annual tonnages of end product at the right. This is a different form of process analysis, horizontal and without symbols. Also, it picks up the problem of yield losses, which is discussed in the following paragraph.

Outflow of waste can be a major part of a layout handling problem. In a sheet-metal shop, for example, not infrequently, trim amounts to twenty or thirty percent of the tonnage involved. This waste material is often awkward to handle. The materials are likely to be dirty, or sharp, unduly bulky, or dangerous, requiring quite a different handling method than that used for the infeeding of materials or components. To overlook this outflow in layout planning can be disastrous. Figure 4–6 shows the conventions used for including these features in the operation-process chart.

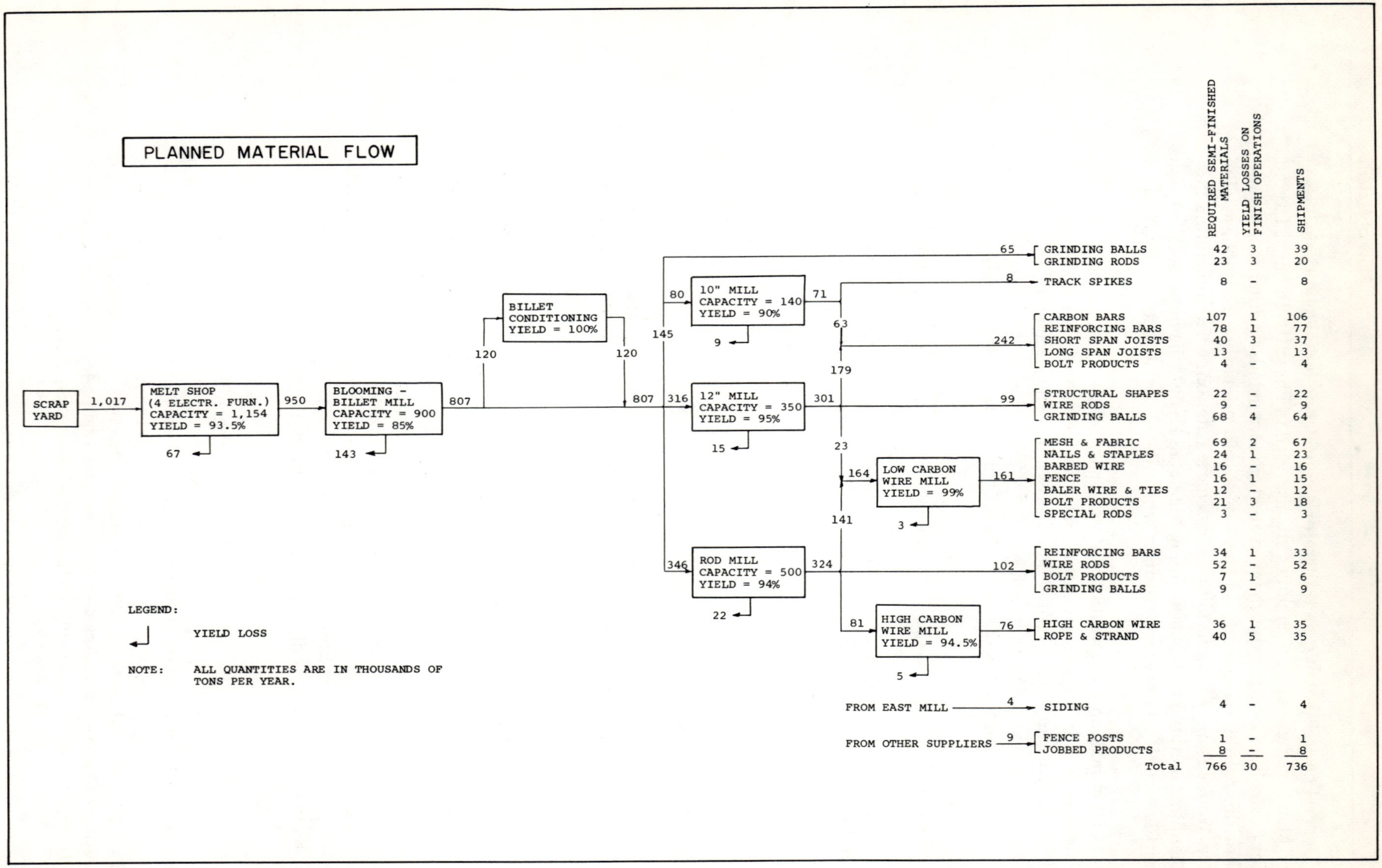


**Figure 4-5.** A form of process chart used in planning major expansion of a steel mill complex. Here each rectangle represents a major activity-area; numbers beside each flow line represent thousands of tons per year; and yield loss is shown by the back-flow arrow. This shows the break-down of a basic material into various derivative products, rather than the build-up into an assembled product as in Figure 4-4.

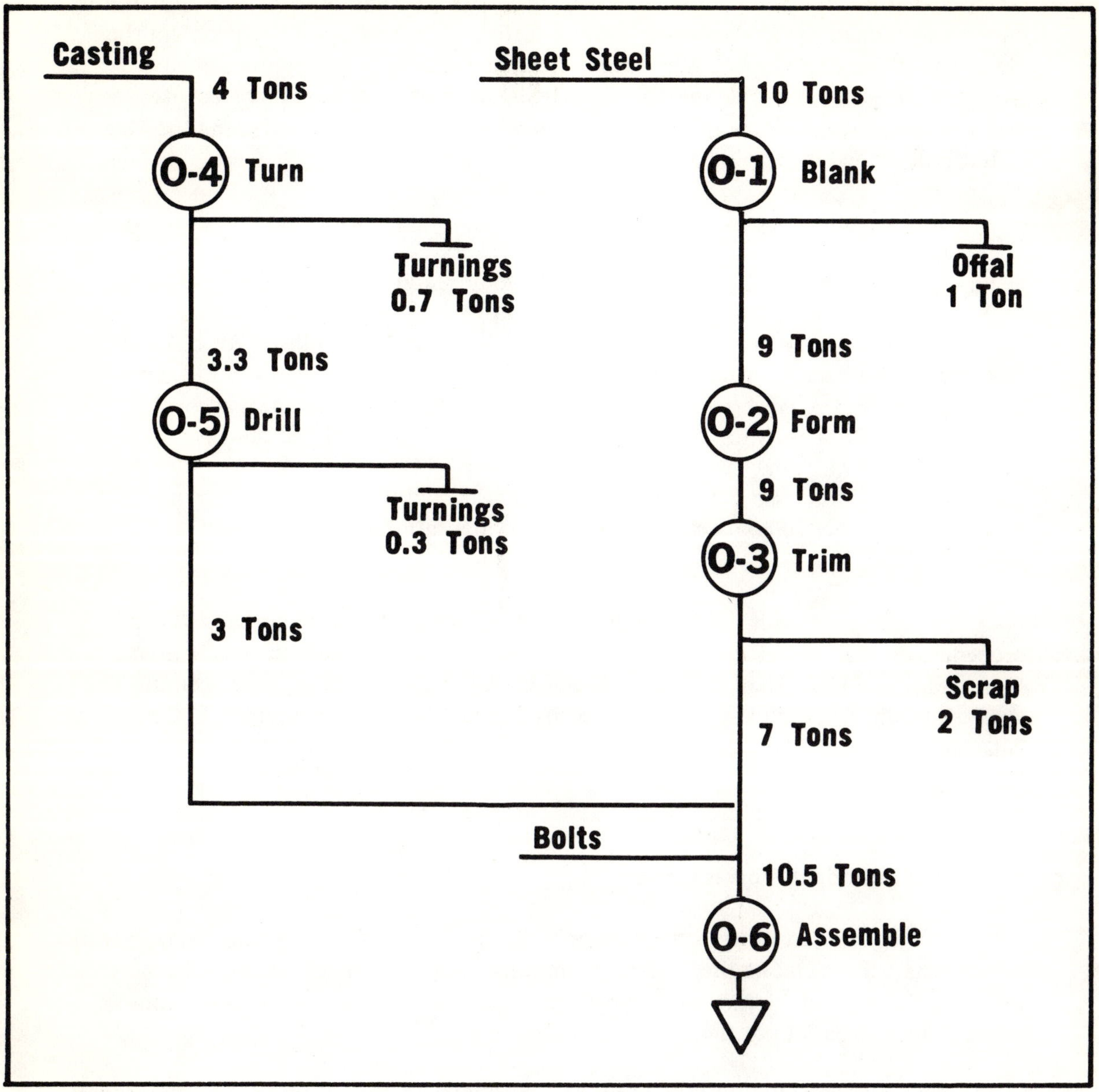


**Figure 4-6.** Example of Operation Process Chart showing intensity of material flow and the out-flow of chips and scrap.

### Measures of Intensity

Where materials are similar or more or less homogeneous, then units like pounds or tons, gallons or other cubic volume, or pallet-loads or tote-boxes are satisfactory measures of the magnitude or intensity of movement. The common calculation is the number of pieces moved per period times the unit of measure per piece. However, when it is necessary to convert from one unit of measure to another, when materials are quite diverse in nature and characteristics, or when there is no common container or handling unit, measurement of flow intensity is more difficult. To help solve this problem the "Mag Count" was developed. Mag Count, the transportability of any item in any condition, is quite a valuable measure, for the conditions of materials change at various stages in the process. (A complete description with tables of values for using the Mag Count is described in Appendix I.)

Mag Count may be a more sophisticated method of measuring flow intensity than is needed, but it—or a simplification of it—is very convenient on many layout planning projects, especially where there is no other readily usable unit of measure.

If the planner's project involves any possible change in methods for transporting materials, and especially if he expects to apply some organized analysis of his materials handling problem(s), he should try to classify his products and materials according to their physical characteristics, quantity, timing, or special control. He can then combine his flow-of-materials analysis for layout planning with his analysis-of-moves for transport planning. This classification of materials is discussed in detail in a companion book, *Systematic Handling Analysis (SHA)*, a condensation of which appears in Appendix XIII.

Remember that it can be costly to get too intrigued with the details of the procedure or analysis being made. Doing so often means that decisions are based on figures far more finite than the future life of the layout warrants or more accurate than the P and Q information itself. Being "precisely wrong" is all too common in the calculation of flow.

### Multi-Product Process Chart

When there are three or four items to be charted, it is best to make an operation-process chart for each one. But when the number of charts becomes many—say six to ten, depending on the nature of the products—it is better to use the multi-product process chart, especially if there is no assembly work.

The multi-product process chart brings all items together on one sheet of paper so they can be pictured easily. It lists, down the left side of a sheet, the *operations* through which the various items pass. Across the top, side by side and each in a separate column, are listed the various *products* or items involved. This pre-establishes each operation line. The routing of each item is then traced through the pre-identified operations. Figures 4–7 and 4–8 show examples of how such charts are made.

By charting these routings side by side, a ready comparison, item by item, of the paths of flow indicated for each can be made. The layout objective is to have a progressive flow with a minimum of back-tracking and to place closest together those operations which have the greatest intensity of flow between them. This means interchanging the horizontal operation lines of the chart until the optimum sequence is obtained.

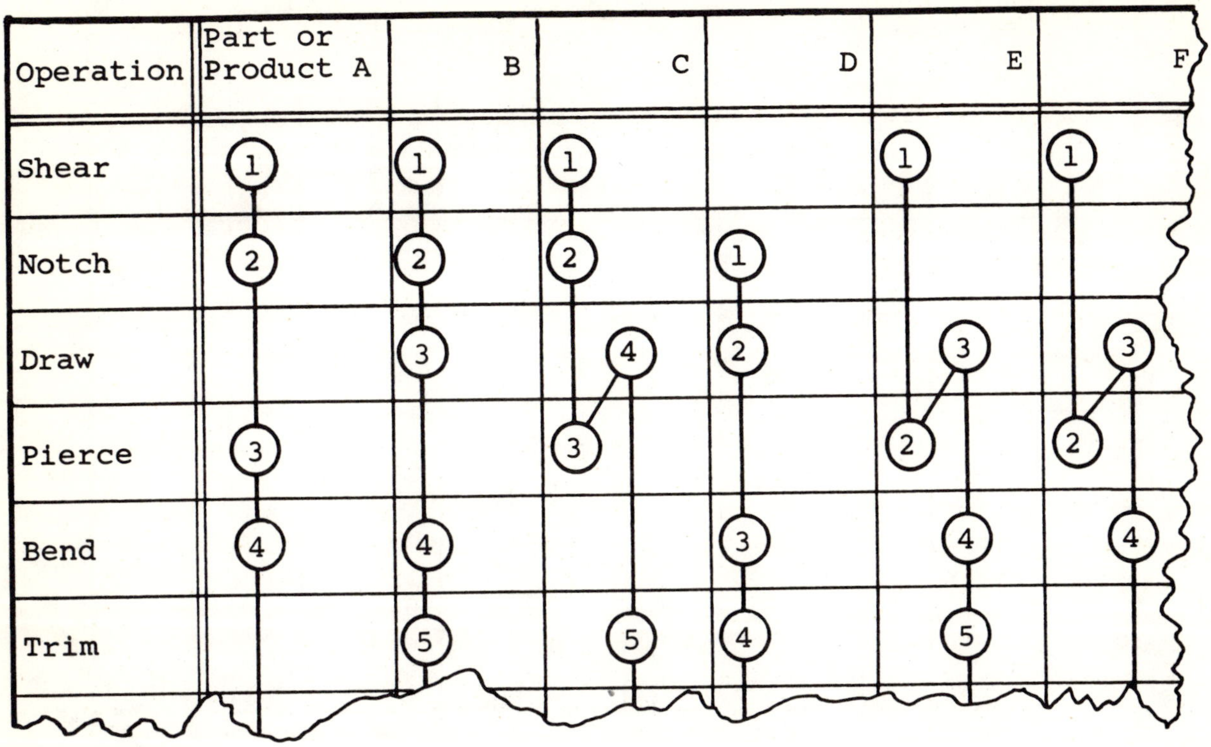


**Figure 4–7.** Multi-Product (Multiple Product or Part) Process Chart.

A simple method of developing the best sequence is merely to calculate the extent of back-tracking. A value of two for each incidence of back-tracking and a value of one for each operation by-passed in the counter-flow affords such a measure. Or if these values are considered negative, they can be combined with a value of plus two for each case where one operation feeds directly to the next and a value of only one if any operation is by-passed in its forward movement. These values are made more meaningful if a measure of the intensity of flow is incorporated into the calculations. More sophisticated mathematical techniques may also be applied. However, be sure not to overlook the benefits of a circular or U-shaped flow pattern when making calculations of this kind.

Note that the multi-product process chart does not show an assembly breakdown of components. Its use is therefore limited to major items and generally to routings listing only two or three lines on the chart for assembly and subassembly.

### Grouping or Selecting

When the number of items involved reaches somewhere around thirty to fifty, some form of grouping or selecting becomes practical.

By combining all or certain items which are *alike in design*, the planner may have a group with a common or reasonably distinct routing sequence. Items which are

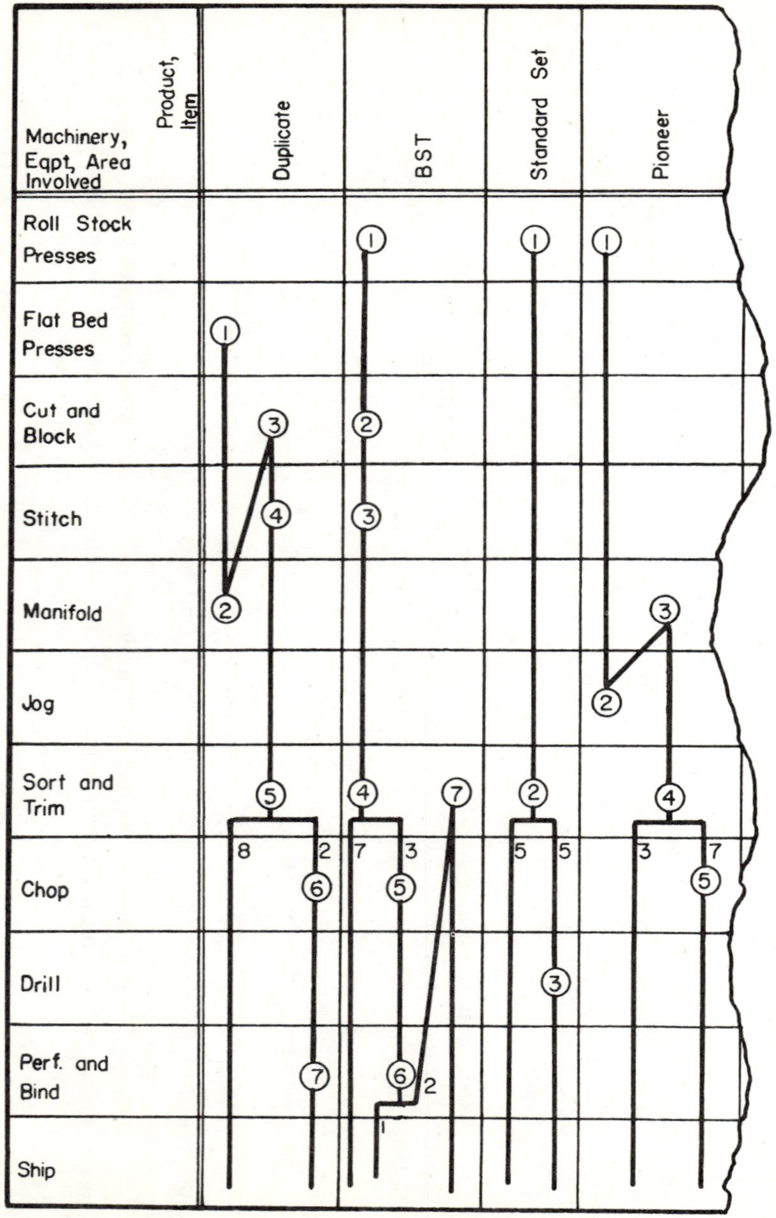


a

| Machinery, Eqpt, Area Involved / Product, Item | D-1 Q=19000 per yr. | D-2 Q=5700 per yr. | D-3 Q=4900 per yr. | D-4 Q=4700 per yr. | E-1 Q=11000 per yr. | G-1 Q=9100 per yr. | S-1 Q=3100 per yr. | Q |
|---|---|---|---|---|---|---|---|---|
| Steel Strip Storage | | | | 1 | | | 1 | |
| Raw (In) Storage | 1 | 1 | 1 | | 1 | | | |
| Extrus'n Storage | | | | | | 1 | | 1 |
| In Process Storage | 4 | | | 3 | | | 3 | |
| Straighten | 2 | 2 | | | | | | |
| Punch | | | | 2<br>4 | | 2 | 2 | |
| Drill | | 3 | 2 | | 2 | | | |
| Counter-sink | | 5 | | | | | 4 | |
| Tap | | | | | 2 | | | |
| Grind Drill | 3 | | | | | | | |
| Polish, Belt Sand | 5 | 4 | | | | 5 | | |
| Thread | 6 | | | | | | | |
| Solvent Clean | 7 | | | | | | | |
| Storage for Assy | 8 | 6 | 3 | 5 | 3 | | | |
| Assembly | 9 | 7 | 4 | 6 | 4 | 3/4<br>6 | 5 | |

b

*alike in process* equipment frequently follow the same routing. Seek out these groups by classifying like designs in the former case (similarities in dimension, shape, chemical, or other characteristics), and, in the latter case, by looking for items which begin or end at the same operation or which pass through certain key operations.

It may then be possible to go back and apply the techniques of the operation process chart or multi-product process chart to the flow problem for the group.

Analyses of this kind lead to group-production layouts. Group production is neither line production nor layout by process, but a combination of the two, which retains the advantages of reduced handling and production supervision and control, yet does not cause too much loss of equipment utilization. See Figure 4–9.

If grouping is impossible, the planner can *select or sample* representative items. Every hundredth part number, every fifth formula, items ordered on various mixes of order number, day, week, and month can be the bases for selecting typical items. A random sampling based on statistical analysis is also valid.

Because there is a chance for error in a small sample, it is often better to select "worst-condition" items. Such a selection rests on the precept that if a layout can handle the worst items, it can handle all of them. Therefore, a selection is made of three to five of the items which rate worst in each or several (not necessarily all) of the following characteristics:

| | |
|---|---|
| Heaviest | Most awkward to handle |
| Largest | Items with most operations |
| Bulkiest | Greatest quantity |
| Most fragile | Worst quality problems |
| Most hazardous | Most customer complaints |
| Costliest | Worst scrap or spoilage record |

Selecting items like this oversimplifies the analysis problem since a few rather than all the items involved in the project are considered. Still, sometimes it is a highly practical way to go. See Figure 4–10.

**Figure 4–8 a & b.** Modifications of the Multi-Product Process Chart. At left (a) is an example of routings split into alternative paths of flow. The percentage of material flowing over each route is shown by a one-to-ten index number. Percentage or actual intensity figures would be more accurate.

At right (b) is an example of how the routing can be charted on a simple cross-section sheet without the formality of circles and connecting lines. The flow is indicated by recording numbers in sequence opposite the appropriate operations. Note the double routing through "Punch" for item D-4, the alternate routing for operation 2 for item E-1, and the double assembly operations (3 and 4) and the return for operation 6 for item G-1. This is a quick way to record routings when no formal operations, process, or routing sheets are available. Connecting lines can be added later according to the actual numerical intensity of flow for each item—by applying the Q data at top of each column and a size, weight, or Mag-Count measure for each item.

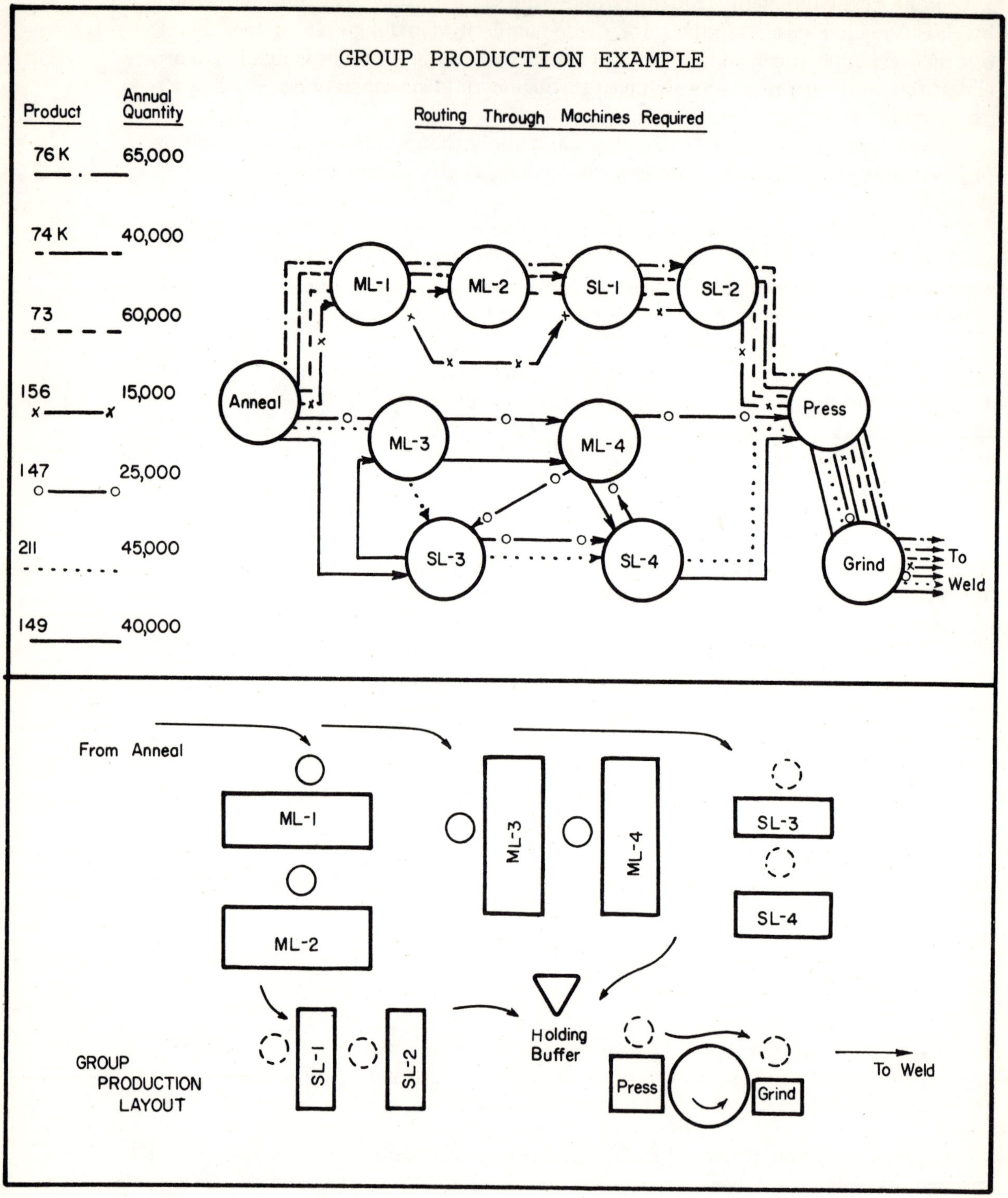


**Figure 4-9.** Group production based on combining certain similar items. This plant produces enamelware utensils for houshold and institutional use. Of some 300 catalog items, the layout planners combined seven reasonably similar teapots and coffee kettles into a group. The flow chart indicated above shows the routing and operating equipment. The actual layout, with the general pattern of flow roughly marked by arrows, is shown below.

### The From-To Chart

Grouping or selecting gives way to the from-to chart when the products, parts, or materials under study are very numerous. The from-to chart is sometimes called a cross chart, and when distance is added to the values, it is termed a travel chart.

Figure 4–11 explains the theory of the from-to chart. The operations or work centers are listed in the same sequence as headings for the horizontal lines and the vertical columns for all operations of the items being charted. For each item, a letter (or strike mark or intensity-of-flow value) is recorded in the appropriate column for each move the item makes. For example, a part moves from Shear to Notch. The appropriate letter is placed in the box "from Shear to Notch." The operation following Notch is Pierce. A letter is therefore marked in the box "from Notch to Pierce."

After all items have been recorded, the letters or quantities in each box are totaled and the total recorded there. This number, then, represents the degree of traffic flow between each pair of operations or work centers.

The box recordings can be checked by adding down each column and across each line. An error has usually been made if the number of moves into an operation or work center does not equal the number of moves out of that operation or work center. However, as shown in Figure 4–12, the line totals may not match column totals when more meaningful measures of flow intensity are actually used.

For accuracy's sake, a relative-value chart, using quantity times size or weight or volume of the item for each incidence of movement, should be made. Intensity of flow figures based on realistic quantitative measures are entered in the from-to chart in the manner described above. Usually, the intensity measures for all moves or transports from one activity to another are tallied and the final figure posted on the chart. With the relative-value chart, the planner can see more clearly which activities should be close to one another and how they should be arranged for optimum flow. In the calculation of this data for each box, separate tally sheets or punched cards can be of real value.

In preparing the chart, it is helpful to list *all* operations or functions involved and to keep the same sequence down the page and across the top. Consistency should prevail, even though nothing may actually come from Shipping, for example, to the operating work centers. It helps to find the correct box and to avoid errors in posting. Moreover, a diagonal line drawn through box 1–1, 2–2, 3–3, etc., will consistently divide the from-to's and the respective to-from's.

If the listing of operations (on horizontal lines and vertical columns) happens to be in best sequence, then the figures above a diagonal from upper left to lower right corners represents moves toward completion. Figures in boxes below the diagonal show counterflow.

The *total* traffic or flow between any two activities or work centers is found by adding both the from-to and to-from values for this pair of activities.

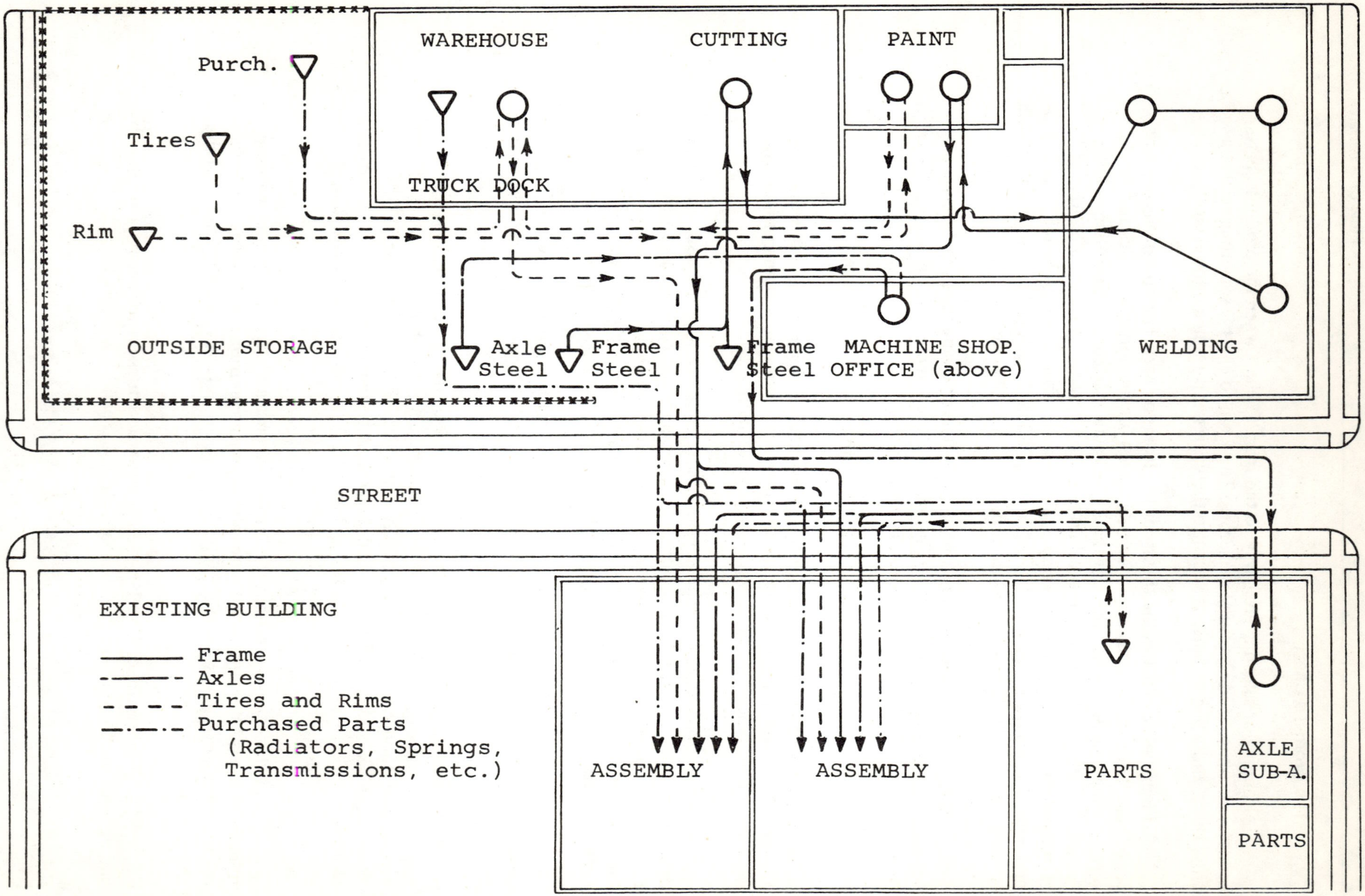
WAREHOUSE
CUTTING
PAINT
Purch.
Tires
Rim
TRUCK DOCK
OUTSIDE STORAGE
Axle Steel
Frame Steel
Frame Steel
MACHINE SHOP. OFFICE (above)
WELDING
STREET
EXISTING BUILDING
Frame
Axles
Tires and Rims
Purchased Parts (Radiators, Springs, Transmissions, etc.)
ASSEMBLY
ASSEMBLY
PARTS
AXLE SUB-A.
PARTS

Always keep in mind that each combination of from-to plus to-from represents a relationship, or relative closeness desired, between the particular pair of operations or activity-areas.

There are various aids to interpreting the chart, once it is made. The boxes are ranked, based on the intensity of flow, starting with the largest and proceeding to the smallest flow. These numbered boxes indicate the order of the most important relationships. Another aid is calculating and recording the percentage-of-total flow intensity that each box accounts for, which yields a measure of the relative importance of the flow between each operation or work center and all others.

### Making the Flow-of-Materials Analysis

We have seen in the discussion above that the type of flow analysis to be made depends largely upon the volume and variety of items involved. And so we return to the P-Q chart.

When the number of items is small but the quantity is large, use the operation process chart—in Phase II, Overall Layout Planning, that is. As the variety increases, the planner needs other means to clarify the determination of flow: the multi-product process chart; grouping or selecting items and the applying of one of the other techniques; and finally the statistical approach of the from-to chart. Of course, the from-to chart itself can be made from selected or sample items; or several can be made, one each for various classes of products or materials.

The easiest way to make any flow-of-material analysis is to work directly from the operations list or process sheets. However, they may *not* show how the material actually flows because of any number of possible operating changes. Moreover, movement of scrap, re-work, and auxiliary materials and supplies, do not show up on the operations list. Customer-returned goods, ends or leftovers sent back to storage, empty containers, and often packing materials are items that do not appear on conventional process or routing sheets. And inprocess storage stops, diversions to weigh scales for counting, and inspections originally not planned, all involve moves of materials (and therefore an influence on relationships), but are not spelled out on available operation planning sheets. This means that a flow of materials based on official company routings more often than not fails to give a complete flow-of-materials analysis.

**Figure 4–10.** Flow diagram of an existing layout, showing the four items involving the greatest handling problems. This plant—divided in the middle by a street—had no process sheets to aid in determining its routings. Because of the need for a quick solution, the production manager selected two items each of the largest, heaviest, greatest quantity, most costly, and most awkward. These reduced to four items which gave him the worst materials handling problem. He diagramed the flow of these on a building plan of the existing layout as shown. The necessity for these particular routings was challenged. The diagram was then used as a basis for the proposed flow of material in planning a general overall layout for a new plant.

| FROM \ TO | Shear 1 | Notch 2 | Draw 3 | Pierce 4 | Bend 5 | Trim 6 |
|---|---|---|---|---|---|---|
| Shear 1 | – | ABC 3 | – | EF 2 | – | – |
| Notch 2 | – | – | BD 2 | AC 2 | – | – |
| Draw 3 | – | – | – | – | BDE F 4 | C 1 |
| Pierce 4 | – | – | CEF 3 | – | A 1 | – |
| Bend 5 | – | – | – | – | – | BDE 3 |
| Trim 6 | – | – | – | – | – | – |

**Figure 4–11.** Making the From-To Chart. The from-to chart is made by listing the operations (or work centers) both down the side and across the top of the chart. Each intersecting box is used to record the movement *from* one operation *to* another. The routing of each product is traced by recording for each move it makes where it comes from and where it moves to. For example, if we take the same information shown on the multi-product process chart of Figure 4-7, we can post the movement as shown in this figure. Product A moves *from* Shear *to* Notch. We can record this by an "A" in the proper square. The material next moves *from* Notch *to* Pierce and again we note it on the chart by recording "A." This is done for each movement of the part. Similarly, each part is recorded.

This example is complete only for the routings shown in Figure 4-7. Actually, letters for items, or even part numbers, are seldom posted to the chart. A simple strike line for each part making a move is the simplest—with the total number added to indicate the total intensity of part numbers moving from each operation or department to each of the others. In practice, a more refined measure (based on P and Q) is tallied and posted in each intersecting box.

It is usually best, therefore, to take at least some measurements of moves actually made on the operating floor—provided existing operations are available. Such measurements can be made by tallying move orders or dispatch tickets and by in-and-out checks of each area.

Besides making a quantitative tally of the items moving over certain routes, the planner must recognize that a qualitative measure for the intensity of materials flow may be even more significant. (See discussion of Mag Count earlier in this chapter and in Appendix I.) Because of this, establishing the intensity measure practically always involves both the quantity of pieces or units to be moved *and* the size or weight or shape or condition of each item.

In many cases, punching our routing sequence on tabulating cards and tallying and extending the intensity of flow on a computer is justified, especially when the ability to printout in sequence the ranked pairs of operations or activities is programmed into it.

### Converting Flow of Materials to Simple Convention

Because of the difficulty and time required to compare and visualize a great many numerical values, SLP converts its flow-of-materials intensities into a common rating system. The rating conventions are:

A *A*bnormally high intensity of flow
E *E*specially high intensity of flow
I *I*mportant intensity of flow
O *O*rdinary intensity of flow
U *U*nimportant moves of negligible intensity

By grouping his quantified flow-of-materials intensities into this vowel-letter scale of values—and its cross-indexed number-of-lines convention, described in Chapter 6—the planner can more readily find and work with his flow data. He can add a minus sign to each vowel letter (A, A–, E, E–, I, I–, O, O–) to give himself half degrees of flow intensity. Mathematically, this gives him a range of accuracy of plus-or-minus six-and-a-quarter percent (± 6¼%), which is usually well within the accuracy of the method or unit of measurement of the meaterials being moved or of the forecast quantities on which the flow-of-materials intensity numbers are calculated (that is, on the P and Q input data).

Converting to the vowel-letter convention is a fairly simple problem of calibrating (see Figure 4–13).

1. Identify each route by activity-areas serving as the origin and destination of the move (always keeping lowest number at left).
2. Complete a common-denominator measurement for the total flow of material (all products or materials in both directions) for each route (or pair of activity-areas).
3. Rank, in descending order of magnitude, the flow intensity for each route.
4. Plot the intensity of each route on a bar chart or graph.
5. Divide the bars at logical break points, recognizing that the A's may bracket perhaps only 10% of the highest routes (but the top 40% of intensity values) and that the O's may bracket intensity values of perhaps only 10% of the largest value (but the lowest 40% of the routes).
6. Draw division lines to indicate the range of vowel-letter ratings, using minus-sign ratings for degrees of flow intensity in between full vowel letters if appropriate.

With these five ranges (or classes), A, E, I, O, U, of flow intensity—nine if half degrees (minus-sign ratings) are used—each route is put into a simple, realistic, order-of-magnitude relationship, ready for subsequent use in comparing closeness desired among the various activity-areas.

## FROM – TO CHART

Plant South (Plastics Division) Project 15697

By R.C.V. With M.R.J.

Item(s) Charted 45 Representative Molded Parts Basis of Values (Q x Size x Care) ÷ 1000 Date 8/10 Page 2 of 5

| Activity or Operation FROM \ TO Activity or Operation | Mold 1 | Normalize 2 | Machine 3 | Assembly 4 | Spray 5 | Machine Spray 6 | Plate 7 | Coat 8 | Buff 9 | Wash or Dip 10 | Tumble 11 | Stamp 12 | Fill & / or Wipe 13 | Pack 14 | Ship 15 | 16 | 17 | 18 | 19 | 20 | TOTALS |
|---|---|---|---|---|---|---|---|---|---|---|---|---|---|---|---|---|---|---|---|---|---|
| Mold 1 | = | 3/26 | 3/160 | 6/232 | 1/2 | 5/631 | 1/884 | - | - | - | 2/400 | 10/468 | - | - | 15/2222 | | | | | @ | 46/5025 |
| Normalize 2 | - | = | - | - | 7/262 | - | 5/376 | - | - | 1/10 | - | 1/180 | 1/2 | 3/129 | 1/120 | | | | | | 19/1079 |
| Machine 3 | - | - | = | - | - | - | - | - | - | - | 2/152 | - | - | - | 1/8 | | | | | | 3/160 |
| Assembly 4 | - | - | - | 1/20 | - | - | - | - | - | - | - | - | - | - | 6/884 | | | | | | 7/904 |
| Spray 5 | - | 1/2 | - | - | 7/414 | - | 5/22 | - | 1/12 | - | - | 1/8 | 2/168 | 1/76 | - | | | | | | 18/702 |
| Machine Spray 6 | - | 9/752 | - | - | - | = | - | - | - | - | - | - | - | - | 4/348 | | | | | | 13/1100 |
| Plate 7 | - | - | - | - | 1/8 | 5/376 | = | 6/910 | - | - | - | - | - | - | - | | | | | | 12/1294 |
| Coat 8 | - | - | - | - | 1/12 | - | - | = | 3/12 | - | - | - | - | 2/886 | - | | | | | | 6/910 |
| Buff 9 | - | - | - | - | - | - | - | - | = | - | - | - | - | 3/22 | 2/10 | | | | | | 5/32 |
| Wash or Dip 10 | - | - | - | - | - | - | - | - | - | = | - | - | - | 1/10 | - | | | | | | 1/10 |
| Tumble 11 | - | - | - | - | - | - | - | - | - | - | = | - | - | - | 4/552 | | | | | | 4/552 |
| Stamp 12 | - | 6/296 | - | 1/92 | - | 3/80 | - | - | 1/8 | - | - | 2/192 | - | 1/180 | - | | | | | | 14/848 |
| Fill & / or Wipe 13 | - | - | - | - | 1/2 | - | 1/12 | - | - | - | - | - | = | 1/156 | - | | | | | | 3/170 |
| Pack 14 | - | - | - | - | - | - | - | - | - | - | - | - | - | = | 12/1320 | | | | | | 12/1320 |
| Ship 15 | - | - | - | - | - | - | - | - | - | - | - | - | - | - | = | | | | | | 0 |
| 16 | | | | | | | | | | | | | | | | | | | | | |
| 17 | | | | | | | | | | | | | | | | | | | | | |
| 18 | | | | | | | | | | | | | | | | | | | | | |
| 19 | | | | | | | | | | | | | | | | | | | | | |
| 20 | | | | | | | | | | | | | | | | | | | | | |
| TOTALS | 0 | 19/1076 | 3/160 | 8/344 @ | 18/700 | 13/1087 | 12/1294 | 6/910 | 5/32 | 1/10 | 4/552 | 14/848 | 3/170 | 12/1459 | 45/5464 @ | | | | | | 163/14,106 |

NOTES: @ 45 parts (jobs) studied but 46 parts actually involved because one finished item involves two detail parts which are assembled together.

**Figure 4-12.** Completed From-To Chart for a plant making a variety of decorated plastic products on a customer-order basis.

Two figures are posted in each box where flow of material is involved. The figure above the slant line represents the number of parts (part numbers) making that particular move. The figure below the slant line represents the actual intensity of flow. The intensity was established in this case by multiplying the annual quantity by the size of each item and the handling care required. Arbitrary scales were used to measure both size and care—similar to those values used in the Mag Count (see Appendix I).

Receiving and Incoming Stores are not shown on the chart, inasmuch as all materials start their first operation in Molding. Also, a selection of items, rather than all items, was made and applied to this chart.

This chart indicates that it is most important to have Molding close to Shipping (2222 units of flow intensity). Next most important is Pack to Ship (1320). Note that while this indicates direction of the flow, the total intensity of movement between any two operations results from going to and coming back. For example, Normalize to Stamp is 1/180 and Stamp to Normalize is 6/296—and therefore the total intensity is 476.

Although the number of *parts* going to an operation should equal the number coming from that operation, the *intensity* figures to and from a given operation will not add up equally. This is primarily because the size and handling care of the item is changed during various operations. Also small, non-molded parts are excluded from this example, yet they go to Assembly where they are added to the molded products. This further contributes to an intensity difference of 344 *to* and 904 *from* Assembly. As recorded in the chart notes, eight parts go *to* Assembly and seven come *from* Assembly. This is because two molded parts are joined there. As a result in this case even the number of items going to are different than those coming from.

Boxes 4-4, 5-5, and 12-12 indicate duplicate operations within the same activity-area.

Note again that each figure in the text having a form number in the lower left corner (as this Figure 4-12 has a form number 136) will be supported by a fresh unused copy of that form in the Working Forms section at back of this book. Quite often, the layout planner will find it convenient or necessary to use a larger-size sheet for his From-To Chart.

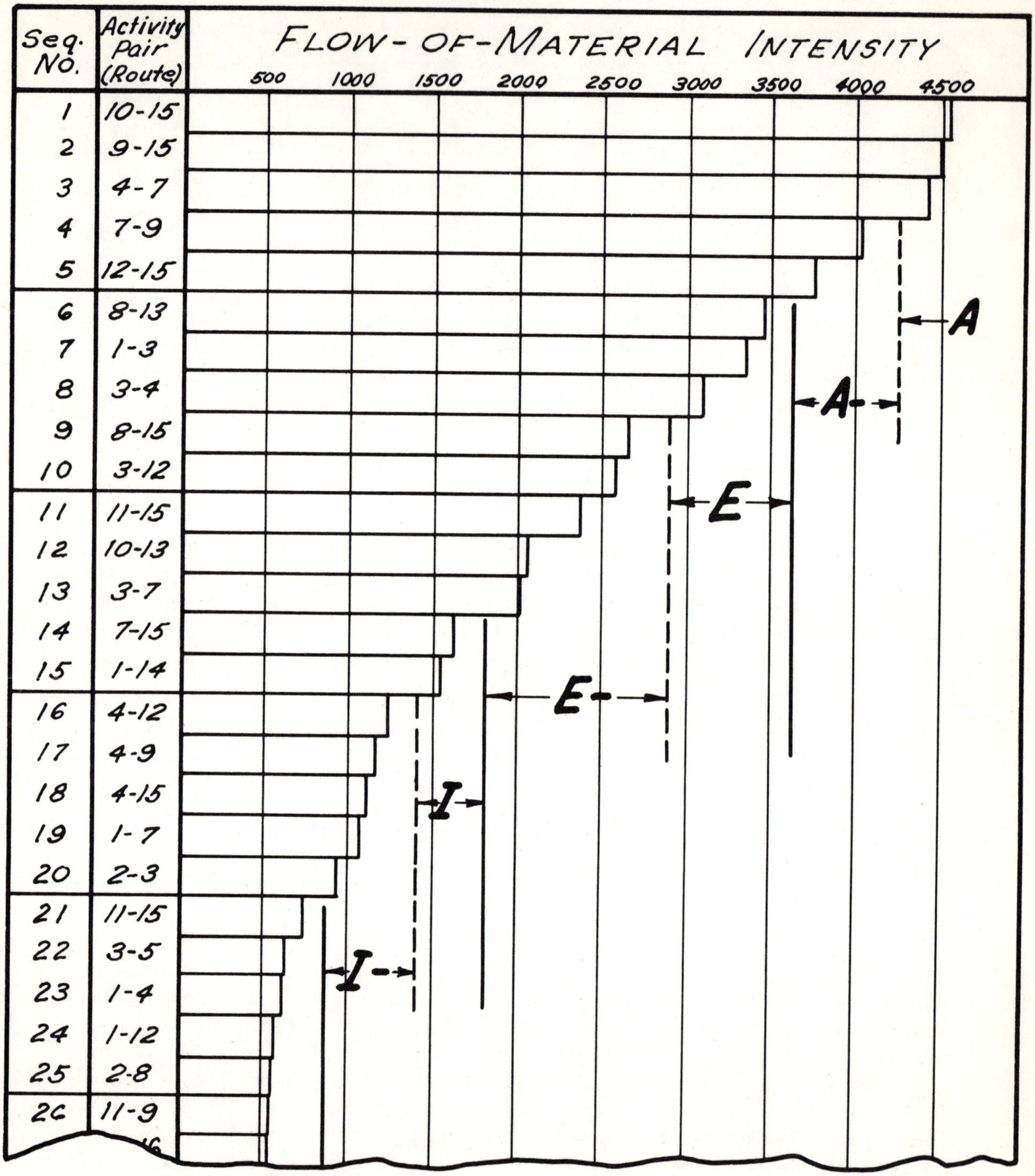


**Figure 4–13.** Calibrating or converting flow-of-material intensities against the vowel-letter rating scale.

# Chapter 5
# Other Than Flow Relationships

In most industrial plants the emphasis is commonly placed on flow of material as the basis for layout arrangement. This has been fully discussed. A flow pattern is determined and diagramed (based on the techniques shown in Chapters 4 and 6) and the rest of the activities are then fitted in and around that pattern. Actually, this is not the best practice as a *general rule* for layout procedure.

## Flow Alone Not Best Basis for Layout

There are several reasons that the flow of material—as determined predominantly by the routing—cannot be the sole basis for layout arrangements.

1. The supporting services must integrate with the flow in an organized way. This integration results from total analysis—analysis of the reasons that certain supporting activities should be close to certain producing or operating areas. The maintenance crib, the superintendent's office, the locker and rest room, and the transformer bank, all have a relatively preferred closeness to each of the producing areas. They are all part of the layout; they must be planned into it, yet they are not part of the flow materials.
2. Frequently, flow of material is relatively unimportant. In some electronic or jewelry plants, only a few pounds of material will be transported during an entire day. In other industries, materials are piped, or one skid-load lasts a worker all week.
3. In completely service industries, office area, or maintenance-and-repair shops, there is often no real or definitive flow of material. Therefore, any general rule must offer a way of relating areas to one another without reference to material flow. And this is true even if we recognize that paper work, equipment, or even people can be considered as the "material" that flows.
4. Additionally, in heavy materials-movement plants, where the influence of material flow will dominate the layout planning, flow will not be the sole basis for arranging the process operations and equipment. Basically, we chart flow to determine the sequence of operations or which departments should be near one another. But flow of materials is only one reason for this closeness. There are many others, which may conflict with or at least cause adjustments in the closeness as based on the analysis of flow. For example, the routing may call for the sequence: form, trim, treat, subassembly, assembly, and pack. For best flow of material, treating should lie between trimming and subassembly. But treating is both a very dirty and a dangerous operation. Therefore, it should be kept away from the delicate subassembly area and its high concentration of workers. The effect of factors like these *and* the distribution of utilities, the cost of controlling quality, product contamination,

and the like must be compared with the importance of material flow and adjustments made as practical.

In any case, some systematic way of *relating service activities to one another* and of *integrating supporting services with the flow of material* is necessary. The Relationship Chart is the best method of meeting this need.

In the SLP pattern of procedures, Flow-of-Materials Analysis is based on P, Q, and R. Activity Relationship Analysis draws on P, Q, and S. T (time is involved in both, as in all analyses to some extent.

### The Relationship Chart

The relationship chart is a cross-section form where the relationship between each activity[1] (or function or area) and all other activities can be recorded. The basis of the form is shown in Figure 5–1.

The relationship chart shows which activities have a relationship to others. Also, it *rates* the importance of the closeness between them and supports the rating with coded back-up reasons. These measures make the relationship chart one of the most highly practical and effective tools available for layout planning. It is undoubtedly the best way of *integrating supporting services* with the operating or producing departments and of planning the arrangement of *office or service areas* having little or no flow of materials.

The chart itself is almost self-explanatory. Where the activity on down-sloping line 1 intersects the activity represented by up-sloping line 3, the relationship between activity 1 and activity 3 is recorded. In this way, there is an intersecting "box" for each pair of activities involved. The basic idea is how to show which activities should be located close together and which far apart, with all in-between relationships also rated relatively and recorded.

The chart can be likened to a from-to chart which has been folded diagonally so that the from-to and the to-from boxes fall on top of each other. The relationship chart thus shows the total relationship, i.e., in both directions.

Further use of the form is illustrated in Figure 5–2. Note that each box is divided horizontally. The upper part is for the closeness-rating value. The lower half is

1. The word "Activity" is a universal term, used throughout this book to designate "things" (other than people and process material) to be located as part of the layout planning. It encompasses, at different levels of planning or in different situations: department, area, function, work center, building, building feature, machine group, operation, and the like.

**Figure 5–1.** The basis of the Relationship Chart. All activities (areas or features) are listed at the left on lines 1, 2, 3, etc. Each activity-line slopes away from the list. Where the down-sloping line 1 intersects the up-sloping line 3, record the required (or desired) relation between Activity 1 and Activity 3. As indicated in b, the vowel-letter rating (c) and number-code supporting reasons for each rating (d) are placed respectively in the top or lower half of each intersecting box.

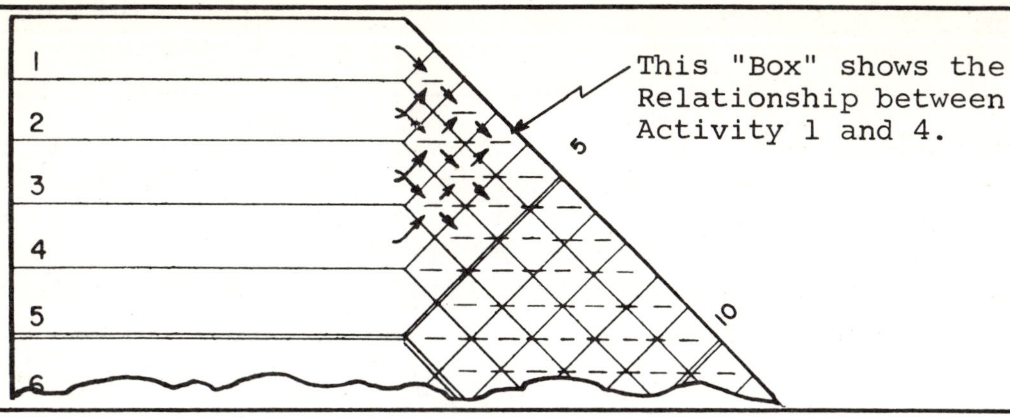


a

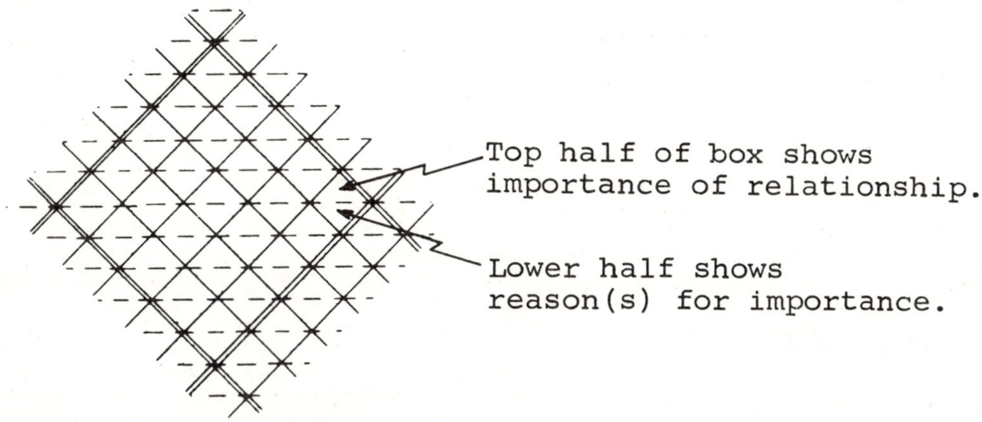


b

| VALUE | RELATIONSHIP |
|---|---|
| A | Closeness absolutely necessary |
| E | Closeness especially important |
| I | Closeness important |
| O | Ordinary closeness OK |
| U | Closeness unimportant |
| X | Closeness not desirable |

Rate the importance of relative closeness required (or desired) between each pair of activities, using the vowel-letter value scale.

c

| Code | REASON |
|---|---|
| 1 | |
| 2 | |
| 3 | |
| 4 | |
| 5 | |
| 6 | |

Support the closeness rating with reasons--listing each reason on a line and entering the respective code number(s) for the reason in lower half of appropriate box.

d

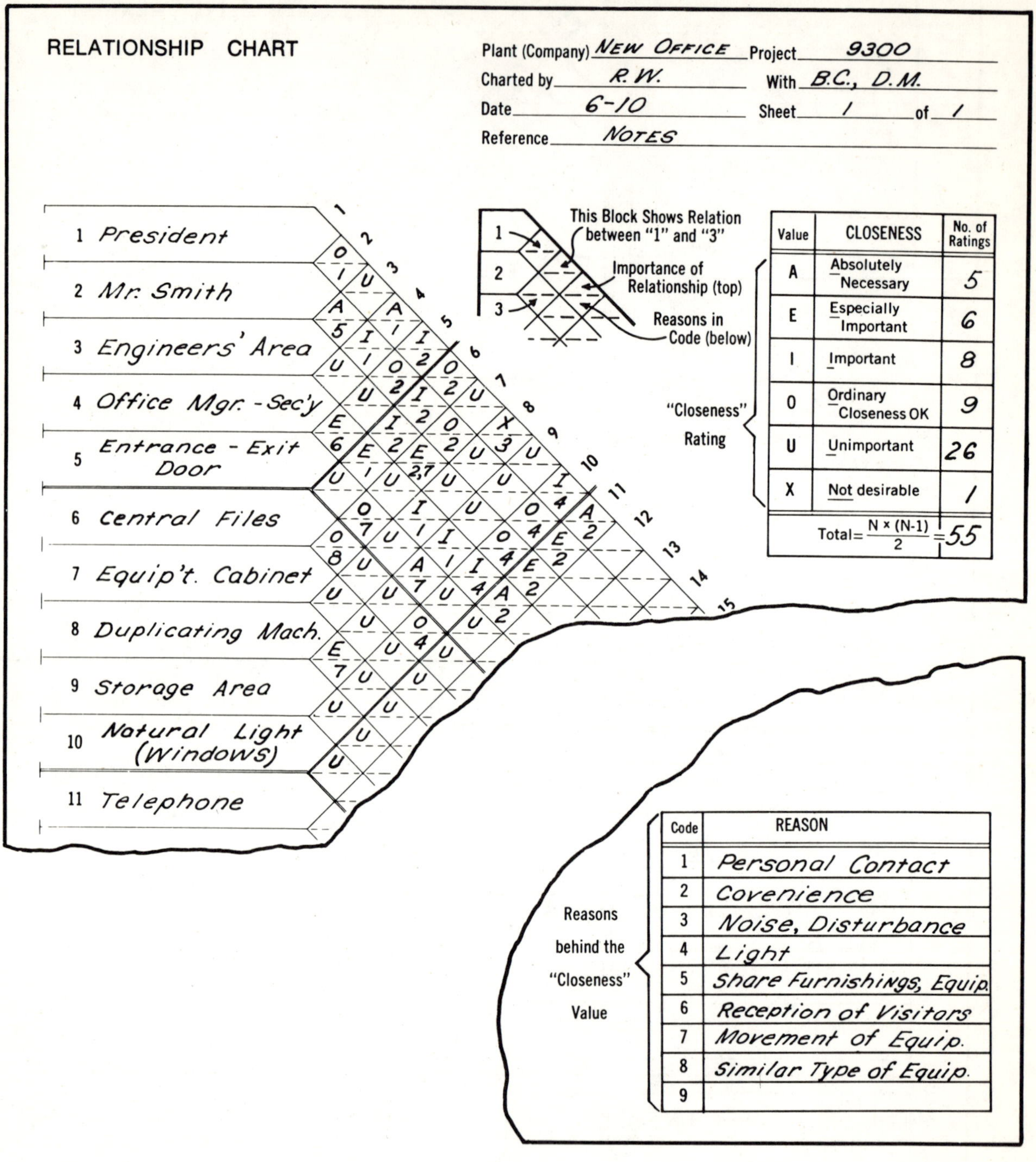


**Figure 5–2.** The relationship chart is extremely effective for planning all activities not tied together with a significant flow pattern. This chart was prepared for an office of consulting soil-test engineers. It indicates that Mr. Smith must be near the Engineers' Area, the telephone to a lesser extent, still less to the office manager—secretary and central files, not at all near the duplicating machine or storage area. Reasons are filled in and recorded in lower half of the appropriate boxes. It is normally not realistic, in charting relationships, to include window or telephone availability. They are included here primarily to indicate the degree of detail to which the planner can go if need be.

for recording the reason(s) for the designated closeness value. A rating and reason(s) for each relationship is thus recorded.

Closeness is rated according to a value scale A, E, I, O, U, and X. A indicates closeness absolutely necessary; X indicates closeness is undesirable and the activities should be kept apart. E, I, O, and U each indicate varying degrees of closeness.

The vowel letters used for closeness rating have been selected with considerable thought.

1. The letters themselves have meaning (in English). A is for *A*bsolutely necessary; E is for *E*specially important; I for *I*mportant; O for *O*rdinary closeness; U for *U*nimportant; and X, the accepted symbol for wrong, stands for a negative degree of closeness–not desirable.
2. A, E, I, O, and U are vowels most people have memorized in sequence and therefore, easy to remember in the closeness-rating context.
3. Numbers were not chosen because they imply an accuracy greater than the rating actually has. Moreover, numbers are used to code the reasons and identify the activities. It would confuse the chart to use them also for rating.

The closeness rating is more meaningful when reasons are tied to the rating. For each reason, a code number is assigned and an explanation is recorded in the "reason block" on the sheet. That code number is then entered in its appropriate box(es). This way, two or three reasons can be marked in a relationship box without too much crowding. With this form, a great deal of information is gathered on one sheet without making it necessary to write notes all over the form.

Most reasons for closeness or distance between activities boil down into eight or ten for any one project. (More room is allowed on the form for additional reasons, or additional reason lines can be added to the form if needed.)

Although many terms may be used and many other reasons are possible, typical reasons supporting relationship ratings include the following:

1. Flow of materials*
2. Need for personal contact
3. Use same equipment
4. Use common records
5. Share same personnel
6. Supervision or control
7. Frequency of contact
8. Urgency of service
9. Cost of utility distribution
10. Use same utilities
11. Degree of communicative or paperwork contact
12. Specific management desires or personal convenience

*As we shall see later on, there is value in saving reason 1 for flow of materials.

### Color Coding

Recording ratings and reasons should be done in pencil. But, because letters or symbols become confusing when used in profusion, the charted information can be more readily applied if the ratings are flagged in color.Therefore, each box on the chart should be re-marked with a coded color. The SLP conventions include the following color code:

| | |
|---|---|
| A – red | O – blue |
| E – orange or yellow | U – uncolored |
| I – green | X – brown |

These colors follow the order of the spectrum so their sequence is easy to remember.

This color coding is done after the original recording in pencil, which saves shifting pencils while making the chart yet allows color to simplify the problem of "reading" the assembled data.

For ease in marking the colors and in reading the chart, and since color represents closeness ratings, not reason(s), the colors are used only in the upper half of each box. Also, in order not to obscure the letter rating already indicated, the colors are drawn in, in three straight lines along the inside edge of each upper-half triangle. . . . See Figure 5–3. The planner may wish to check back on the letter rating; or he may have explanation reference-note letters, asterisks, or other information in the upper half of the box that should not be obscured. And from a practical standpoint, it is much easier to read a photo-reproduced chart if the original pencil letters are clear.

In most projects, nearly half the boxes will have a value of U. To save time and maintain clarity, the planner does not color the "unimportants," although it can be done. However, so as not to overlook any box, always indicate it has been considered by marking a U in the upper half of the appropriate box. This ensures that all relationships have indeed been considered.

### Refinements in Charting

Layout planners not familiar with vowel-letter rating have a great tendency to overassign A ratings. Operating people fall into this same trap. To prevent this, don't assign A to any pair of activities unless it truly is "absolutely necessary" that the two be adjacent each other. Actually, it is desirable to have an increasing frequency of rating occurrences from A through U–say, 2 to 5% A, 3 to 10% E, 5 to 15% I, 10 to 25% O. The frequency of X depends on the nature of the project.

The range from A through U involves four rating changes. Each letter, thus, has a range of 25%. To get greater accuracy, there must be less spread between ratings. Therefore, a minus sign following the letter rating (A– or E–) represents halfway between one rating and the next below it. To color code a minus rating on the chart, use a dotted or broken line of the respective color. The addition of minus signs then allows ranges of accuracy of 12½%.

Another refinement can be used when a rating of extremely undesirable is called for: Use double X (XX) and color the chart black, as indicated in Figure 5–3.

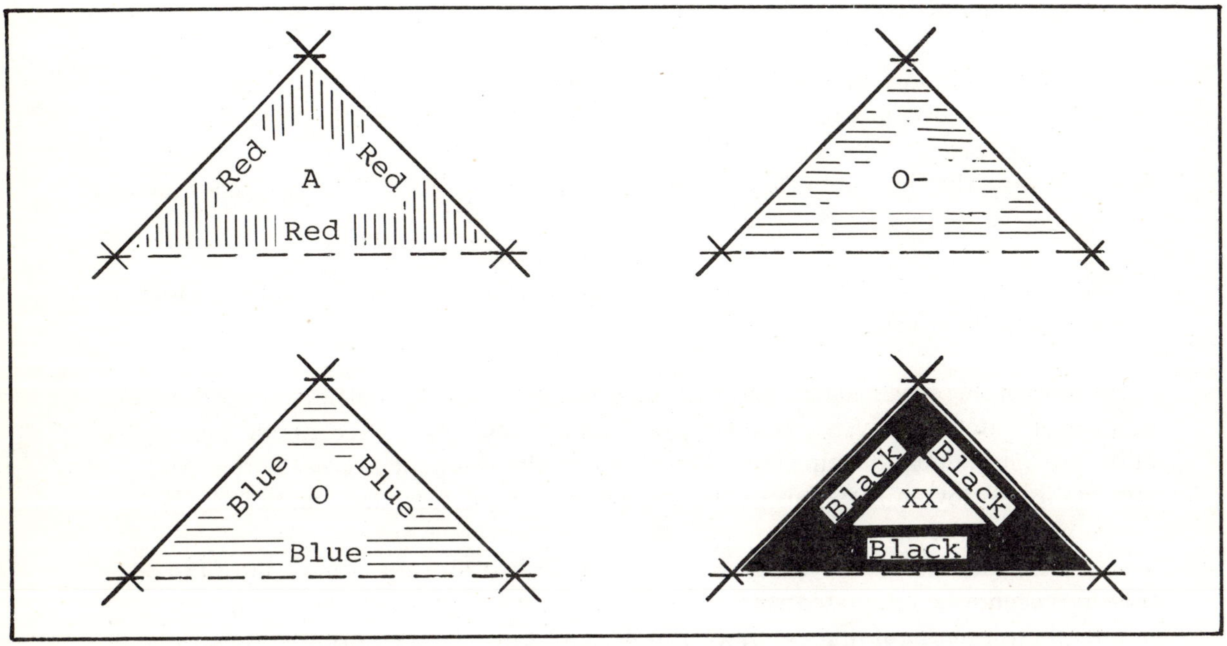


**Figure 5–3.** Preferred method of coloring the relationship chart. By marking three colored lines inside the top half of each box, we flag the various relationships with easy-to-find colors. The triangular arrangement of the three lines retains the clarity of the original letter rating. Note, in-between ratings are marked with minus and coded with a broken line in its respective color. The double XX—with black color—can be used for a rating of Extremely Undesirable.

### The Procedure

The procedure for making a relationship chart varies, among other things, according to whether or not productive operations are included with service or supporting activities. Both the flow-of-materials chart and the relationship chart can be used separately, or they may be combined.

If there is no service support, that is, no other-than-flow considerations, there is little need for the relationship chart. Flow analysis determines the relationships. But in most cases, it is practical to incorporate the operating activities and the supporting services on the same relationship chart. Thus the relationship of services is integrated with the operating or producing activities they support.

In practice, the SLP procedure most often calls for the following general sequence of steps:

1. Develop the intensities of flow for the operating activities.
2. Rate or classify the intensities between each pair of activities into groupings:
   Abnormally high intensities . . . . . . . . . . . . . . . . . . . . . . . . . . A
   Especially high intensities . . . . . . . . . . . . . . . . . . . . . . . . . . E
   Important intensities . . . . . . . . . . . . . . . . . . . . . . . . . . . . . . I
   Ordinary intensities . . . . . . . . . . . . . . . . . . . . . . . . . . . . . . . O
   Unimportant or negligible intensities . . . . . . . . . . . . . . . . . . . U
3. Then develop a relationship chart for all the service or other-than-flow factors.
4. Join the flow and other-than-flow ratings and make a *combined* relationship chart.

This sequence of *combining flow and other-than-flow relationships* is explained in detail in Appendix XII.

When there is no significant flow of material, as in many offices or service areas, no flow-of-material analysis is necessary, and all activities are related to one another only through the relationship chart. The method followed depends on the relative importance of flow of materials versus other-than-flow factors. See Figure 5–4.

The relationship chart is especially helpful when the largest portion of the relationships cannot be calculated; and when the planner is faced with only opinions—frequently conflicting at that—with wishful thinking or wrong assumptions, or with other people's desires for status or self-betterment. With the relationship chart the people involved can participate; they see the qualitative approach being used; they are made to explain why. Thus, opinions are sifted and evaluated, and a decision is based on each relationship having a specific closeness value. In planning offices and service areas, the relationship chart is undoubtedly the most practical and important technique available.

In preparing the chart, when productive or operating activities are to be included, the planner, converts the flow-of-materials intensity figures (developed by

---

**Figure 5–4.** Type of work determines procedure to follow. Flow of Materials and Activity Relationship are the two basic procedures to establish the desired (or required) closeness of various areas, activities, or functions to each other.

When the materials are large in size or quantity, flow of materials becomes the primary basis for determining the relationship of areas—with activity relationship used to tie-in the service and supporting areas. When there is no problem of size, quantity, or difficulty of moving the product or material, activity relationship becomes the chief procedure—with little if any flow-of-material (or flow-of-papers) study being justified.

In between these extremes, both procedures should be used—integrated from the start or independently with subsequent tie-in. In this former method, "flow of materials" becomes just one of the elements or reasons in the construction of the activity relationship chart.

TYPE OF WORK DETERMINES PROCEDURE TO FOLLOW

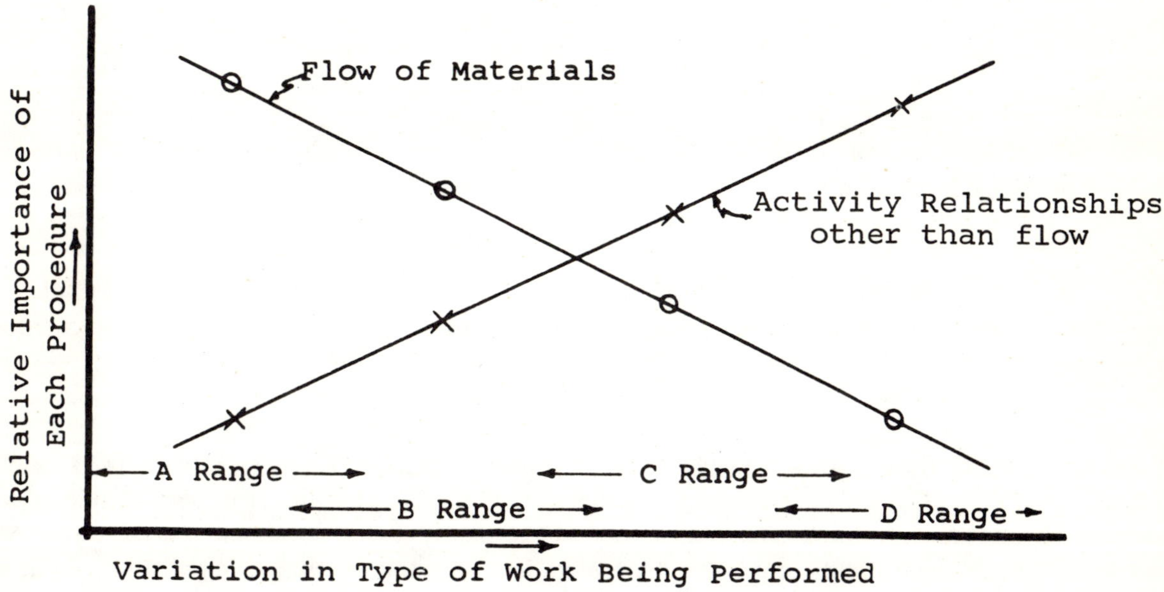


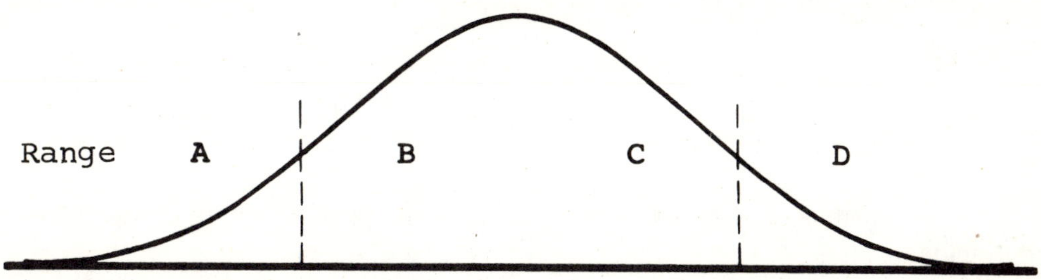


| | |
|---|---|
| Range A. | Plant layouts involving heavy or awkward products or materials, or relatively large quantities of parts or materials. (Flour Mill; Steel Fabricator; Mass Production Manufacturing) |
| Range B. | Job shop layouts--having no major clear-cut pattern(s) of material movement. (Tool Making; Custom Plating; General or Specialty Manufacturing) |
| Range C. | Service Shops with significant amounts of material sequencing or office areas involving heavy flow of papers. (Overhaul and Maintenance Shops; Test Laboratories; High-Volume Office Operations) |
| Range D. | General Office Areas. (Experimental Labs; Drafting Room; Most Office Areas) |

flow-of-materials analysis) to the vowel-letter rating. This may seem to dilute the accuracy of these figures. But this is good since flow can be overemphasized and frequently is. Flow intensity calculations are based on anticipated estimates for P, Q, and R—which estimates are probably accurate only to plus or minus 10% at best. Figures unsoftened by adjustment to vowel-letter rating suggest greater accuracy than is realistic. Moreover, as pointed out earlier in this chapter, there are factors other than flow of materials that influence the closeness of even the productive or process activities to each other, and the flow figures should be modified anyway by these other factors to establish suitable relationships among the process departments.

These very acts—adjusting the relationships from flow intensities to vowel-letter ratings and relating each service to each operating department—establish the closeness-desired ratings for all relationships on the same basis. That is, a single, common-denominator scale of measure for closeness is used for all relationships.

In preparing the chart, identify the activities (see Chapter 3, Input Data) and list them on the chart. Group together activities which are similar or come under the same supervision. This will make it easier to mark, get approval, and use the chart. Also, it may be difficult to use the chart when there are too many activities listed. About 40 or 50 is a practical limit (the working form supplied with this book has room for only 45). It may be necessary to consolidate, segregate, or temporarily overlook certain activities, rather than including too many.

Utility sources and site or building features are sometimes significant and should be listed when they are. North light, neighboring hazards, even passing highways, may be involved as activities.

There are a number of ways to establish the relationship ratings. Generally these include

1. The planner, knowing the operations, does his own rating.
2. Calculations, like those for flow intensity or work sampling studies, serve as the basis for selected relationships.
3. Opinions can be solicited from others involved with the operation or its support in face-to-face discussion. This is frequently done at the same time information on which to base space requirements is gathered. When trying to get others to understand the relative degree of closeness which each vowel letter represents, the planner is wise to establish typical "benchmark relationships," or benchmark ratings similar to those shown in Figure 5–5.
4. Questionnaires can be distributed to persons in charge of the activities involved. The planner cross-compares the responses and posts them on the relationship chart. See Figure 5–6.
5. A small group of managers responsible for the area can meet to decide jointly on each relationship, the planner serving as recorder. (One technique that helps the group keep in mind the specific pair of activities under consideration is to have a set of cards with the number and name of an activity printed clearly and boldly on each card and to have a pair of cards always on the discussion table.)

BENCHMARK RELATIONSHIPS

| Ltr. | Pairs of Activities | Reasons for Closeness |
|---|---|---|
| A | Steel Storage and Shear Area | Qnty. of material moved; Similar handling problems. |
| | Final Inspection and Packing | Damage to unpacked items; Inspection slip loose until packing completed. |
| | Clean and Paint | Use same personnel, utilities, supervision; Similar type building. |
| E | Reception and Visitor Parking | Convenience; Security. |
| | Metal Finish and Welding | Quantity and shape of material moved. |
| | Maintenance and Sub-Assembly | Frequency and urgency of service. |
| I | Shear Area and Presses | Qnty. of material moved. |
| | Sub-Assembly and Final Assembly | Volume of material moved; Share same personnel. |
| | Vault and Accounting | Movement of records; Security; Convenience. |

| Ltr. | Pairs of Activities | Reasons for Closeness |
|---|---|---|
| O | Maintenance and Receiving | Movement of Supplies. |
| | Salvage and Tool Room | Share same equipment. |
| | Mail Room and Plant Office | Frequency of contact. |
| U | Maintenance and Cafeteria | Service is minor. |
| | Welding and Purch.Prts. Stor. | Slight contact. |
| | Engineering and Shipping | Infrequent contact. |
| X | Welding and Paint | Dirt; Fire hazard. |
| | Incinerator and Main Office | Smoke, odors, dirt; Appearance. |
| | Press Shop and Tool Room | Vibration. |

**Figure 5-5.** Benchmark relationships aid consistency and speed when establishing relationship for all pairs of activities. These are benchmarks used in a project involving the relayout of a light manufacturing company.

6. Establish with each supervisor a ranking of his activity's relations to each other activity and then convert the ranking to ratings. (It is often easier for operating managers to force things into a sequence than to make the distinction of specific ratings.) Cards, with the names of each activity for developing the ranking and benchmark relationships for converting to ratings are of help here, too.

### Check and Endorsement

Regardless of which method(s) is used to establish and record the rated relationships, the final chart should be checked by, or with, the people who will actually have to make the layout work.

Several methods of checking, or combinations of methods, can be followed:

1. Take the chart to the department heads, or others in charge of each activity involved, and have them review their relationships.
2. Circulate the chart for approval and have it returned signed "OK" or with suggestions for change.

To: All Department Heads
Subject: Systematic Facilities Planning

In our efforts to establish the best practical layout, we would like the benefit of your experience and thinking.

We are after the best arrangement of each activity (or area or office) to each of the others. Activities that relate to each other should be close; unrelated activities should be apart.

For each of the activities that relate to your operations, please fill out one of the attached recording sheets. Mark the number and name of your activity at the top and fill in the date and your name. Use one sheet for each activity under your responsibility. Then, indicate the relationship of this activity (marked at the top) to each of the other activities listed. Do this by marking (at left) the closeness relationship value and the reasons therefor (coded). Use the values and code numbers explained at the meeting and shown again below.

When marking these sheets, assume a brand new, theoretically ideal condition exists and that you are completely unlimited by existing conditions, building, or arrangement.

Undoubtedly new ideas will occur to you as you work this out. Be sure to list them at the bottom of the recording sheet so every potential good idea is captured.

This is a method of getting the best thinking of each of us set down in specifics. Thank you for your help.

B. G. Edwards

| VALUE | RELATIONSHIP |
|---|---|
| A | Closeness absolutely necessary |
| E | Closeness especially important |
| I | Closeness important |
| O | Ordinary closeness ok |
| U | Unimportant to be close |
| X | Closeness not desirable |

Example: E/2,5 indicates especially important relationship because of equipment common to both activities and because of frequent personal contact.

| CODE | REASON |
|---|---|
| 1 | Sequence of work flow |
| 2 | Use common equipment |
| 3 | Share same personnel |
| 4 | Share same space |
| 5 | Degree of personal contact |
| 6 | Paperwork contact |
| 7 | Use common records |
| 8 | Perform similar work |
| 9 | Noise, vibration, fumes and hazards |
| 10 | Other________ |

a

3. Have some independent reviewer check your closeness-rating values and the posting of them in the correct boxes.

Figure 5–7 shows a completed relationship chart with signatures for approval and colors added, though coloring the full upper triangle is not recommended.

Suggested alterations or corrections are best recorded on a sheet similar to the modification record shown in Figure 5–8. Activities should be identified by line number on the chart as well as by description in order to speed the work and avoid

**Activity - Relationship Recording Sheet**

Activity No._____ Name ____________________ Date ____________ Supplied by ______________

| Val. | Code | No. | Activity | Val. | Code | No. | Activity |
|---|---|---|---|---|---|---|---|
| | | 1 | Body Shop | | | 30 | Parts Storage |
| | | 2 | Trim-Upholstery Shop | | | 31 | Inventory & Purchasing |
| | | 3 | P.M. & Unit Repair | | | 32 | Accnt'g. & Audit Office |
| | | 4 | Radio Repair | | | 33 | Records Storage |
| | | 5 | Paint Shop | | | 34 | Stationery-Supplies |
| | | 6 | Bus | | | 35 | Ex Offi |

Notes of explanation ________________________________________

New ideas that suggest themselves ______________________________

b

**Figure 5–6 a & b.** An example of how relations between activities can be established by questionnaire-type of recording sheet. Here the memorandum (a) instructs the various department heads to record the relationship of each of their activities (or areas or offices) to each of the others on the preplanned Recording Sheet (b)—doing so in the same recording conventions as will be used for consolidation on the relationship chart. The example is from a city taxi-cab garage—itself a service, rather than production-type of facility. Figure 5–7 shows the chart made from the consolidation and reappraisal of the returned recording sheets.

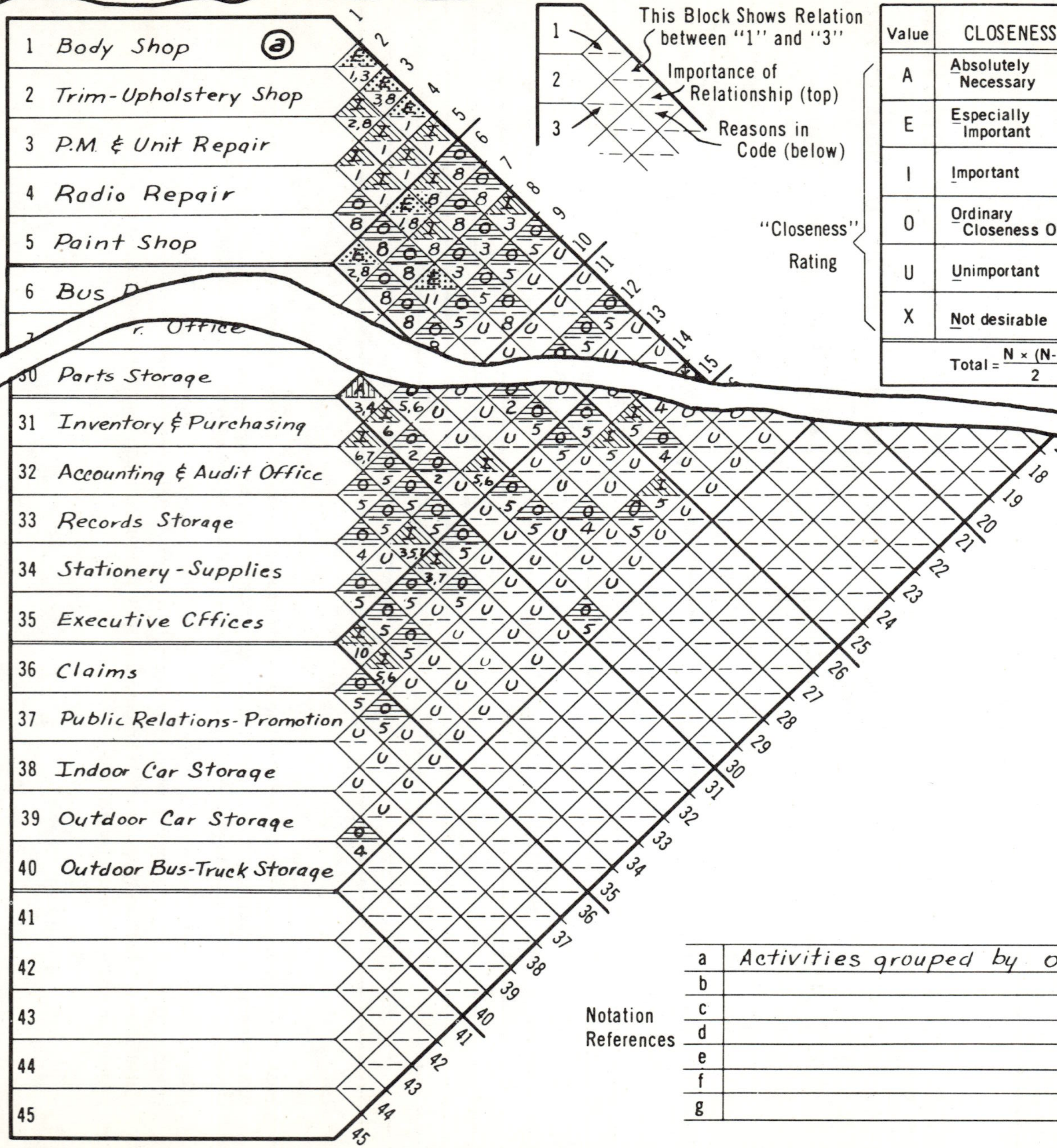


"Closeness" Rating

| Value | CLOSENESS | Color | No. of Ratings |
|---|---|---|---|
| A | Absolutely Necessary | RED | 4 |
| E | Especially Important | OR-YEL | 10 |
| I | Important | GREEN | 29 |
| O | Ordinary Closeness OK | BLUE | 52 |
| U | Unimportant | U | 680 |
| X | Not desirable | BROWN | 5 |
| | Total = $\frac{N \times (N-1)}{2}$ = 780 | | = 780 ok |

| Code | REASON |
|---|---|
| 1 | Sequence of Work Flow |
| 2 | Use same Equipment or Utilities |
| 3 | Share same Personnel |
| 4 | Share same Space (Alternating Requirements) |
| 5 | Degree of Personal Contact |
| 6 | Degree of Paperwork Contact |
| 7 | Use Common Records |

| Dept. / Date | Sign "OK" | Dept. / Date | Sign "OK" |
|---|---|---|---|
| Shops 3/2 | A.T. Halvorsen | | |
| Warehse 3/2 | REW | | |
| Purch. 3/2 | M. T. Frank | | |
| Acctg 3/5 | H. Y. Britt | | |
| 3/6 | B. T. Locke | | |
| 3/9 | B D E | | |
| 3/9 | [illegible] | | |

Notation References

a Activities grouped by organizational responsibility
b
c
d
e
f
g

errors. Also show the name of the person who is responsible. This, together with any further listing of activities found OK (on bottom of modification record sheet), helps ensure checking complete.

Since each box involves two activities, any changes in a box's data should be endorsed by representatives of (or those responsible for) both activities (in many cases it is only one person). This means keeping track of who is responsible for each activity. To aid the analyst and the approver, it is easiest if the list of activities is grouped by person responsible for the activity, as in Figure 5–7.

In most cases, the two persons will agree on the relationship desired between each other's activities. Disagreements can usually be best resolved by calling both parties together, although sometimes a full meeting of representatives from all activities is faster and more helpful.

In seeking this check and endorsement, remember that the operating people probably see today's conditions and relationships only, or at least predominantly. It may be that they are not as well informed as you about plans for future product design, production engineering, and marketing programs of the company.

In any event, a sign-off or other official agreement with the relationships recorded on the chart should be considered a must. The data is then formally established so it can be used to begin diagraming the arrangement of the various activities. Getting approval forces the planner to involve the responsible people in his project. Early participation of those who will later operate the areas being laid out gets them into the act, convinces them of the soundness of the approach, places relationships on a more objective basis with reasons to substantiate them, and goes a long way towards

---

**Figure 5–7.** Completed relationship chart for a taxicab office-garage. In such a facility there are actually several different flow patterns: cabs coming in for gas-up and meter control, cabs going to or coming from the storage lot, cabs moving to and from the garage for either preventive maintenance work or emergency or extra work; flow of trip record and turned-in cash; and in this case additional flow of repair parts, advertising signs, buses, baggage trucks, and rental cars. The relationship chart is an ideal technique for untangling in a fairly simple way this type of interrelationship.

This chart results from, and summarizes the data reported by, each department head written to in Figure 5–6a. Where the two persons responsible for each pair of activities reported the same relationship and reasons, this rating was accepted as is. Any differences were resolved by the layout planner contacting the people involved. Unresolved differences called for a face-to-face get together with the layout planner of the two people concerned.

See inside front cover for color reproduction of this same illustration. Note how the color makes the chart easy to read. In actual practice, colored straight lines in triangular shape will be used, rather than coloring the full top half of each box which obliterates the rating vowel letter. No color and no supporting reasons are used for the unimportant relationships.

Sign offs are used to be sure the data is correct before spending further time planning and then finding errors in this underlying data.

MODIFICATION RECORD

Dept./Activity No._____ Descpt._______________ Person____________

| Relation between This | and That | Now is | Should be | Reason(s) |
|---|---|---|---|---|
| | | | | |
| | | | | |
| | | | | |

Other depts. or activities examined
and found OK include:

Signed________________

**Figure 5-8.** Sheet used for recording changes on the relationship chart. This keeps the chart free of erasures and allows a record of the suggested changes. A modification record like this can be taken along or sent out with a copy of the summary relationship chart to obtain approvals in an organized way.

selling the eventual layout plan developed from the charted data. Here is one of the most effective ways of getting participation into layout planning long before the physical plans take shape.

Experience indicates it is essential to total all the A's then the E's . . . including the U's. Every relationship must be accounted for later, when adjusting the layout, or the layout becomes distorted. To check for total verification, the sum of all letter totals should equal the number of potential relationships on the chart, namely,

$$\frac{N \times (N-1)}{2}$$

where N equals the number of activities listed.

#### Conclusion

The rated and reason-supported relationship chart is a systematic means to get data into usable form. It yields a set of relationship requirements to which the layout planning can be keyed—And all this on one sheet of paper. Moreover, it provides the planner with ready reasons for departments' being located closely or far away—sound backing for the data and highly valuable when challenged at approval time.

The chart insures that a decision is made for the relation of each activity to each other activity. Thus, it serves as a check sheet. A chart of only ten activities means 45 relationships, and even the best layout man cannot keep that many decisions in mind. Figure 5-9 summarizes the actual procedure for making the chart.

PROCEDURE FOR ACTIVITY RELATIONSHIP ANALYSIS

1. Identify all activities involved.

    a. Make a list of departments, areas, operations or features and go over this with department managers or supervisors involved to check coverage and terminology.

    b. Keep like activities or those reporting to same individual grouped together--as in an organization chart.

    c. Do not use more than forty-five activities on any one chart. Consolidate as necessary in advance and re-chart secondary activities.

2. List the activities on a relationship chart.

    a. Set down productive operations first, supporting services thereafter.

    b. Include site or building features (rail siding, elevator shaft, transformer).

3. Determine or establish the desired relationship for each pair of activities and the reason(s) therefor. This can be done:

    a. by your knowledge of the operating practices.

    b. by making a series of calculations for each major consideration, or reason, just as you would do for flow of material. (Don't overlook the combining of flow and other-than-flow ratings--in Appendix XII.)

    c. by personal visit and discussion with department managers or area supervisors involved.

    d. by group explanation and individual recording sheets.

    e. by group discussion among a few key managers.

4. Enter data on a relationship chart--consolidating all notes, tallies, and calculations--so you have one set of approved, specific relationships and reasons from which to plan your layout.

    a. The chart itself acts as a check sheet to make sure the relationship between every activity and every other activity has in fact been considered.

    b. Get approval of the chart.

**Figure 5-9.** Procedure for making the Activity Relationship Chart.

Chapter 6

# Flow and/or Activity Relationship Diagram

After the relationship of activities has been charted—either by flow-of-materials analysis, activity relationship charting, or a combination of the two—the next step is to diagram this information. This step is Box 3 in the SLP pattern of procedures.

At this stage, the planner is seeking a visual picture of the data gathered up to this point and of the calculations and analyses made from them. Now the tallied or charted information, which shows the sequence of activities and the relative importance of the closeness of each activity to each other activity, is transferred and translated into a geographic arrangement. This geographic arrangement should in fact now locate the activities according to the degree of closeness charted.

Bear in mind that we are dealing here with Phase II, Overall Layout, not with diagraming detailed layouts. Detailed layouts are covered in Part Three.

## Diagraming

There are many techniques available for diagraming. The general sequence is to work from the charted information to a visual diagram. The planner usually starts with the most important relationships and subsequently, diagrams the less important relationships. Later on, the activities themselves will be expanded according to the actual space allotted to each, to develop a space relationship diagram (Box 6 of the SLP pattern).

The essentials for diagraming involve (1) convenient and simple set of symbols for identification of each activity (area or feature); and (2) some method of indicating the relative closeness between activities and/or the direction and relative intensity of materials flow.

When diagraming the relationships from an activity relationship chart, the planner should already have incorporated materials flow with the supporting services and other-than-flow relationships involved. If so, the quantified flow-of-materials and other-than-flow relationships are diagramed as combined relationships.

## Diagraming Flow

Diagraming materials flow without the integration of service and other activities, can be done in one of two ways: Either start with the normal material in-point and continue on through the flow sequence; or, start with the two departments or activities having the greatest intensity of flow and continue in decreasing order of flow intensity. Be sure to note that the total intensity is the sum of the flow in both directions.

The final form of the flow diagram normally is something like that shown in Figure 6–1. The flow is diagramed in such a way that the supporting services can be fitted to it as seems practical.

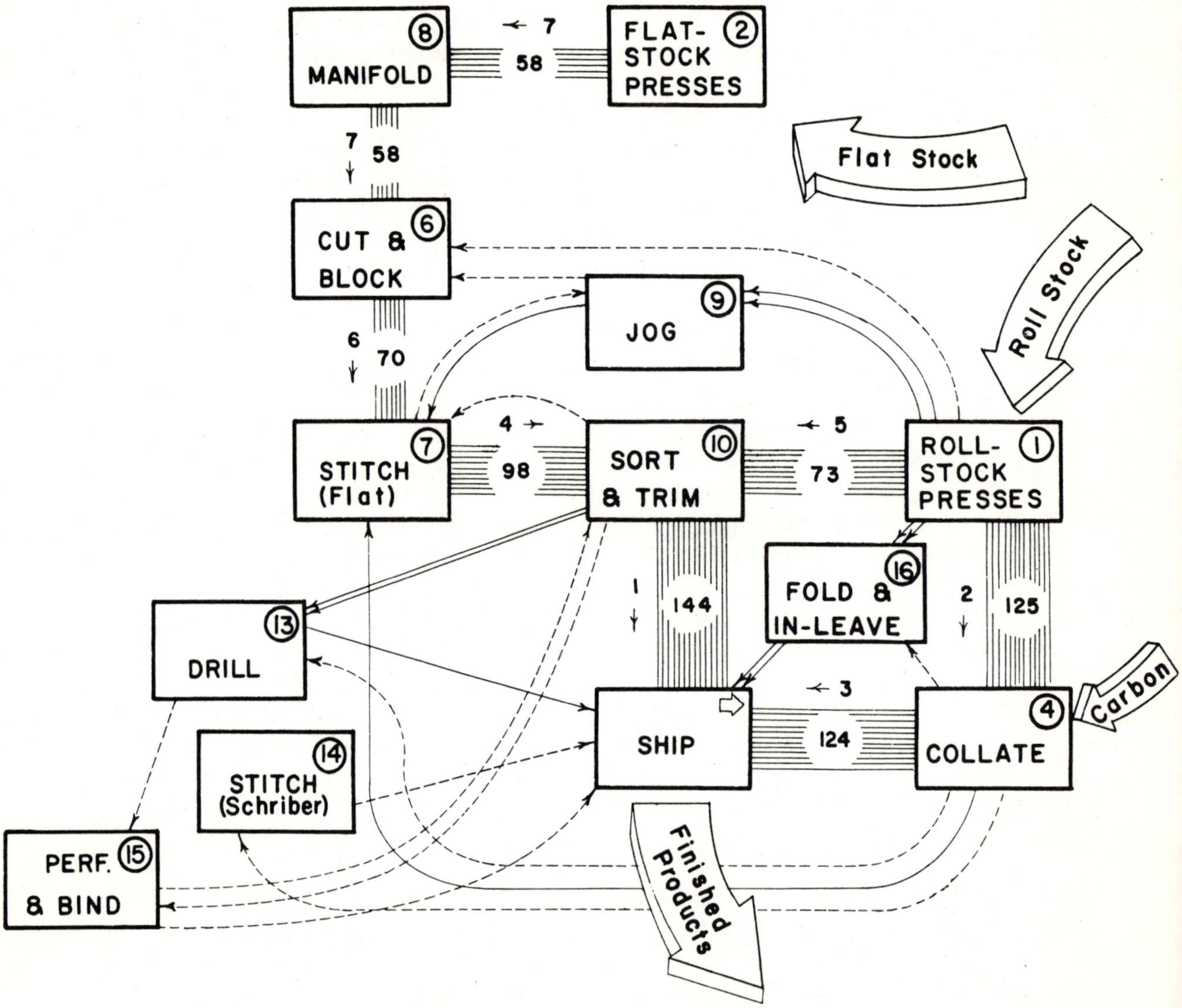


**Figure 6–1.** The flow diagram for a printing plant. Products are business forms and sales-books. Each departmental area is represented by a rectangular block, and flow intensity, by the number of lines connecting each pair of departments. Each connecting flow-line represents 10 "units of flow." The actual number of flow units is shown inside the cluster of flow lines, and the direction is indicated by an arrow. Beside each arrow is marked the order of magnitude of flow intensity—starting with the largest intensity first. The number of flow lines connecting each pair of departments are shown to the closest 10 units; dotted lines represent five or less units.

Normally, diagraming is done as a series of trials and refinements. That is, the diagraming is roughed; then as more information is added various arrangements appear or occur. If many trials are made on one sheet, the diagram is confused. Therefore, it is best to begin a second arrangement on clean paper or to rework the diagram on a tracing-paper overlay. It is not uncommon to go through six or eight trials before an acceptable diagram is developed.

In their initial trials diagrams need not be complicated. In fact, the simpler their construction, as long as they can be understood, the better.

### Flow for Various Products

It is not uncommon to have several products or classes of materials to be diagramed. There are two ways. (1) Several digrams can be made—one for each product, group of products, or class of material; or (2) one diagram can be made with different colors letters, numbers, or symbols representing different products, groups, or classes. Figure 6–2 shows a method by which this can be done.

Generally speaking, however, it is better to consolidate all this information at the charting stage rather than at the time of diagraming. That is, analyze the flow of various products or classes of material sufficiently to derive a net, composite analysis of flow. Then the diagram can be converted directly from the flow chart.

### Diagraming to Determine Flow

In some cases, it is practical to diagram the existing flow pattern right on a plan of the existing layout and so get a picture of the existing overall relationships. This can then be subjected to analysis for improvements in a way that is easier for operating supervisors or untrained layout planners to understand than figures or charted information. In other cases, to better visualize the problem, it is helpful to rough out a "present" flow diagram before making an analysis of "proposed" flow. The "present" flow diagram then can be compared with the planned-for diagram.

Often a company has no operations list, process sheets, or written-down routing to work from. This means the planner has to get out in the plant or area involved and actually trace the flow of materials. The easiest way to do this is on a sketch or floor plan of the existing area. The resultant tracing of flow is really a flow diagram. (This is discussed in Chapter 4, and an example of flow diagraming for an existing plant is shown in Figure 4–10.)

### Diagraming Activity Relationships

When the planner is not concerned about the direction of material movement or when there is no significant flow of materials involved in the relationships, he can prepare his diagram using only the closeness ratings recorded on the relationship chart.

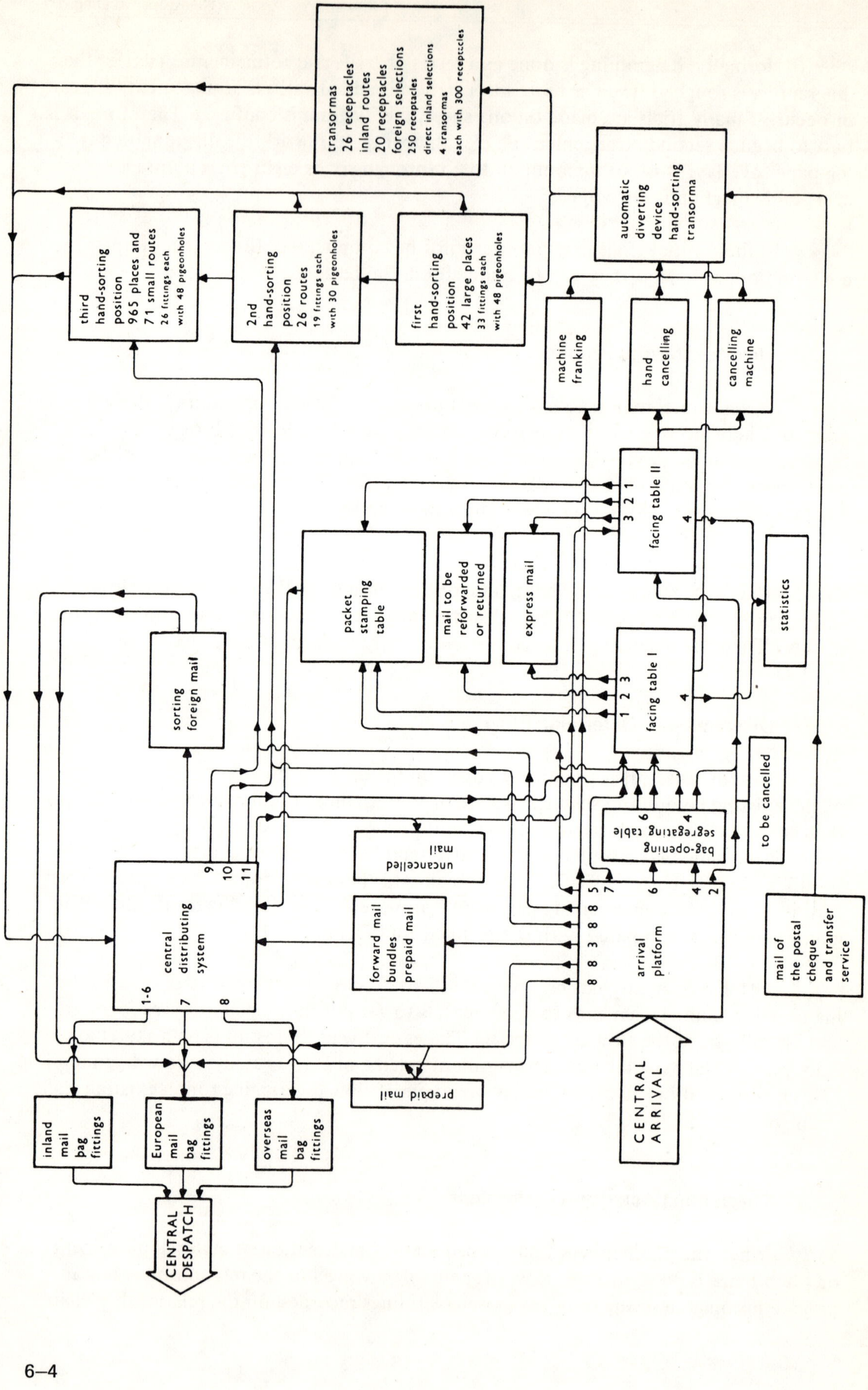
transormas
26 receptacles
inland routes
20 receptacles
foreign selections
250 receptacles
direct inland selections
4 transormas
each with 300 receptacles
automatic diverting device hand-sorting transorma
third hand-sorting position
965 places and 71 small routes
26 fittings each with 48 pigeonholes
2nd hand-sorting position
26 routes
19 fittings each with 30 pigeonholes
first hand-sorting position
42 large places
33 fittings each with 48 pigeonholes
machine franking
hand cancelling
cancelling machine
facing table II
facing table I
packet stamping table
mail to be reforwarded or returned
express mail
statistics
sorting foreign mail
to be cancelled
bag-opening segregating table
uncancelled mail
central distributing system
forward mail bundles prepaid mail
arrival platform
mail of the postal cheque and transfer service
CENTRAL ARRIVAL
prepaid mail
inland mail bag fittings
European mail bag fittings
overseas mail bag fittings
CENTRAL DESPATCH

The specific procedure includes a set of conventions for diagraming. These conventions are used both to save time and to aid in understanding and interpretation. A simple diagram using the conventions is described in Figure 6–3.

Cartographers have long known the value of conventions in making maps. They use a host of symbols, letters, numbers, and colors. Diagraming flow and activity relationships is, essentially, applied cartography, so it is not surprising that the SLP procedure utilizes the same types of codes.

### Conventions for Diagraming

The conventions used in Systematic Layout Planning are illustrated in Figure 6–4. They include the following:

1. A symbol for the type of activity
2. A number (or letter) identification for each activity
3. A number-of-lines code for the intensity of flow or closeness value
4. A color code, also for the intensity or closeness value (Color use is optional, but recommended for the final diagram.)
5. A color for each type of activity (optional here, but used later when converting to a space relationship diagram)

The *symbols* are both self-descriptive and readily available on any die cut rule having the standard process chart symbols. In certain cases, symbols must be reoriented in order to get a distinct identification. But no new symbol is added which would make it necessary to obtain still another rule in order to use the conventions.

The meaning of each symbol is adapted directly from that established in the A.S.M.E. standard for process charting. Although two new symbols have been added, each of them is merely a 90-degree reorientation of a conventional symbol. These two include (1) the *turned arrow*, resembling a house, and therefore used to designate building features or office areas, and (2) a *turned D*, or delay symbol, representing service areas and auxiliaries such as maintenance, salvage, compressor room, and the like.

The activity *number*—usually the line number on the activity relationship chart—is inserted inside each of the symbols.

**Figure 6-2.** Flow diagram of the incoming, picked-up mail within a post office. This illustrates a method of showing the movement for different types of products. Each type of letter is indicated by a number. For example: 2 indicates uncancelled mail collected from post boxes, 3 is bundles of prepaid printed matter, 8 is postage-paid mail and cancelled mail for foreign countries originating in branch post offices.

Because of the complexity of flow patterns, the diagraming will have to be done in two or three stages. A diagram like this would be followed by one indicating the intensity of flow as well as the type of item, before actually placing the various function or activity blocks in geographic relation. From booklet describing post office operations at The Hague, Netherlands.

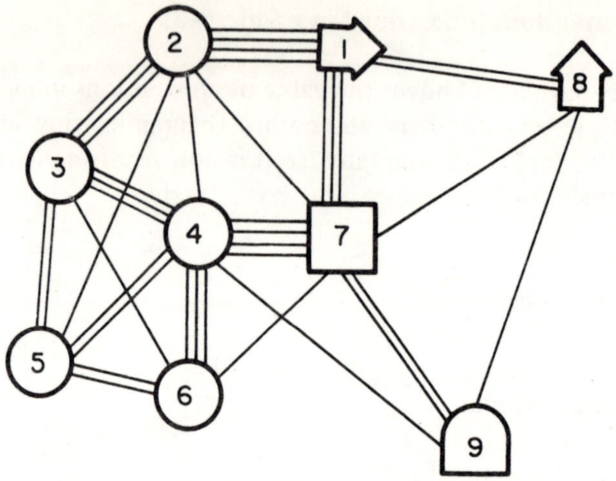


**Figure 6-3.** The method of diagraming relationships involves connecting the activities by a number-of-lines code. The shape of each symbol indicates the type of activity; the number inside is the activity identification; the number of connecting lines indicates the rated closeness. Here, the shipping-receiving dock (1) relates to the first production area (2) by four lines (closeness absolutely necessary), to final inspection (7) by three lines (closeness especially important), and to the front office (8) by two lines (closeness important). Other activities are similarly related.

In connecting each pair of activities, a number-of-lines coding is used to indicate *closeness.*

When making trial diagrams, it is quickest to use a medium-hard lead pencil. This penciled record can be duplicated and eliminates changing colors as the diagram is developed. On the other hand, the color code is the same as that used in charting the relationships and therefore has a direct relation with the activity relationship chart. Inasmuch as colors aid visualization, it is best practice to go back (or have someone else go back) and color over the black lines when the chart is completed or being made ready for interpretation. Applying color also makes it easier to assure that all relationships are diagramed.

For a negative number of lines, use a wiggly line (or two wiggly lines) to represent an X (or XX) closeness rating. The wiggly line resembles a spring pushing the two activities away from each other, in contrast to straight lines which can be likened to elastic-rubber bands.

A dotted line can be used for a half degree of closeness value and is consistent with the minus sign and dotted triangle used in coding half degrees on the relationship chart.

Sometimes the department number, rather than the chart line number, is used inside the symbol to identify the activity. And where it seems helpful, both the chart line and the department number or meaningful letter(s) can be used.

| Process Chart Symbols & Action* | Symbols Extended to Identify Activities & Areas | Color Ident. | Black & White** |
|---|---|---|---|
| ○ Operation | ○ Forming or Treating Areas | Green** | |
| | ○ Assembly, Sub-Assembly, Dis-Assembly | Red** | |
| ⇨ Transportation | ⇨ Transport-related Activities/Areas | Orange Yellow** | |
| ▽ Storage | ▽ Storage Activities/Areas | Orange Yellow** | |
| D Delay | D Set-down or Hold Areas | Orange Yellow** | |
| □ Inspection | □ Inspect, Test, Check Areas | Blue** | |
| | Service & Support Activities/Areas | Blue** | |
| | ⇧ Office or Planning Areas, or Building Features | Brown** (Gray) | |

*A.S.M.E. Standard
**I.M.M.S. Standard
(Adopted as basic to SLP procedure)

| Vowel Letter | No. Value | No. of Lines | Closeness Rating | Color Code |
|---|---|---|---|---|
| A | 4 | //// | Absolutely Necessary | Red** |
| E | 3 | /// | Especially Important | Orange Yellow** |
| I | 2 | // | Important | Green** |
| O | 1 | / | Ordinary | Blue** |
| U | 0 | | Unimportant | Uncol-ored** |
| X | -1 | | Not Desirable | Brown** |
| XX | -2, -3, -4, ? | | Extremely Undesirable | Black |

**Figure 6–4.** Conventions used for diagraming activity relationships. A minus sign, indicating half a degree of closeness, is diagramed as a broken line. Planners should work primarily in black and white; color for the relationship lines is optional. Later, when adding space to his diagram, the planner will use color to emphasize type of area.

Although the shape of the symbol identifies the type of activity, a color code is also set up for these symbols. This color, in contrast to that indicating closeness, is for type of activity and will be particularly helpful in Box 6 when arranging space relationship diagrams.

Some of these conventions may seem a bit superfluous, but it is best to have the full range of them available and standardized so that they can be used when wanted. Enough alternatives or options are included in the conventions so that diagraming can be done quickly and in a meaningful way, yet, at the same time, the work sheets can be dressed up for presentation to others.

The color codes for closeness rating and for type-of-activity identification are identical to those approved by the International Material Management Society—after a seven-year study—and adopted as its standard. See Figure 6–4 and Appendix V.

### Procedure For Making Activity Relationship Diagram

Working from a relationship chart or form for combining flow and other relationships (see Appendix XII), the planner starts with the A relationships. These are shown in red on the relationship chart. The significant symbol for the type of activity is drawn and the activity number is inserted within it. This is then connected by four lines to the A-related activity with its significant symbol and its activity number placed therein. See Figure 6–5. Following down diagonal-sloping row 1 for any A's, then row 2 . . . to obtain an orderly sequence of transferring the relationships.

**Figure 6–5.** Relationship diagraming procedure for a general overall layout of a small office area. This office layout project involved newly leased space for a firm of consulting soil-test engineers.

The charting of these activity relations was shown back in Figure 5–2. The charted relationships are then diagramed—to give a geographical arrangement of the activities. Each activity is diagramed in symbolic shape as well as by number identification. These conventions are an inherent part of the SLP methodology. Note that in the office area, the circle symbol is used for an office operation.

First diagramed are the most important relationships: the A's with four connecting lines. Then the E's are added, with three connecting lines. They are spaced farther apart—theoretically twice as far apart as four-line relationships. In Diagram 3, the A and E relationships are rearranged and the I-rated relations added. Here also, activity "11"—telephone—is spread out, for phones are recognized as decentralized and in fact can be put almost anywhere. Finally in Diagram 4, the O's and X's are added and the diagram rearranged for final "best fit". The space is already leased and activity "10"—natural light—is accepted as being available along one side of the arrangement rather than as one concentrated "spot". Note also that the telephones (11) have been excluded from Diagram 4 to avoid unnecessary confusion.

It usually takes from three to eight diagrams—each successively increased and refined—to get the most satisfactory arrangement. Once diagramed, the space allowed each activity can then be applied to the activities, as will be discussed in Chapter 8.

DIAGRAM 1

5 A's (4 Lines - Red)

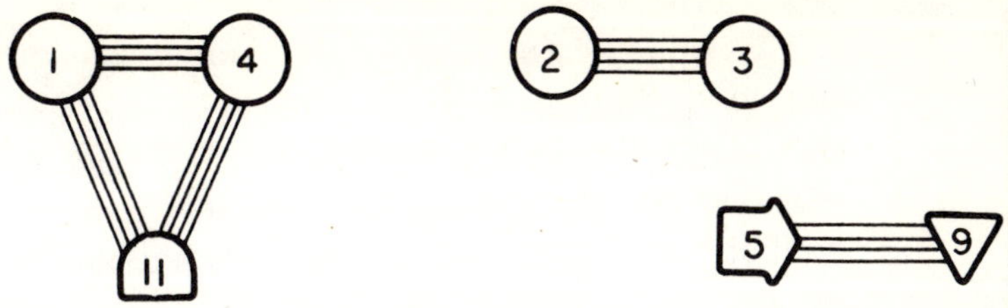


DIAGRAM 2

5 A's (4 Lines - Red)
6 E's (3 Lines - Or.-Yel.)

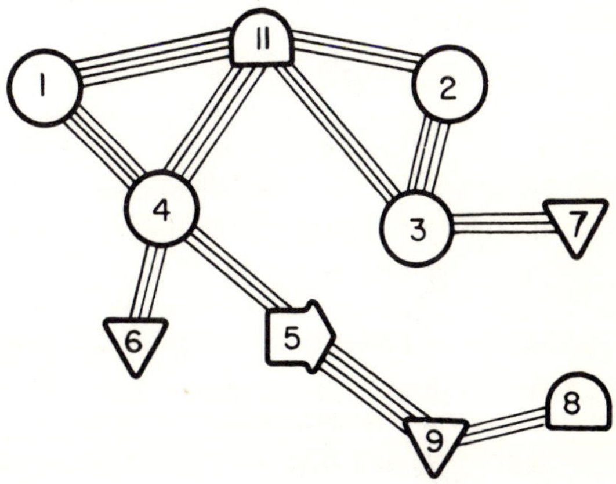


DIAGRAM 3

5 A's (4 Lines - Red)
6 E's (3 Lines - Or.-Yel.)
8 I's (2 Lines - Green)

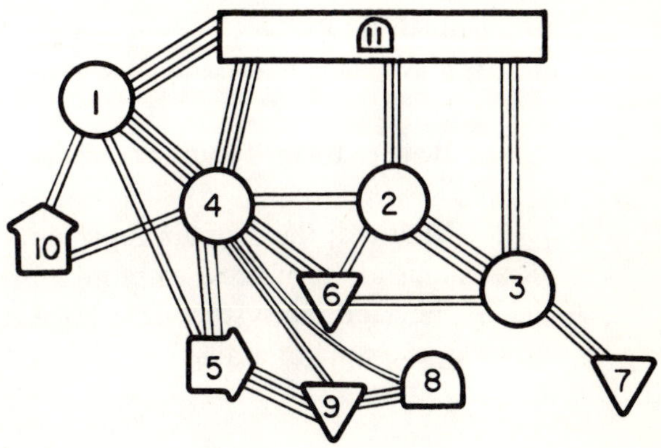


DIAGRAM 4

5 A's (4 Lines - Red)
6 E's (3 Lines - Or.-Yel.)
8 I's (2 Lines - Green)
9 O's (Single - Blue)
1 X (Wiggly - Brown)

(11) Not included

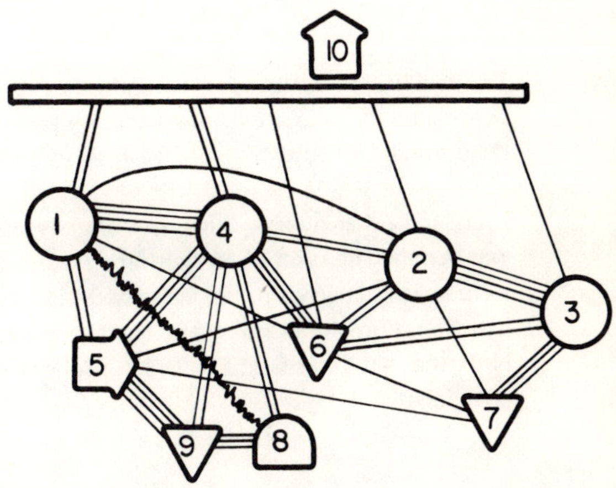

After all A relationships are diagramed, generally by spreading them appropriately on a blank sheet of paper, the next most important relationships are added. These are the E's, or three-line relationships. This usually means rearranging the four-line relationships before adding the three-line ones. Again, the pertinent type-of-activity symbol is used with the activity number inside.

By this time, ordinarily the diagram must be rearranged because the pattern drawn did not give a correct geographic diagram. What the planner seeks, of course, is an arrangement where the distance between E-rated activities (length of three lines) is approximately twice that between A-rated activities (length of four lines).

Then I relationships are added. Again it is usually desirable to rearrange the diagram, using a fresh sheet or a fresh position on the same sheet, to make an improved pattern or orientation of the A, E, and I activities.

The same is done for the O relationships and for the X and the XX relationships. The double wiggly lines, when they are used, are usually considered to have a negative value of approximately minus three, or about as bad as an E relationship is good. Therefore, any double wiggly lines should usually be added at the same time the I values are diagramed.

The planner may draw six or eight diagrams—successively increasing and refining as he goes—before being satisfied with the arrangement. The diagram is considered complete when all the boxes on the chart have been diagramed, and when the diagram's arrangement best meets the desired conditions of closeness. That is, with short four lines and long wiggly lines and relatively appropriate lengths for the other relationships.

Be sure to verify your count of each vowel-letter's frequency.

Finished, the diagram represents the theoretically ideal relationship of the activities, independent of the area required for each, and before modifying considerations, such as the handling system or storage equipment, have been incorporated. Figure 6–6 shows an example of a diagram, before color coding, based on the above procedure.

**Figure 6–6.** A combined flow and other-than-flow relationship diagram for a new ball-and-roller-bearing plant. This shows the diagram carried only to the "Important" (2-line) relationships. This diagram will be adjusted on a fresh sheet and then the "O" and "X" relationships added.

Flow and other relationships have already been combined on the relationship chart before diagraming was begun. The symbol conventions are used to identify the type of activity, although the numbers inside each activity symbol agree with line number of the previously made relationship chart for full identification. A rail siding and road access are shown merely to help identify the activities further. Note that activities 5 and 6 were consolidated into one activity—probably during the relationship charting.

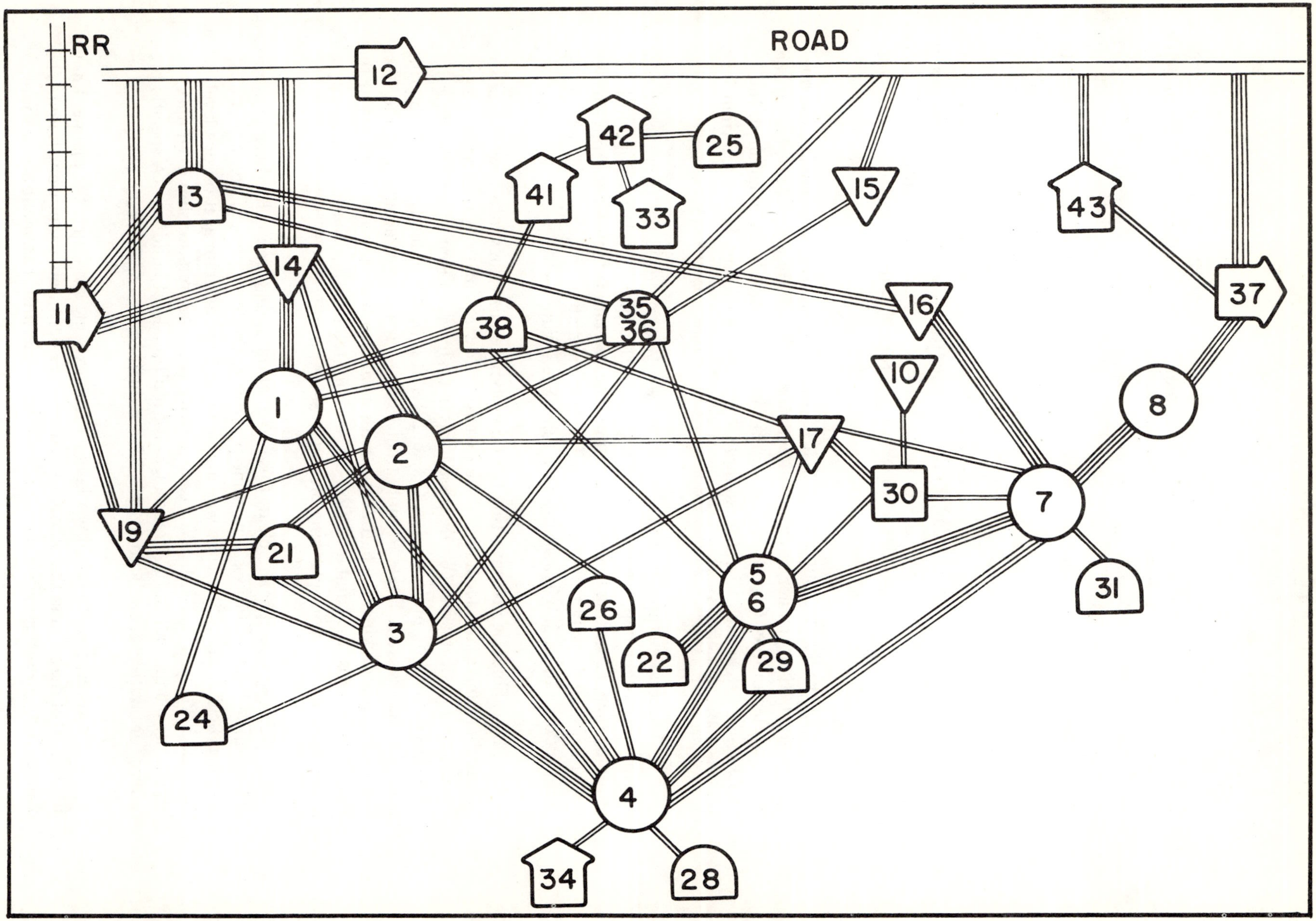
RR
ROAD
12
42
25
13
41
33
15
43
14
11
38
35
36
16
37
10
1
8
2
17
30
7
19
21
5
6
31
26
3
22
29
24
4
34
28

### Refinements

In making the diagram, the planner can avoid unnecessary obscuring or tangling of connecting lines by recognizing, yet withholding from the diagram, certain activities and relationships. The telephones can be run to almost any point and so they were not diagramed when the O's and X's were added to the diagram in Figure 6–5, for example.

Also, to illustrate that an activity has to be close to many other activities, the shape of the normal symbol can be distorted to aid in visualizing the relationships diagramed. Figure 6–5 shows this also—first for the telephones and in Diagram 4 for the windows.

Whenever one activity has connecting lines to a great many other activities, it is a pretty good sign that the activity might better be divided or decentralized. This is especially true of such activities as toilets or rest rooms, foremen's offices, files, storage, and various utilities.

We have pointed out already that in preparing the diagram, one can use one of two methods. The planner can work directly from the relationship chart, taking one activity after another to pick off the significant rating color being diagramed. He checks his work at each trial, for example, by counting the number of red triangles on the chart and matching this with the number of four-line connections.

A second method, useful when some twenty or more activities are involved, interposes another step: a separate recording is posted of each relationship on the chart, grouping them according to closeness rating. The diagram is then made from this listing rather than directly from the chart. This intermediate step helps avoid errors and on complex diagraming it will actually save time. See Figure 6–7.

By calibrating or converting the flow-of-materials values into vowel-letter values before diagraming, the accuracy of the extensive flow calculations that have been made is weakened or decreased. (We discussed this in Chapter 5.) Remember, the flow of materials is based on projected P's, Q's, and R's. It is highly unlikely that these projections themselves are accurate to more than 12½%—the limits of accuracy of the vowel-letter conventions when the half degree with its dotted-line indicator is used. For these and other reasons, the combined flow-and-activity-relationship chart is most acceptable in the majority of layout-planning projects. (See Appendix XII.)

To keep it simple and clear, this explanation of diagraming has been concentrated on single-level operations. The same practice generally applies in *multi-level* diagraming. Begin exactly the same way and complete the diagram as if on one level. When the areas have been established for each activity the space relationship diagram(s) begun, the planner adjusts and rearranges for multiple-level buildings. (A specific procedure for multi-level applications of SLP is explained in Appendix XV.

However, sometimes, in working with an existing structure, the planner knows in advance of diagraming that certain fixed features definitely cannot be altered. The planner recognizes these, lists the significant ones (elevator, basement boiler room,

ACTIVITY RELATIONSHIPS GROUPED BY CLOSENESS RATING

Activities Rated A--Absolutely Necessary--(4 Lines--Red)

| | | | | |
|---|---|---|---|---|
| 1. | Punch Press | and | 3. | Lathes |
| | | | 14. | Steel Storage |
| 2. | Screw Machine | and | 14. | Steel Storage |
| 3. | Lathes | and | 4. | Heat Treat |
| 4. | Heat Treat | and | 5. | Grinding |
| 5/6. | Grinding (5) | and | 7. | Assembly |
| | Grinding Washer (6) | | 22. | Coolant Cleaner |

**Figure 6–7.** Recapping the relationship chart prior to diagraming. When there are many relationships to be diagramed, a recap of these relationships will help avoid errors and save repeated scrutiny of the chart. This is done by listing each pair of activities, grouping them by rated closeness—all A-rated (red) relationships first, then E-rated relationships. This lists all activities in the order in which the diagraming will be done. The example here is for the same ball-and-roller bearing plant diagramed in figure 6–6. Actually, the diagram was made from this recap.

second floor show room, and the like) on the activity relationship chart, and designates that fixed activity when diagraming.

### Location Considerations

Diagraming relationships can be done with or without considerations of the actual location to be used. Generally speaking, the planner attempts to diagram the ideal arrangement of the various activities and so should use a theoretical diagram based on the data gathered. For this reason, we prefer to make the diagram completely independent of any existing buildings or existing site plans. This makes it possible to develop any arrangement without preconceived ideas, without forcing the planners into practical limitations too early. If this "arranging" is approached with a free and open mind and the theoretically best relationship is diagramed, then there is the possibility of capturing real contributions for the layout plans. There is ample time later on to adjust the theoretical or ideal arrangement to the practical limitations of fixed site and building restrictions.

On the other hand, it may take somewhat longer to go through this approach. Moreover, in some cases it is known that an existing building structure or a certain floor of a specific building must be used with no possibility of changing building features. In these situations, it may be practical to do the diagraming directly on a floor plan of the existing site or building.

## PROCEDURE FOR DIAGRAMING RELATIONSHIPS

1. Identify by number and name the activities (areas or features) to be diagramed. Code these by type-of-activity symbol--preferably right on the relationship chart.

2. If not already done in making the relationship chart, calibrate or otherwise relate flow-of-material intensities to the vowel-letter closeness-rating values, combine them with other-than-flow relationships, and record the combined relationship ratings.

3. Diagram the A (4-line) relationships first, as diagram one.

4. Rearrange to get approximately equal lengths of all 4-line relationships, then add the E (3-line) relationships. This will usually be diagram two. Add any XX (double zig-zag or double wiggly line) relationships at this time.

5. Rearrange, trying to get approximately equal lengths of the 3-line relationships but making them approximately twice as long as the 4-line relationships. Then add the I (2-line) relationships and rearrange as necessary.

6. Add the X (one wiggly line) relationships when adding the O (1-line) relationships. Then rearrange, perhaps by making two or three more diagrams to get the most acceptable fit--the arrangement having the single lines approximately four times as long as the four lines, and the other lines in similar reverse proportion.

7. Check and recopy the final diagram. It will become the basis for the layout when space is added and adjustments are made for modifying considerations and practical limitations.

**Figure 6-8.** The SLP procedure for diagraming relationships.

At this point, the planner is diagraming only the activities, not the space they require. That is, he is not attempting to fit certain areas together according to their size, but interested only in visually relating the activities to each other according to the relative closeness established. Later on, the space to be allowed each activity is incorporated with the diagram. But at this stage the planner works only with the activities themselves.

## Conclusion

There are many methods for diagraming flow of materials and activity relations, but the simple orderly procedure described in this chapter is the fastest and most practical way to proceed. Many further refinements can be made on complicated or sophisticated layout projects, but for most cases this technique is adequate. In fact, this procedure is the most universally applicable approach to arranging in a geographical pattern the relationship of activities to one another, for it is usable in planning industrial plants, service industries, and offices or service areas. And it has a specific convention for diagraming negative relationships, often overlooked in less universal methods of diagraming.

Figure 6–8 summarizes the procedure for diagraming flow and activity relationships.

## Chapter 7
# Space Determinations

Up to this point in Phase II of the SLP pattern of procedures, we have ignored space. But once the geographic arrangement of the various activities involved is worked out, the planner must establish the space for each activity. The space, or area, is then fitted to the activity relationship diagram in a Space Relationship Diagram, Box 6 of the pattern.

The space relationship diagram is actually a crude layout. Refined and re-arranged, and incorporating the modifying considerations and their practical limitations, the diagram becomes the layout. This chapter deals with Boxes 4 and 5: *Space Requirements* and *Space Available.*

The omission of space considerations up to this point does not mean that space calculations have to wait until the activities have been diagramed. Actually, space calculations can be made any time after the operating and the supporting activities have been established. However, generally the planner has a much better idea of the break-down or division of activities, and therefore of the division of space involved, if he does not dive into space determination too early. Wait at least until the input data has been analyzed, preferably until the flow of materials and activity relationship chart has been prepared.

Obviously, this rule doesn't hold for establishing the total space in order to solve a Phase I (Location of the Area to be Laid Out) problem. To decide which location is suitable, one must know how much space is required. Such a figure, however, is generally a rough estimate or an overall figure, calculated without necessarily extensive analysis.

What is covered in this chapter is the area determinations necessary to effective Phase II planning (General Overall Layout).

Because, for best results the phases overlap, in practice, the planner may have to consider Phase II space requirements to satisfy a Phase I solution. Similarly, the planner frequently must consider fairly detailed data on machinery, equipment, services, and the like (normally dealt with in Phase III) in order to determine the overall space requirements suitable for solving the general overall layout of Phase II.

### Space Requirements

Basically, there are five ways in which to determine space requirements. Each has its place; all may even be used on the same project. Different methods of determining space requirements tend to check each other, thereby giving more credence to the figures. Moreover, very precise methods of space determination may be suitable or required when fixed investment is large. Conversely, in the planning of warehouses or general office areas, there is somewhat less at stake, and accurate details may be more of an exercise in mental gymnastics than of practical use.

The five chief ways of determining space requirements include

1. Calculation
2. Converting
3. Space standards
4. Roughed-out layout
5. Ratio trend and projection

These are arranged generally in the order of greatest accuracy, and probably also in the order of most frequent use.

### Machinery and Equipment Inventory

Before the calculation method of determining space requirements can be used, the machinery and equipment involved in the project must be identified. If no inventory of equipment is on hand, the planner may have to take a physical inventory in the existing layout areas–then modify it for future conditions. We suggest accounting for each piece of equipment by department or activity so that the information will conform with the activity relationship diagram and the other preceding work.

One can also work from a file of equipment location records. It may be necessary to work from an accounting department asset record or from a professional appraiser's list. Or maybe, equipment-history cards are available in the maintenance department.

In small plants, service industries, and office areas, maintaining an equipment inventory on an up-to-date basis may not be practical. Large companies, on the other hand, especially those with frequent rearrangement of facilities or a major investment in machinery and equipment, usually find such a record almost a necessity.

There are various ways in which companies can keep their equipment inventories. We suggest the form shown in Figure 7–1, a simple and convenient form for

**Figure 7–1.** Machinery and Equipment Layout Data sheet used for space planning and inventory record. Here is a standard sheet or card with sketch, photographs, specifications, and service requirements. Note that one master sheet can be made for several duplicate pieces of equipment.

Such a record can save valuable time in working out plant layout problems. For equipment currently in use, a photocopy of its original sheet can be filed according to the activity or area where it is located. And when the equipment is moved, the sheet can be refiled for the new location. This way we can keep a complete inventory by type of equipment and at the same time by area where equipment is in use. The same applies for areas in their planning stages when equipment *is to be* used. Data sheet copies for equipment held in storage and available for use can be held in a "free and available" file. When the machine is disposed of, the record is closed out and sheet is filed away in past records.

## MACHINERY & EQUIPMENT LAYOUT DATA

Plant *Eatalot Milling Co.* Project *66532*

Prepared by *M. Jolly* With *J. Brinse* Date *10-2*

Water ____ Steam *✓* Drains ____

Comp. Air ____ Gas ____

Foundation ____ Pit ____ Level/Lag *Lag*

Exhaust ____ Sp'l. Elect. ____

Max. Height *90 ins.*

Weight (less attach.) *5100 lbs.*

| Electrical | H.P. | Volts | Cycle | Phase | Amps. | | |
|---|---|---|---|---|---|---|---|
| Drive Motor | *100* | *220* | *60* | | | | |
| Auxiliary Motor | *20* | *220* | *60* | | | | |
| Auxiliary Motor | *3/4* | *220* | *60* | | | | |

Name/Type *Pellet Mill* File *M-35*

Manufacturer *Wenger Mixer Mfg. Co.* Size/Model *Multi-Duty*

Speed/Capacity *1800 RPM* Signif. Ident. *Serial 4362*

| | |
|---|---|
| Left-Right | *84 inches* |
| Front-Back | *82 inches* |
| Net Floor Area | *48 sq. ft.* |
| Worker/Maint. Area | *56 sq. ft.* |
| Material Set-down | *—* |
| Area for Aisles | *28 sq. ft.* |
| Service/Other | *@* |
| Gross Area (plan for) | *132 sq. ft.* |

Co. Mach/Eqpt. Ident. Numbers Covered by this Sheet

| | | |
|---|---|---|
| *M-35* | | |
| *M-38* | | |
| | | |
| | | |

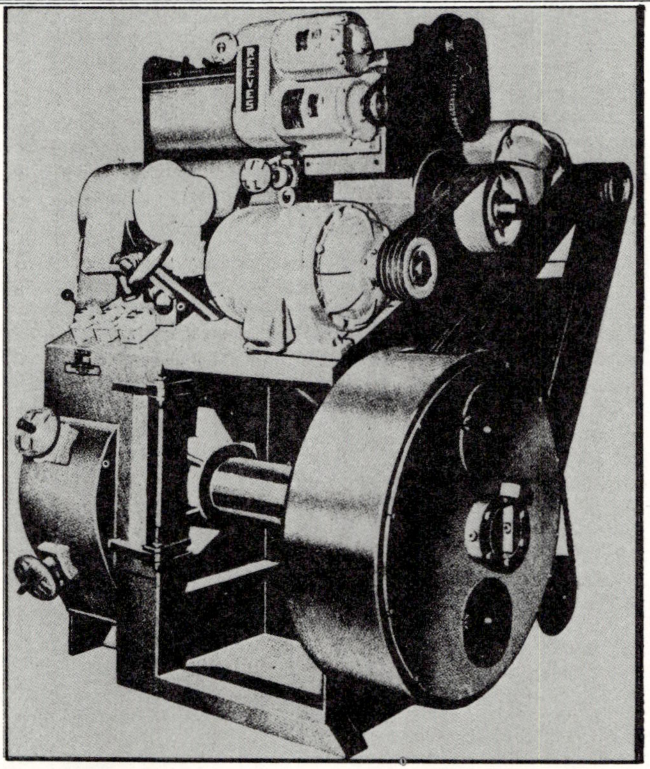

ELEVATION SKETCH OR PHOTO SOURCE *Catalogue* DATE *9-58*

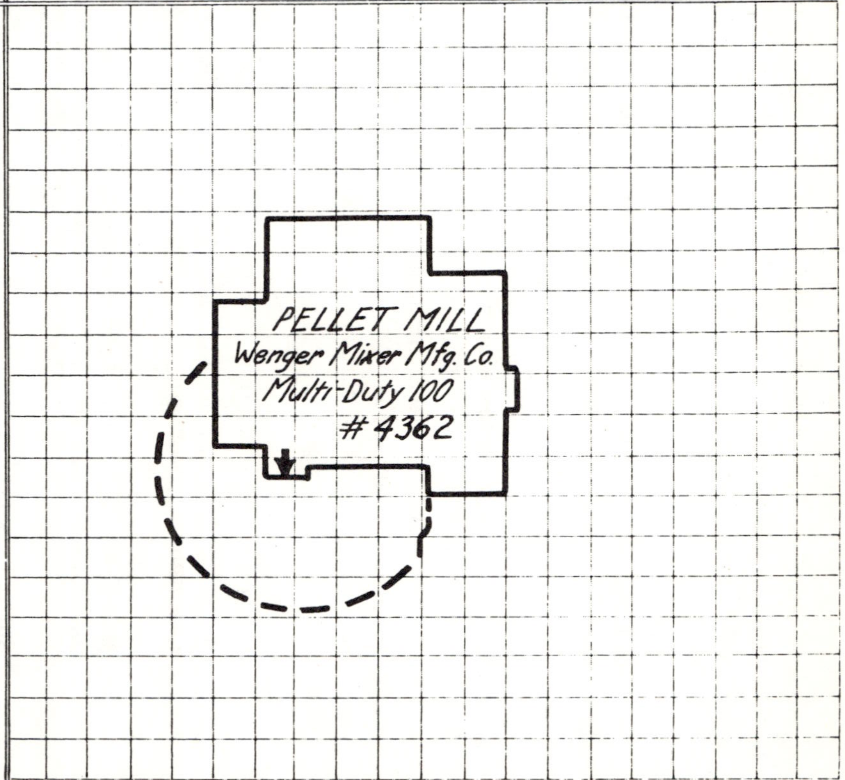


PLAN VIEW SCALE *1/4* INCH = 1 Ft.

Reference Notations/changes *@ Service area included "Worker/Maint. Area" figures*

recording and filing the necessary information—especially that needed for plant layout work. Although all of the detail may not be needed for a Phase II layout, it will in all likelihood be needed later on when the detail layout is being planned.

For initially gathering inventory information, it may be practical to use a shop worksheet and then transfer the information to the Machinery & Equipment Layout Data sheet. Such a worksheet is shown in Figure 7–2.

In addition to a record of machinery and equipment, it is highly useful to have a classification system for the equipment. This is generally set up by equipment type, thereby establishing a filing system. At the same time, this allows cross-filing by departmental areas or activities. There are many systems of classification available, and no simple one is universally used. We recommend a letter prefix indicating the class of equipment followed by a numerical system, with suffix letters as appropriate. For example:

> Prefix letter(s) indicating the general class or type of machinery or equipment: W—welding equipment; P—positioners; C—cutting torches and equipment; L—Lathes
>
> Numbers indicating subtype of equipment: W1—DC arc welders floor mounted; W6—spot welders; W8—gas welding equipment
>
> Suffix letter(s) indicating still further breakdown or accessories

An excellent classification of materials handling equipment developed by the American Materials Handling Society, is available in the *Materials Handling Handbook.*[1] See Appendix III for a partial breakdown of this classification. It needs only the prefix H (for handling) to tie in with the above system.

The classification described above refers only to type of equipment, not identification of a particular piece of machinery or equipment. To identify specific items of equipment an identification or "tag" number should be assigned each. This is usually done from a master log—not infrequently by the accounting group responsible for fixed asset records and depreciation schedules. Tag numbers should match the number tagged to or otherwise marked on the equipment. While there are many systems, perhaps the easiest is simply to start with 10,001 and keep a numerical sequence, using the same

1. Bolz, Hagemann et al, *Materials Handling Handbook* (New York: The Ronald Press Company, 1958).

**Figure 7–2.** Worksheet to gather information concerning details of each piece of machinery or equipment. Data will be used to develop space requirements, to establish utilities and services needed, and as a basis for template or model preparation. This is the working or shop sheet, from which the machinery and equipment layout data sheet (like Figure 7–1) is prepared. This latter sheet then becomes a permanent record in the layout office.

## MACHINERY & EQUIPMENT LAYOUT DATA WORK SHEET

File ____________

Plant St. Louis — Bldg. 34 — Dept. Wire Mill
Area Wire Drawing — Nearest Column Location — as of —
Compiled by R C Kneeley — With — Date 10-5

Company Identification No. 5
Size/Model/Style
Speed 1550 FPM — Capacity Finished size ⓐ
Sgnif. Identfn. (Serial No., Code) 7B-7H

Name/type 16" Wire Drawing Machine
Manufacturer Federal Wire Rope Corp.

**Utilities and Service Reqmts:**

Water Yes — Steam No
Drains Yes - Front of machine - opposite operator side
Comprsd. Air Yes - Front of machine — Gas No
Other Piping
Foundation No — Pit No
Level Yes — Lag No
Exhaust No
Elect. Outlet Yes — Direct Conctn. Yes
Spcl. Electr. Yes - To connect spooler

| Motors | HP | Volt | Cycle | Phase | Amp |
|---|---|---|---|---|---|
| Drive | 50 | 440 | 60 | 3 | |
| Auxiliary | 3/4 | 440 | 60 | 3 | |
| Auxiliary | | | | | |

**Dimensions and Area Reqmts:**

Height at max. point 7'2"
Weight (less attachments)
Left-Right (base) 21'-0"
Left-Right (max) 21'-6"
Front-Back (base) 2'-8"
Front-Back (max) 4'-0"
Net Floor Area Req'd. 56 sq. Ft.
Worker/Maint. Area 122
Material Setdown Area 72
Area for Aisles ⓑ
Support/Other Areas 27
Gross area allocated 277 sq. Ft.
(on current ☒ basis
projected ☐ basis)

Elevation Sketch (Scale 1/4" = 2 Ft)

Template Sketch (Scale 1/4" = 2 Ft.)

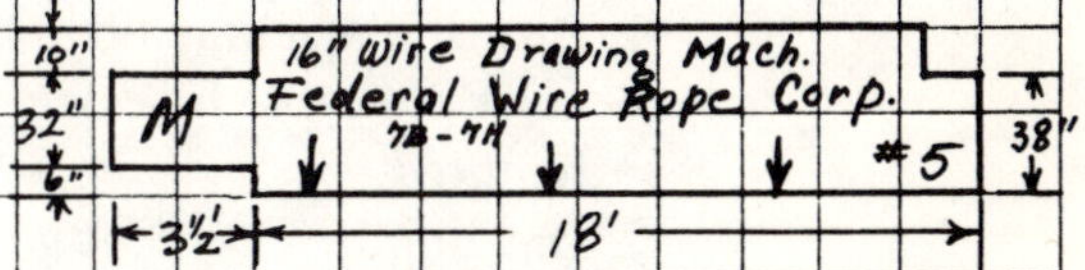


1/4"

Reference or Explanatory Notes:
ⓐ 0.034" min. diam. to 0.060" max. diam. wire
ⓑ Aisles are incl. in work/maint. Area.

File ____

letter prefix mentioned above to tie class of equipment to each tag number. See Appendix II for details of a simple identification system.

### Calculation Method of Determining Space

The *calculation method* of determining space requirements is generally the most accurate. It involves breaking down each activity or area into the subareas and individual space elements that make up its total space.

First, determine the *amount* of area for each space element, then multiply by the *number* of elements required to do the job, and add any *extra* space not generally apportionable to any of the elements.

For industrial plants this calculation can be compiled as shown in Figure 7–3. Here, each element of machinery and equipment is listed and data recorded for the area occupied by the equipment itself, for the operator's work and maintenance area, and for material set-down area—all built into a total area per machine or piece of equipment.

To calculate the number of machines or pieces of equipment needed on any project, the operation times for each part number, the number of pieces per year or per period, and allowances for downtime, scrap, and the like, must be known. (See Appendix IV for more on calculating machine requirements.) The number of machines required—ignoring scrap, downtime, and the like—equals:

$$\text{Number of machines required} = \frac{\begin{array}{c}\text{pieces per hour to meet}\\ \text{production requirements}\end{array}}{\begin{array}{c}\text{pieces per hour}\\ \text{per machine}\end{array}} = \frac{\begin{array}{c}\text{time per piece}\\ \text{per machine}\end{array}}{\begin{array}{c}\text{time per piece to meet}\\ \text{production requirements}\end{array}}$$

In calculating machines required, certain precautions must be taken.

1. Part of a machine obviously can not be purchased. Generally the planner must go to the full, or next full, machine, when a fraction of a machine is required.
2. One hundred percent good work is not possible. We must decide how much we expect in spoilage and make an allowance in determining the number of machines.

**Figure 7–3.** Calculation of space requirements by measuring individual units of machinery and equipment. Equipment is listed by identification and name at left. On the same line are entered the dimensions and the space calculations. At the bottom are added requirements applicable to the activity in question but not assignable to any specific piece of equipment. Note that at the time the equipment is examined for dimensions, its physical features are also recorded (at right). This information may not seem necessary, but undboutedly will be needed later on and this is a logical time and way to get it.

## MACHINERY & EQUIPMENT AREA & FEATURES SHEET

Company/Plant *Nortot* Bldg/Dept/Area *Cafeteria - Kitchen*

Project *73513*
By *O.M. Cramer* With *D.F. Addis*
Date *8-3* Sheet *2* of *2*

| Machine or Equipment Identification Number | Name and/or Description | Left – Right *inches* | Front – Back *inches* | Height *inches* | Area for Machinery or Eqpt. in *sq ft* | Operator(s) Work & Maintenance Area | Material Set-down Area | Total Area each Machine or piece of Eqpt. | No. of Mchs/Eqpt. | Total Net Area* (in *sq. ft.*) | 110-A.C. | 220-A.C. | Other Power | Ampere Rating | Water | Steam | Drains | Compressed Air | Other Piping *Gas* | Foundation/Pit | Exhaust/Hood | Dust Collector | Comments ----- Shape/Configuration Special Requirements |
|---|---|---|---|---|---|---|---|---|---|---|---|---|---|---|---|---|---|---|---|---|---|---|---|
| *Area Total* | *Brought Forward* | | | | | | | | | *1218* | | | | | | | | | | | | | |
| *Bake* | *Area* | | | | | | | | | | | | | | | | | | | | | | |
| *D-1* | *Sink* | 24 | 24 | – | 4 | 6 | 4 | 14 | 1 | 14 | – | – | – | – | ✓ | – | ✓ | – | – | – | – | – | |
| *D-2* | *Food Prep. Table* | 120 | 30 | – | 25 | 111 | – | 136 | 1 | 136 | ✓ | – | – | – | – | – | – | – | – | – | – | – | |
| *D-3* | *Steamer (stag)* | 19 | 30 | 25 | 4+ | 4+ | – | 8 | 1 | 8 | ✓ | – | – | – | ✓ | ✓ | ✓ | – | ✓ | – | ✓ | – | |
| *D-4* | *Proof Box* | 21½ | 28 | 63 | 4+ | 10(a) | – | 14 | 1 | 14 | ✓ | – | – | – | – | – | – | – | – | – | – | – | |
| *D-5* | *Bake Oven* | 60 | 40 | – | 17- | 23 | – | 40 | 1 | 40 | – | – | – | – | – | – | – | – | ✓ | – | ✓ | – | *Use Tray 18×26* |
| *D-6* | *Portable Mixer, Hobart A-200* | 18 | 18 | – | 2+ | 5- | 4 | 11 | 1 | 11 | ✓ | – | – | – | – | – | – | – | – | – | – | – | |
| *Cook* | *Area* | | | | | | | | | | | | | | | | | | | | | | |
| *F-7* | *Range* | 29 | 34 | – | 7- | 17(b) | – | 24 | 1 | 24 | – | – | – | – | ✓ | – | – | – | ✓ | – | ✓ | – | |
| *F-8* | *Range* | 29 | 34 | – | 7- | 17(c) | – | 24 | 1 | 24 | – | – | – | – | ✓ | – | – | – | ✓ | – | ✓ | – | |
| *F-10* | *Portable Slicer, U.S. Model 806* | 26 | 20 | 20 | 3½ | 10- | 6 | 19 | 1 | 19 | ✓ | – | – | – | – | – | – | – | – | – | – | – | |

* Required space for main or delivery aisles and service areas not included.

Total Net Area Required* (in *Sq. Ft.*) *1508*

Aisles *720*
Services *305*
Other *–*
Total Area Rq'd. *2535*

Reference Notes:
a. *Door Swing Is 21½ Inches*
b. *Door Swing Is 17 Inches*
c. *Door Swing Is 17 Inches*
d.
e.
f.

RICHARD MUTHER & ASSOCIATES – 153

3. Known or anticipated capacity-reducing delays—whether due to the operator (avoidable delays) or to the operating practices of the plant (unavoidable delays)—must be incorporated, if they have not been built into the figures as a delay allowance when the operating times per piece were established.
4. Machine utilization, due to chang-over or set-ups, failure of utilities, repair, or maintenance, is an important factor in all plants and must be compensated for on an individual machine basis.
5. Peak conditions of quantity—if not already built into the requirement figures—should be allowed for. Actually, these may vary from machine to machine, depending upon the critical nature of the operation, whether or not the work is tightly scheduled, and whether or not the work can be put on alternate machines or sent outside to subcontract shops. If the machine is already operating on two or three shifts, the possibility of solving this problem by overtime is unlikely.
6. In balancing production lines, the planner must account for the fact that extra machine capacity may be available for use by other areas, but that because it is segregated for use in the production line, it may not be practical to move other materials in and out of the machine.
7. Where only a small portion of an additional machine is required, an attack on methods improvement or simplifying the job may reduce operating time sufficiently to avoid the investment in an additional machine which would be used only a small portion of the time.

No specific standard form has been developed for calculating storage and service-area space because of the wide diversity of activities that go on in these types of areas. Therefore, we should make our own worksheets to suit the particular requirements of the layout planning project.

Examples of such worksheets are shown in Figure 7–4. Not infrequently, one must dig much deeper than merely gathering the data for these forms. We are trying to keep this presentation simple, but the planner should not hesitate to use more detailed or more sophisticated methods of space calculation whenever it is warranted.

For calculating space for office areas, the working form shown in Figure 7–5 is highly practical.

### The Converting Method of Space Determination

A second method of determining space requirements is known as the *converting*, or factoring, *method*: establishing what space is being occupied at the present time and

**Figure 7–4.** Work sheets to gather facts for calculating space requirements. Form (a) was used to accumulate requirements for a relocated plastic-molding department. The type of press and the number of each were supplied by the engineering department. Salesmens' catalogs and special prints of new presses gave other dimensions. Use Forms (b) and (c) similarly for storage and service-space requirements. (From R. Muther, *Practical Plant Layout*, New York: McGraw-Hill Book Company, 1955).

Department or Activity SALVAGE CRIB

SERVICE - SPACE REQUIREMENTS

Sheet 1 of 4

Compiled by RL date 2-9

| Type of Service | Locations of Service areas of this type | Factors used to calculate area | Area Req'd. per Service area | Total Area | Height, Pit, Foundation | Special Requirements of Service Space | |
|---|---|---|---|---|---|---|---|
| | | | | | | Utilities | Shape |

(c)

Department or Activity RAW STORES

STORAGE - SPACE REQUIREMENTS

Sheet 1 of 6

Compiled by RL date 2-4

| Mat'l.or Product Stored | Unit or Container Stored in | Quantity to be stored in this area | Probable Method of Stacking | Area Req'd. per Unit Stored | Total Area Req'd. | Height of Storage Stack | Special Requirements as to shape of storage space |
|---|---|---|---|---|---|---|---|

(b)

Department or Activity MOLDING

OPERATION - SPACE REQUIREMENTS

SHEET 1 of 14

Compiled by RL date 2-6

| Type of Machine or Equip. | No. of Each Req'd. | Area Occupied by Equip. | Working Area, (Men and Material) | Access Area, (Handling & Service) | Aisle or Other | Area per Mach. | Total Area Req'd. | Height of Machines or Equip. | Special Requirements as to Utilities or Shape of Workplace |
|---|---|---|---|---|---|---|---|---|---|
| W.S. 32 | 1 | 18×3 | 18×8 | 18×4 | 0 | 270 | 270 | 7'-8" to hopper | Room needed for draw bars and die changes |
| W.S. 16 | 2 | 16×3 | 16×8 | 16×4 | 0 | 240 | 480 | 7'-4" to hopper | Room needed for draw bars and die changes |
| R.P. 8 | 8 | 16×3 | 16×4 | 16×3 | 16×2 | 192 | 1,536 | 6'-6" to hopper | Room needed for draw bars and die changes |
| R.P. 4 | 2 | 14×3 | 14×5 | 14×3 | 0 | 154 | 308 | 6' to hopper | Room needed for draw bars and die changes |
| Impc. 16 | 2 | 12×3 | 12×5 | 12×4 | 12×2 | 168 | 336 | 12'-6" to hopper | Die end on aisle |

(a)

converting it to what will be required for the proposed layout. This conversion is usually a matter of logic, best estimate, or educated guess.

In using the converting method there is a great temptation merely to jump directly from "what we have now" to "what we anticipate needing." Do not be drawn into this trap. The layout planner must be more discriminating. Adjust the existing space to what is actually needed now and then convert it for each individual area. The form in Figure 7–6 is recommended for it allows sound decision-making at each stage.

Converting is a highly practical method, especially applicable:

1. When the project involved must be done in a hurry
2. When space requirements for Phase I only are being planned
3. When the nature of the work done in any activity or area is so diverse and complicated that detailed calculations are not warranted
4. When key elements required for calculation, such as the product or quantity information, are too general or too indefinite to justify using the calculation method.

It is not uncommon to calculate space requirements for manufacturing areas and then use the converting method for establishing supporting service and storage areas.

### Space Standards

The use of pre-established space standards is a practical way to determine requirements for many projects. Basically, once the area requirements for a given machine or space element are established, the planner should be able to use that figure over and over again. See Figure 7–7 for an example of space standards.

In actual practice, however, this does not always work. In fact, there is great danger in adopting the standards established by someone else unless the planner understands what is involved in the space element, what the working conditions are—or are anticipated to be—and how to refer to and get access to the back-up data underlying the establishment of the standards.

For example, a space standard of 300 square feet gross area per automobile for parking lots is acceptable for plants in the United States. However, this same standard does not apply in Europe where smaller cars predominate.

Similar discrepancies exist in various publications. It is tempting to use certain published standards. But these were established for specific companies under specific

**Figure 7–5.** Calculating space requirements for office areas. This example shows the gathered information for an existing office being relocated. If the operations, organization, or size of staff are to be different at the new location, another sheet with projected space and equipment requirements should be made up.

## OFFICE LAYOUT REQUIREMENTS DATA

☒ Present ☐ Projected

Company *Cader* Bldg. *A* Floor *1* Page** *1*

Sq. ft. now Assigned** Dept. *Marketing* Sect'n *B* Sheet *1* of *2*

| Room or Area Number | Identification: Personal Name or Work Group. (List each individual — — arranged by work group— or list name of work group or function. If individual's name is used, check appropriate columns; if working group, enter numbers.) | Job title or Description | Space Req'd: Male | Female | Type of Space # | Square Feet* (Net Area) | Equipment Involved (Fill in Quant'y, Size & Description as req'd.) — Regular Equipment: Standard Desk *34x60* | Typist Desk | Bench or Table | Size | Chair or Stool | Letter File | Legal File | | Telephone | Special Light | Dict./Trans. Eqpt. | Tel./Dict. on Stand | Elec. Mach on Desk | Elec. Mach. on Stand | Other Equipment: No. | Description | Left to Right | Front to Back | Height | Special Utilities or Requirements (Describe) |
|---|---|---|---|---|---|---|---|---|---|---|---|---|---|---|---|---|---|---|---|---|---|---|---|---|---|---|
| | *Department Heads* | | | | | | | | | | | | | | | | | | | | | | | | | |
| | *J. Jones* | *Mkt. Research* | ✓ | | *P* | *80* | *1* | | *1* | *30x60* | *2ᵃ* | | | | *1* | | | | | | | | | | | |
| | *S. Smithson* | *Marketing* | ✓ | | *P* | *80* | *1* | | *1* | *30x60* | *2ᵃ* | | | | *1* | | | | | | | | | | | |
| | *H. Wolfer* | *Promotion* | ✓ | | *P* | *80* | *1* | | *1* | *30x60* | *2ᵃ* | | | | *1* | | | | | *1* | | | | | | |
| | *L. Charles* | *Field Sales* | ✓ | | *P* | *80* | *1* | | *1* | *30x60* | *2ᵃ* | | | | *1* | | | | | | | | | | | |
| | *Supervisor* | | | | | | | | | | | | | | | | | | | | | | | | | |
| | *B. Bushler* | *Research* | ✓ | | *O* | *38* | *1* | | | | *2ᵃ* | | | | *1* | | | | | | *1* | *Bookcase* | *36* | *12* | *30* | |
| | *D. Dolligsen* | *Education* | ✓ | | *O* | *35* | *1* | | | | *2ᵃ* | | | | *1* | | | | | | | | | | | |
| | *Secretary* | | | | | | | | | | | | | | | | | | | | | | | | | |
| | *B. Portier* | *Secretary* | | ✓ | *O* | *38* | | *1* | | *34x54* | *1ᶜ* | *3* | | | *1* | | *1* | *1* | | | *1* | *Supply Cabinet* | *36* | *18* | *72* | |
| | *R. Trotter* | *Secretary* | | ✓ | *O* | *38* | | *1* | | *34x54* | *1ᶜ* | | *3* | | *1* | | | | *1* | | *1* | *Office Valet* | *48* | *16* | *89* | |
| | *City Sales* | | | | | | | | | | | | | | | | | | | | | | | | | |
| | *E. Able* | *Manager* | ✓ | | *O* | *80* | *1* | | *1* | *34x54* | *2ᵃ* | | | | *1* | | | | | | | | | | | |
| | *G. Howard* | *Cashier* | | ✓ | *O* | *65* | | *1* | | *34x54* | *2ᵇ* | | *1* | | *1* | | | | *1* | | *1* | *Coin Sorter & Stand* | *39* | *36* | *49* | |
| | | | | | | | | | | | | | | | | | | | | | *1* | *Safe* | *22* | *22* | *31* | |
| | | | | | | | | | | | | | | | | | | | | | *1* | *Coin Wrapper & Stand* | *18* | *24* | *48* | |
| | | | | | | | | | | | | | | | | | | | | | *1* | *Storage Cabinet* | *36* | *18* | *72* | |
| | | | | | | | | | | | | | | | | | | | | | *1* | *Counter* | *36* | *14* | *42* | |
| | *Conference Room* | | | | *P* | *115* | | | *1* | *36x96* | *10* | | | | | | | | | | *2* | *Hall Trees* | *18* | *18* | *66* | |
| | Totals | | *7* | *3* | | *729* | *7* | *3* | *6* | | *28ᵈ* | *3* | *4* | | *10* | | *1* | *1* | *2* | *1* | *10* | *Assorted* | | | | |

# Type of Space: P-Private E-Semi-Private Enclosure O-Open or Semi-Open S-Service or Special Areas (For special areas furnish sketch and dimensions) G-General areas not specifically assigned. (Shared areas not considered service or spcl.)

*Net Area: Excludes stairways, restrooms, central corridors, etc. (except where specifically listed), but includes working and access area or apportioned share thereof.

**Filled in by layout engineer (or office layout analyst)

Reference Notes: *a. Includes: Executive and Side Chair*
*b. Includes: Steno and Side Chair*
*c. Includes: Steno Chair*
*d. Includes: 7 Executive Chairs, 3 Steno Chairs, 18 Side Chairs*

Data Recorded by *O.M.C.* Date *8/3*

RICHARD MUTHER & ASSOCIATES – 151

conditions and are not necessarily adoptable generally. Therefore, we suggest using pre-established space standards developed by others as a guide only. Of course, once you develop your own standards and keep a record of how they were established, they can be used over and over again in your company so long as the same conditions prevail.

### Roughed-Out Layout

In some layout projects, calculating or converting is impractical, and no standards are available. If a scale plan of the area is available, if templates or models of the equipment involved are already on hand, and particularly if certain activities are critical or represent very high investment, it may be advisable to rough out detail layouts of certain areas and use them for space requirements. This preliminary detail planning is a perfect example of phases' overlapping—here Phase II into Phase III. This method of space determination is expected to be used for critical areas of high investment, relatively fixed equipment, large machinery, or multiple work stations that should line up (as a conveyorized assembly line). What's more, both management and operating supervisors have more confidence in the planner's space requirements when they can see an intended layout.

For areas where equipment can be easily moved, it is unnecessary to be too specific in the arrangement. That is a Phase III job. Only sufficient attention is given to the roughed-out layout to assure that it is an acceptable arrangement that can be made to work. Everyone should understand that this roughed-out layout is not the layout which will eventually be selected. In fact, the eventual shape of the area may be quite unlike the roughed-out version. However, this technique is adequate for establishing the space requirements for use in planning Phase II, general overall layout.

**Figure 7-6.** The conversion method of space determination. Activities are listed at the left with the space each currently occupies shown beside it. Note that the space unit at the top of each column can be varied, depending on the area (square feet, square meters, acres, etc.) or whether an amount or a percentage figure is used (columns c, e, and h).

Before converting the present areas to the plan-for areas, several steps should be taken. First, present space should be adjusted from what it "happens to be" to "what it should be" to do the current job satisfactorily. This results in the figures in column d. Then, based on the expansion or contraction plans, the basis of which is identified at the top of the form, an increase or decrease figure is assigned for each activity. This is much more accurate and discriminating than using a blanket figure for all activities. The conversion is then made to the required areas (column f). The final step is to compare the required area with the area available. Adjustments between the two will then result in the plan-for areas shown in column g.

The example shown here is for the expansion of a small company producing special electronic tubes and nuclear detectors. Note how they have used the form to plan an immediate expansion, columns e, f, and g based on sales two years in the future, and have projected their space planning for a second period ahead. With this form the desired plan-for periods or quantities are keyed to the appropriate columns by identification at the top of the form.

## SPACE REQUIREMENTS — — CONVERTING

Plant *Roberts Laboratories* Project *35638*
By *R. S. Linton* With *J. C. Betts*
Date *9-1* Sheet *1* of *1*

Basis (year, period, quant'y) of Cols. e, f, g *Plus 2 Yr. Sales* Cols. h, j, k *Plus 4 Yr. Sales*

| a Activity — — Area or Dept. (Unit →) | b Area Now Occupied (Sq. Ft.) | c + or − Adjstmt. (Sq. Ft.) | d Should Have Now (Sq. Ft.) | e Increase Decrease (%) | f Req'd Area Determined (Sq. Ft.) | g Plan-For Area (Sq. Ft.) | h Increase Decrease (%) | j Req'd Area Determined (Sq. Ft.) | k Plan-For Area (Sq. Ft.) |
|---|---|---|---|---|---|---|---|---|---|
| 1. Purchased Parts Storage | 30 | +70 | 100 | 0 | 100 | 75 | +20 | 120 | 120 |
| 2. Machine Shop | 400 | +600 | 1000 | −15 | 850 | 850 | +10 | 1100 | 1050 |
| 3. Brazing | 60 | +60 | 120 | +50 | 180 | 150 | +75 | 210 | 200 |
| 4. Plating Uranium | 50 | +50 | 100 | +20 | 120 | 110 | +50 | 150 | 150 |
| 5. Plating Other | 40 | +110 | 150 | +10 | 165 | 165 | +30 | 195 | 200 |
| 6. Welding | 190 | +210 | 400 | −5 | 380 | 380 | +10 | 440 | 440 |
| 7. Sand Blast | 45 | +45 | 90 | +5 | 95 | 90 | +10 | 100 | 100 |
| 8. Seal Making | 190 | +210 | 400 | +20 | 480 | 440 | +30 | 520 | 500 |
| 9. Leak Test | 20 | +20 | 40 | +25 | 50 | 50 | +30 | 50 | 50 |
| 10. Vacuum Furnace | 25 | +75 | 100 | +10 | 110 | 100 | +50 | 150 | 150 |
| 11. Tube Exhaust | 165 | +205 | 370 | +20 | 445 | 400 | +25 | 460 | 450 |
| 12. Gas Preparation | 15 | +35 | 50 | +30 | 65 | 65 | +40 | 70 | 70 |
| 13. Receiving | 150 | 0 | 150 | +10 | 165 | 165 | +20 | 180 | 180 |
| 14. Shipping | 200 | 0 | 200 | +10 | 220 | 200 | +20 | 240 | 240 |
| 15. Office and Rest Rooms | 300 | −20 | 280 | 0 | 280 | 260 | +30 | 365 | 300 |
| 16. | | | | | | | | | |
| 17. | | | | | | | | | |
| 18. | | | | | | | | | |
| TOTALS | 1880 | +1670 | 3550 | — | 3705 | 3500 | — | 4350 | 4200 |

NOTES:

Page 4 GR-1 to GR-9

| Code | Description | Min. Area Required (square feet) |
|---|---|---|
| **Grinders--GR-1 thru GR-12** | | |
| GR-1-B | B & D 10" Bench | 4 |
| GR-1-B | B & D 6" Pedestal | 4 |
| GR-1-B | Black & Decker H-4700 | 7 |
| GR-1-B | Blount 14" Heavy Duty | 18 |
| GR-1-C | Cinn. Pedestal | 7 |
| GR-1-S | Sterling 14" Tool | 18 |
| GR-1-S | Standard | 16 |
| GR-1-V | Van Norman 39 Spcl.Rad. | 45 |
| GR-1-W | Wickes 2-Spdl. Tool | 12 |
| GR-2-B | Badger 220 | 64 |
| GR-2-H | Hanchett 121 | 100 |
| GR-3-G | Gardner 125 | 120 |
| GR-3-H | Ha[illegible]t 230 | [illegible] |

**Figure 7-7.** Estimating space requirements by use of pre-determined space standards. Here is a sample sheet from space standards developed by a multi-plant manufacturer operating several hundred machine tools.

"Minimum area required" in square feet is established as left-right and front-back dimensions including table travel, overhang, and service doors open; plus 18 inches on three sides for cleaning or adjusting the machinery plus two feet on the operator's side.

These figures are then multiplied by a factor based on the anticipated space required for set-down space, access ways, and departmental service areas. The factor ranges from 1.3 for normal layouts to 1.8 where handling and stocking of material in the department is a real problem. The total estimated area for the department equals the sum of the minimum areas required for all the machines times the concentration-or-dispersion factor.

Almost any plant or office can set-up this kind of space standards. The organization of the data for ready reference and the estimating of the factor value are the important practical limitations to this method.

### Ratio Trend and Projection

The *ratio trend and projection method* is limited to general space requirements. It is perhaps the least accurate of any of the five methods. Still, in terms of long-range planning, it may be fully adequate, especially in areas of offices and general storage, where equipment is movable, fixed investment is relatively low, and the property can be used for more than one purpose.

The ratio trend and projection method establishes a ratio of square feet to some other factor, square feet per person employed, for example, or square feet per man-hour per year.

First establish what the ratio was in each of various periods of time in the past. By comparing these ratios for past periods, one establishes a trend for the ratio. Then project into the future what the ratio is likely to be. From this and a projection of the companion portion of the ratio, the space to meet those projections can be derived. An example of the method applied is shown in Figure 7–8.

Square foot per unit produced is perhaps the best ratio to use. Other ratio factors not illustrated which may be used include square foot per dollar of investment and square foot per operation.

### Activities Area and Features Summary

In this chapter, we have concentrated, thus far, on the amount, or quantity, or size, of space required. Actually, space for planning purposes involves *amount, kind,* and *shape* or configuration. In determining space requirements for each activity, it is logical to consider also the physical features or special requirements that distinguish its kind and shape.

And when the planner documents and summarizes space requirements for each subactivity (subarea), he should indicate all the pertinent space data. The Activities Area and Features sheet is such a combined form. It is illustrated in Figure 7–9.

This sheet can be used as a recap sheet—with the total area required listed on the top line and each of the subactivities or subareas listed on a separate line thereunder. Or one sheet can be used for each of the departments or subareas involved—with the subareas listed on each line. Further, several sheets can be used, one for the various activities in the plant, another for the office areas, and still another for the outdoor or yard areas, with a covering recap sheet on the front. A larger form, with lines spaced to match the number of activities on the relationship chart, is most frequently employed.

In any event, the activity is identified down the lefthand column—by number and by name. The third column contains the area required in square feet or square meters (as determined by one of the five methods described above). The total figure can be entered at the top, beside the name for the activity to which the sheet applies.

On the same sheet, the physical features and the shape or configuration required for the area in question are noted, in the center and righthand portions of the form. The lefthand center section has six columns where the major characteristics of each subactivity's space can be recorded. These usually affect or pertain to the building itself. Enter the unit of measure above the double line so that numbers only need be entered on each line.

When it comes to utilities and auxiliary services, the planner does not have full information on them when he determines the space. The pipe diameter and water pressure necessary, for example, are unknown. Still, the planner knows for certain types of operations, water and drains are a vital part of the installation. Therefore, a system of rating the relative importance of such features is necessary, but at this stage the planner is not concerned with questions of detail capacities or specifications for the services.

Here again, the A-to-U value rating system is used. The appropriate vowel is placed where called for under each auxiliary or utility on the line involved to indicate the relative importance to the layout planning it is likely to play. Note that where no utility or auxiliary service is required a dash is entered. The letters serve as red flags, telling where the planner must be concerned with certain features.

In the far right column, requirements for shape or configuration of the space are recorded. For example, if the rail dock is one subactivity, its long and narrow nature should be indicated. Or in planning a steel rolling mill, which must by its very nature have so many stands in line, indicate the area as so-and-so long by such-and-such wide. In general storage or bench-assembly areas, the requirements for shape or configuration are of less importance and need not be recorded.

At the bottom of the sheet is a place for notation references. Any additional comments are marked here and coded back to the point they refer to by marking the respective lowercase a, b, or c, at that point on the sheet. This is an effective way to keep working papers clean and free of notes which confuse or obliterate the data.

Whether the activities area and features sheet is used or not, some recap summary of the amount and nature of the space is needed. When doing such a recap, list the *same* activities used in the materials flow and activity relationships analyses, the same numbering identification, with the same order of listing, even though it may mean regrouping the activities for a space summary.

### Space Requirements Versus Space Available

As often as not, a layout planning project is more restricted by space limitations than any other factor—except investment money. Reduced investment funds themselves usually result in reduced space available whenever modernization of buildings or new construction is involved.

**Figure 7-8.** Ratio trend and projection method of determining area required. Figures are established for past and present sales, production, man-power and area allocated (A). From these, ratios are calculated (B). Then sales and production figures are forecast for future periods of time (C) and ratios are projected based on trends and anticipated operating conditions (D). Using forecasts and the projected ratios in various combinations, calculations are made for the square feet probably required (F). These calculated area requirements are then modified by judgment and the likely space available to program expansion plans.

| Group | | | 4 Yrs. Ago | 2 Yrs. Ago | This Year | Two Years From Now | Five Years From Now | Group |
|---|---|---|---|---|---|---|---|---|
| A | 1 | $ Net Sales | 385M | 855M | 1,300M | 1,800M | 2,600M | C |
| A | 2 | No. of pieces (units) produced | 805 | 1,720 | 2,660 | 3,800 | 5,500 | C |
| A | 3 | No. of Shop Employees | 15 | 35 | 51 | (1 ÷ 7) 67 | (1 ÷ 7) 93 | E |
| A | 4 | No. of Office Employees | 8 | 11 | 16 | (1 ÷ 8) 20 | (1 ÷ 8) 24 | E |
| A | 5 | Square feet in Service Shop | 5,250 | 11,000 | 17,000 | (1 ÷ 9) 22,500<br>(3 x 11) 23,400<br>(2 x 13) 24,700 | (1 ÷ 9) 30,600<br>(3 x 11) 32,600<br>(2 x 13) 35,800 | F |
| A | 6 | Square feet in Office | 750 | 1,600 | 2,100 | (1 ÷ 10) 2,840<br>(4 x 12) 2,700 | (1 ÷ 10) 3,900<br>(4 x 12) 3,200 | F |
| B | 7 | $ Net Sales per Employee (Shop) | $25,600 | $24,400 | $25,500 | 27,000 | 28,000 | D |
| B | 8 | $ Net Sales per Employee (Office) | 48,200 | 77,600 | 81,300 | 90,000 | 110,000 | D |
| B | 9 | $ Net Sales per sq. ft. (Shop) | 73 | 78 | 77 | 80 | 85 | D |
| B | 10 | $ Net Sales per sq. ft. (Office) | 514 | 535 | 620 | 635 | 665 | D |
| B | 11 | Sq. ft. Shop per Shop Employee | 350 | 315 | 334 | 350 | 350 | D |
| B | 12 | Sq. ft. Office per Office Employee | 94 | 145 | 131 | 135 | 135 | D |
| B | 13 | Sq. ft. Shop per piece (unit) produced | 6.5 | 6.4 | 6.4 | 6.5 | 6.5 | D |

A. Taken from company records or estimates of officials.
B. Ratios derived from figures in "A".
C. Forecast sales and production.
D. Ratios projected--based on trends and anticipated products and methods.
E. Manpower calculations based on "C" forecasts and "D" ratios.
F. Square foot calculations based on "C" and "E" figures, and "D" ratios.

Regardless of cause, the planner seldom has—and probably should not be permitted to have—all the space he wants. This means compromising or balancing what is determined as space *required* with what can logically be made *available.* The balancing is aptly termed the shoehorn-and-paring-knife stage of the SLP pattern, for it directly involves Box 4 and Box 5.

The problem of balancing space requirements against space available is really three problems:

1. Will the total amount of space available be adequate?
2. Will the divisions of space available (buildings, floors, rooms) match in amount the various areas (departments, activities, organizational groups) required?
3. Will the character or condition of the available space or space divisions be suitable for the work required to be done in the various areas?

Balancing total amount is usually a simple matter of adding and comparing. If the area requirements will not fit, the requirements must be trimmed or squeezed. Rather than making a single percentage reduction for all areas involved, make the required reductions in the areas where the least hurt to total company operation will result. This means rating (preferably with vowel-letter values) each area in order to establish which must retain its amount and which can afford to be reduced.

Usually, general, open, flexible, multipurpose areas without fixed equipment are the ones that can be reduced. After all, one can always somehow manage to find additional storage or office space when it is needed. But, this is also one reason that many layouts "go in" with inadequate storage and service areas.

A harder job than balancing totals is matching the various divisions of available space with the individual areas in both amount and condition. The more honeycombed or divided the physical space available, and the more diversified its physical features or characteristics, the more difficult matching is and the greater is the likelihood of waste or idle space.

**Figure 7–9.** Summary recording of area amounts and characteristics. This form is divided into several sections: a) the areas for each activity, b) the physical features pertaining to building structure, c) the features pertaining to auxiliaries or utilities, and d) any requirements as to shape or configuration of the areas involved. Note space at the bottom is for explanations or comments—with letter-code reference to the point needing explanation. With this form, we can consider and record the characteristics of the space for each activity at the time we determine its area.

In using the section on physical features, the left portion shows the physical requirements of the space that relate to the structure of the building. At the right, specific dimensions for the utilities and auxiliaries will not yet be known or it will not pay to engineer them at this time. Instead of size of water lines or pressure of steam, a relative-importance vowel-letter rating is used. Each activity requires relative degrees of attention being given to its various utilities and supporting auxiliaries. Thus, the rating letter identifies, or signals, the relative importance of these features in the same way as closeness was rated on the activity relationship chart.

## ACTIVITIES AREA & FEATURES SHEET

Plant Chicago
Project 472-C3
By J.O. Hook With
Date 7/1 Page 3 of 4

| | Activity: No. | Activity: Name | Area in Sq. Ft. | Physical Features Required: O'Head Clearance | Max. Overhead Supported Load | Max. Floor Loading | Min. Column Spacing | Water & Drains | Steam | Compressed Air | Foundations –or Pits | Fire or Explosion Hazard | Special Ventilation | Special Electrification | Other | Requirements for Shape or Configuration of Area (Space) |
|---|---|---|---|---|---|---|---|---|---|---|---|---|---|---|---|---|
| | | | | Enter Unit and Required Amount under each | | | | Relative Importance of Features: A – Absolutely Necessary; E – Especially Important; I – Important; O – Ordinary Importance; — · Not Required | | | | | | | | Enter Requirements for Shape or Configuration and Reasons therefore |
| | | Airlines Food Service | Total: 2580 ⓐ | Ft. | | P.S.F. | Ft. | | | | | | | | | |
| 1. | A | Office | 210 | 9 | No Overhead Loads | 100 | 10X 10 | — | — | — | — | — | — | — | — | |
| 2. | B | Food Testing | 245 | 9 | | 100 | 10X 20 | I | — | — | — | — | — | — | — | |
| 3. | C | Wash & Clean-up | 280 | 9 | | 100 | 10X 20 | A | E | — | — | — | I | — | — | |
| 4. | D | Packaging | 300 | 9 | | 100 | 10X 20 | — | — | — | — | — | — | — | — | |
| 5. | E | Pastry/Dessert | 320 | 12 | | 100 | 20X 20 | A | — | — | — | — | E | — | — | Use Existing Hood |
| 6. | F | Hot Food Prep. | 240 | 12 | | 100 | 20X 20 | A | A | — | — | — | E | — | — | Use Existing Hood |
| 7. | G | Salad and Sandwich | 280 | 9 | | 100 | 20X 20 | A | — | — | — | — | — | — | — | |
| 8. | H | Raw Food Storage | 525 | 12 | | 100 | 20X 20 | — | — | — | — | — | — | — | ⓒ | |
| 9. | J | Ready Food Storage | 120 | 12 | | 100 | 20X 28 | — | — | — | — | — | — | — | — | Rectangular - 4½ x 27' Hot & Cold Storage Boxes |
| 10. | — | Beverage | 50 | 10 | | 100 | 10X 20 | A | — | — | — | — | I ⓓ | I | — | |
| 11. | | Unassigned | 10 ⓑ | 9 | | 100 | — | — | — | — | — | — | — | — | — | |
| 12. | | | | | | | | | | | | | | | | |
| 13. | | | | | | | | | | | | | | | | |
| 14. | | | | | | | | | | | | | | | | |
| 15. | | | | | | | | | | | | | | | | |

Sub-Activities or Areas

Notation References:
- a Does Not Include 360 Sq. Ft. For Rest Rooms
- b Aisle Space To Outside Entrance
- c Need Walk-in Refrigeration
- d Fan Now Being Used

No. ______ Activity ______ Sheet 1 of 1

RICHARD MUTHER & ASSOCIATES – 150

In practice, some form of visual balancing is helpful. The chart in Figure 7–10 is an example. This chart, coded with colors, numbers, and symbols (representing condition of space, degree or priority of permissible reduction, or fixity of location) can be used for more complicated and space-matching problems. In complex cases, the planner may have to accept certain space figures tentatively and adjust them later when actually manipulating the areas into a layout.

### Conclusion

We should point out that not only is there more than one way to skin a cat—there is more than one way to solve the problem of limited space. The easiest is to ask top management for more space or for money to construct more. In many, if not most, cases, this simply isn't practical. Other courses of action, ways of finding additional "equivalent space" without going to new construction, are listed below.

1. Increase working hours—third shift, weekends, overtime.
2. Improve methods, processes, equipment.
3. Redesign products or simplify product line or components.
4. Readjust inventory policy, possibly with revised plan of distribution.
5. Overhaul production planning and control to get more from existing facilities.
6. Initiate a housekeeping campaign and scrap drive to capture wasted space.
7. Rearrange the existing layout for better space utilization—even though something else may have to be sacrificed.
8. Go up or overhead—with mechanical handling, stacking, service-storage balcony.
9. Lease nearby space and move storage, offices, or customer service areas there, thus freeing space to expand operations at the existing plant. (The use of public warehouses falls into this category.)
10. Buy, rather than make, items that are marginal; subcontract certain products or components.
11. Sell the existing building, rather than expand, and buy or lease another.
12. Decentralize. Divide the operations into two or three groups and spread to leased or available buildings in other communities.
13. Buy out or merge with another company and integrate the work of both.

**Figure 7–10.** Balancing space requirements against space available. By using a bar-chart for area measurements, various space combinations can be visualized. The requirements are on the left; available (or possibly available) space amounts on the right.

This sheet-metal fabricating and assembly plant producing vending machines wants to expand. It needs 27,900 square feet for a new wall-type series of designs. Space required is measured off at left. Space possibly available is matched at right in various combinations. In this case there is no fit practical for the space available is in widely separated buildings—even though there is a fairly close match in area totals with combination C.

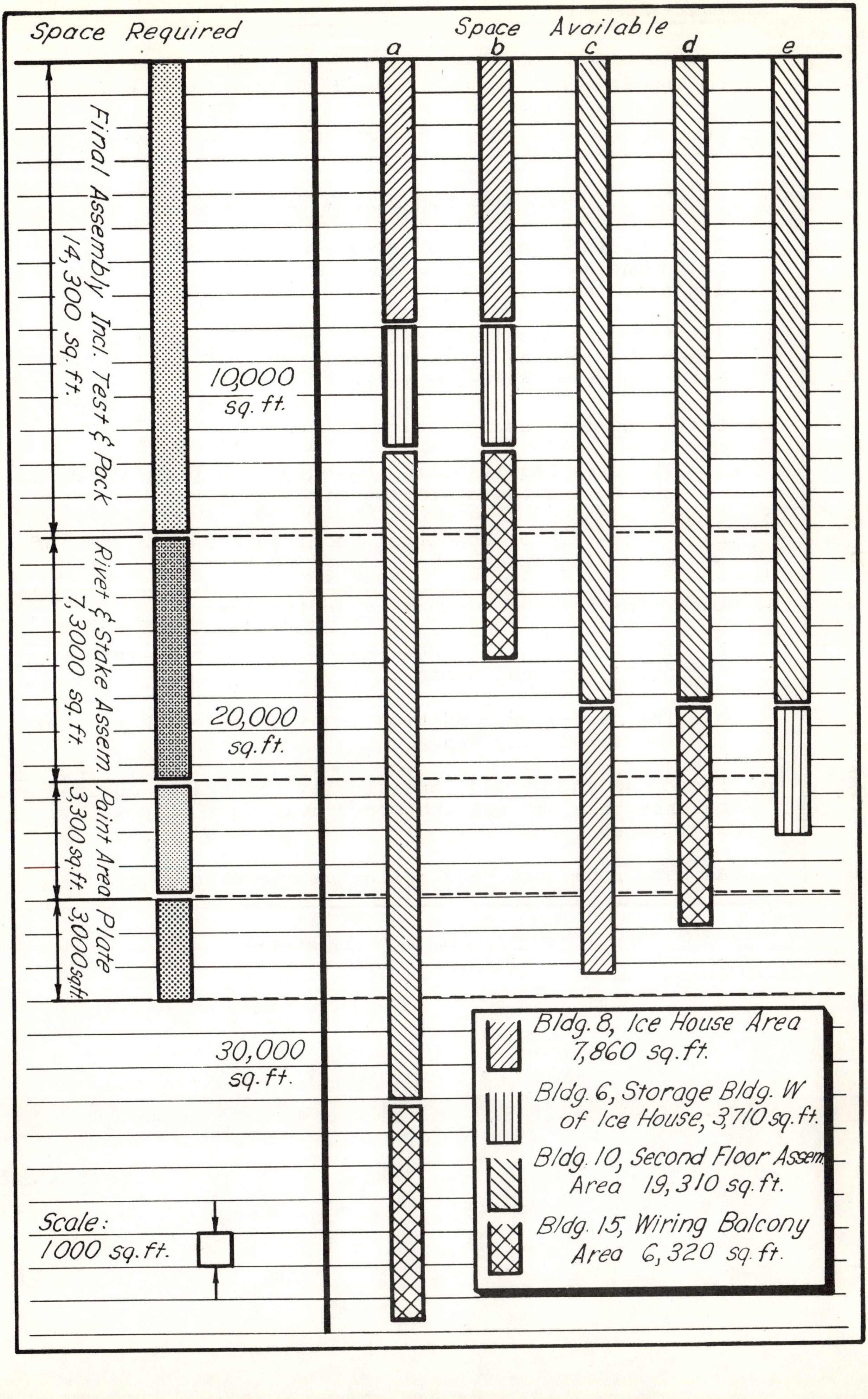

Space Required
Space Available
a
b
c
d
e
Final Assembly Incl. Test & Pack
14,300 sq.ft.
Rivet & Stake Assem.
7,3000 sq.ft.
Paint Area
3,300 sq.ft.
Plate
3,000 sq.ft.
10,000
sq. ft.
20,000
sq. ft.
30,000
sq. ft.
Bldg. 8, Ice House Area
7,860 sq.ft.
Bldg. 6, Storage Bldg. W
of Ice House, 3,710 sq.ft.
Bldg. 10, Second Floor Assem.
Area 19,310 sq.ft.
Bldg. 15, Wiring Balcony
Area 6,320 sq.ft.
Scale:
1000 sq.ft.

For a summary of this chapter, refer to Figure 7–11. It outlines the procedure for determining the space to be allowed or allocated for layout-planning projects.

PROCEDURE FOR DETERMINING SPACE

1. Identify the activities (areas or features) involved, using the same numbering and terminology as for charting and diagraming.

2. Identify the machinery and equipment involved or at least the general type of machinery and equipment-- both operating and supporting.

3. Determine for the operating activities:

    a. The area requirements, based on the plan-for P, Q, and R, and the operating times involved.

    b. The nature or condition required for each area of operating space.

4. Determine for the supporting activities:

    a. The area requirements, based on the plan-for P, Q, and S, and the times involved.

    b. The nature or condition required for each area of the supporting space.

5. Recap the amount and condition of the space required and balance this against the space available or possibly available.

6. Adjust, rebalance, and refine as necessary.

**Figure 7–11.** Procedure for determining space allocation or allowance.

## Chapter 8
# Space Relationship Diagram

We are now ready to tackle Box 6 of the SLP pattern of procedures: the Space Relationship Diagram. Flow and activity relationships having been determined and diagramed into a geographical arrangement, the space requirements for each activity having been established and balanced against the space available, the planner applies the space to the diagram.

### Fitting Space to Diagram

In fitting space to the diagram, the planner again has the flow and/or the activity relationship alternatives. That is, he can

1. combine the space with the flow diagram only;
2. combine the space with the activity relationship diagram; or
3. combine the space with a combined flow and other-than-flow relationship diagram.

The method chosen, again, depends upon the relative importance of materials flow and the relationship of supporting services—as emphasized in Chapter 5.

When using the flow diagram as the basis, convert each activity designated on the diagram from its identifying symbol to its allowed size. Work to some convenient scale to designate the areas. Generally, this is much easier to do on a sheet of paper printed with cross-section grid lines.

Each activity is still identified by symbol, number, and, possibly, name, but now—in addition to diagraming to scale—the actual square feet may be written in. This way, the space is recorded both in actual numbers and visually, in relative size. An example of this is shown in Figure 8–1.

When working from the activity relationship diagram—either service only or combined flow and other-than-flow—proceed in a similar way. Enlarge each activity symbol into its specific size, on cross-section paper, at a convenient scale. Retain the same geographical arrangement of the activity relationship diagram. Each area is adequately identified to refer to the previous charting and diagraming. An example is shown in Figure 8–2.

Many refinements of space relationship diagrams can be made to show particular information pertinent to the layout planning project at hand: Existing buildings versus new construction; number of employees; need for, or profit potential of, expanding or relocating any activity: fixity or difficulty of relocating: required condition

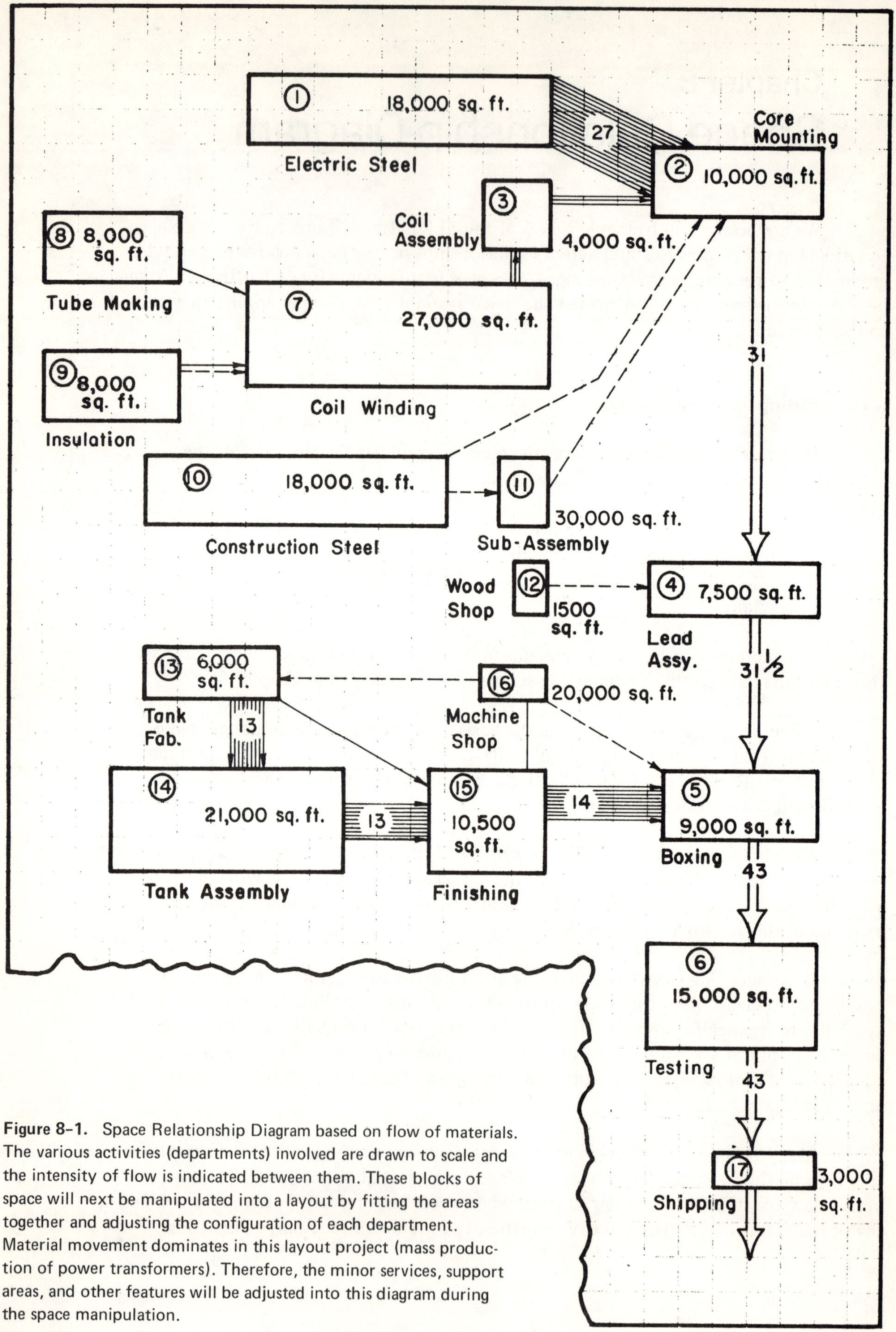


**Figure 8-1.** Space Relationship Diagram based on flow of materials. The various activities (departments) involved are drawn to scale and the intensity of flow is indicated between them. These blocks of space will next be manipulated into a layout by fitting the areas together and adjusting the configuration of each department. Material movement dominates in this layout project (mass production of power transformers). Therefore, the minor services, support areas, and other features will be adjusted into this diagram during the space manipulation.

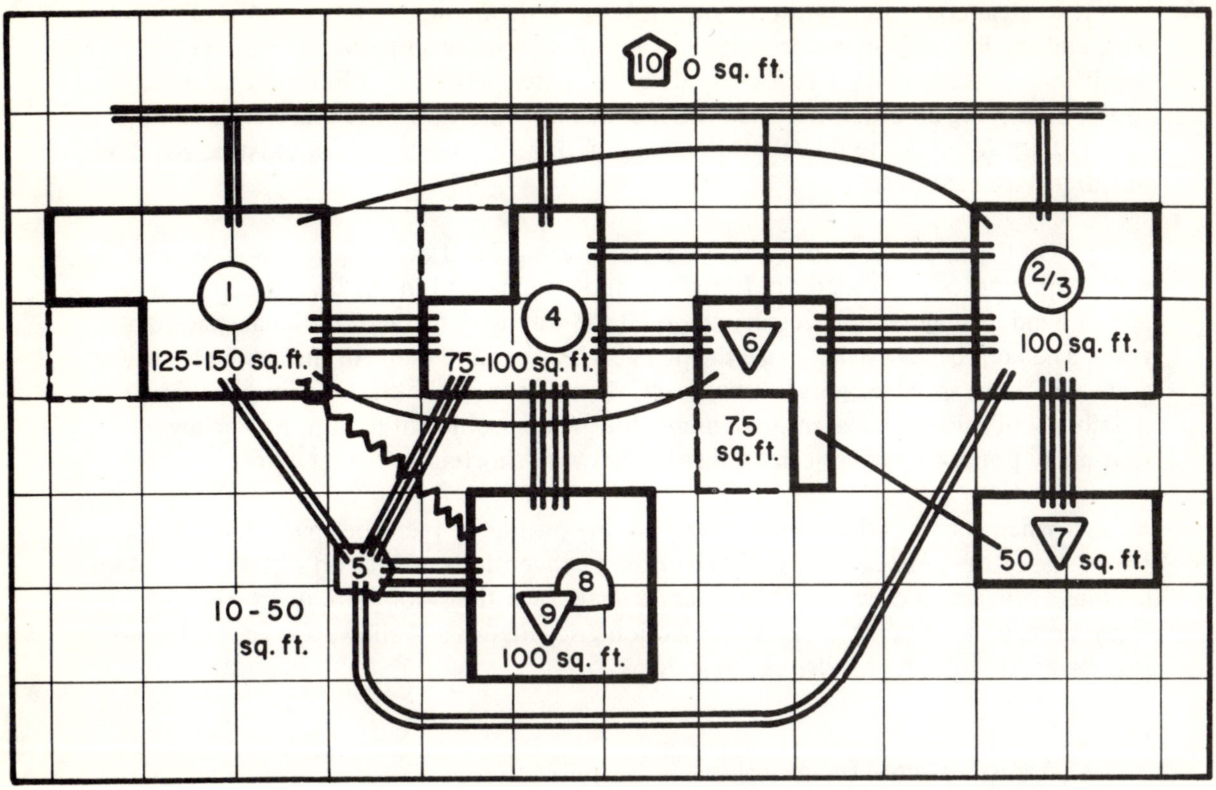


**Figure 8–2.** Space Relationship Diagram based on activity relationship chart and diagram. This example is for planning of the same engineering office discussed in Figures 5–2 and 6–5. The space allowed is merely added to its proper designated symbol on the activity relationship diagram.

There are ranges in the amounts of space, established when the requirements were determined. These ranges are handled by diagraming the minimum figure in solid lines and dotting-in the additional amount. Note also that some of the activities are combined. When this occurs, the rated-closeness lines are added together—as activities 2 and 3 in relation to 6. In practice, each relationship for combined activities should be re-examined for two one-line relationships don't necessarily equal one two-line relationship.

of the area; suitability of existing location—all be coded into the diagram by use of colors, symbols, letters, and the like. An example of refinements built into the space relationship chart is shown in Figure 8–3. In complex layout problems, showing such refinements may make the diagram too confusing. In such cases, it may be better to prepare the diagram on a reproducible print and, using several copies, to show different significant information on each copy. (See coding for multi-level layouts in Appendix XV.)

If a specific location for the layout has already been definitely decided, it may be practical to draw the space relationship diagram directly on floor plans of the designated building. This can lead to short-cuts in planning. However, it can also lead to jumping to conclusions and missing real opportunities to make major improvements or contributions to the layout. Therefore, to realize the full potential of planning the layout to ideal conditions without the limitations of existing columns, walls, rail sidings, and the like, it is better to work out the space relationship diagram to meet the conditions of the activity relationship diagram. Later, when adjusting the space diagram to the modifying considerations and their practical limitations, the planner has ample opportunity to bring the theoretical diagram within the constraints of existing buildings or other fixed features.

On the other hand, if it is known that certain fixed building features, such as walls, columns, or floor loads, definitely cannot be changed in the layout planning project at hand, it may be unnecessary to go all the way toward the fully ideal space diagram. Then it is better to recognize the physical features and shape of the available space and to place the space diagram in this specific location from the start. So long as there is no possibility of missing a real improvement, follow this course of action. But be aware that many potential savings can be lost by this very shortcut.

When starting with the existing building outline to plan the space diagram, take advantage of the cross-section grid and coding conventions, just as on a clear sheet. This is done by placing a clear cross-section grid sheet over the floor plan, or by placing a transparency of the print over a cross-section grid. Figure 8–4 shows a space relationship diagram built on an existing floor plan.

### Adjusting Space Relationships

As we can see from the above text and illustrations, we are practically at the layout stage. A space relationship chart is, in effect, a layout, when the space is joined together and fitted in the appropriate way. This joining and fitting of the spaces involved can be done in two ways:

1. By *sketching* to scale, on cross-section paper, various alternative combinations of fits and configurations of the activities involved
2. By *moving unit-area blocks* of space—the actual areas being represented by an equivalent number of movable blocks—and making various arrangements of them.

The advantages of sketching on cross-section paper are: immediate availability; a record of each alternative tried; and a plan that can be reproduced conveniently.

The advantages of the unit-area block method, on the other hand, are: the space need be counted into blocks only once and not for each sketch; time is saved because only one coding need be marked on the blocks; and the physical replicas of

**Figure 8–3.** Space Relationship Diagram with added refinements. This was used in planning a rearrangement of a structural steel fabricating plant. The activities (or areas) are drawn to scale, based on the area requirements anticipated. Activity identification is in the center and other information as explained is indicated in each of the four corners. The relationships are shown in number-of-lines code. However, the reason for the relationship is partially indicated by the four different types of lines—three for material movement and one for service relationships.

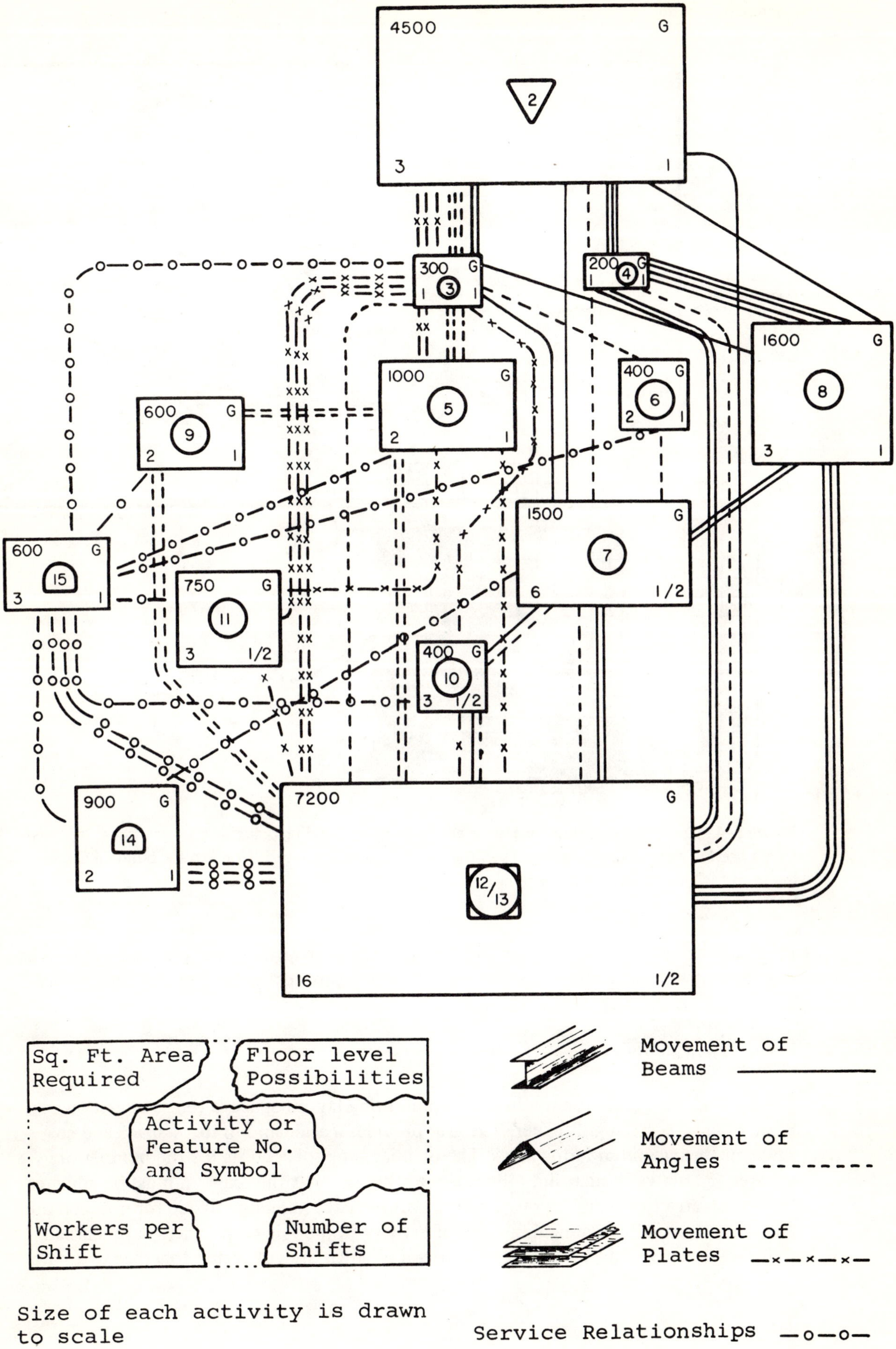
4500 G 2 3 1
300 G 3 1 1
200 G 4 1 1
1000 G 5 2 1
400 G 6 2 1
1600 G 8 3 1
600 G 9 2 1
1500 G 7 6 1/2
600 G 15 3 1
750 G 11 3 1/2
400 G 10 3 1/2
900 G 14 2 1
7200 G 12/13 16 1/2
Sq. Ft. Area Required
Floor level Possibilities
Activity or Feature No. and Symbol
Workers per Shift
Number of Shifts
Size of each activity is drawn to scale
Movement of Beams
Movement of Angles
Movement of Plates
Service Relationships

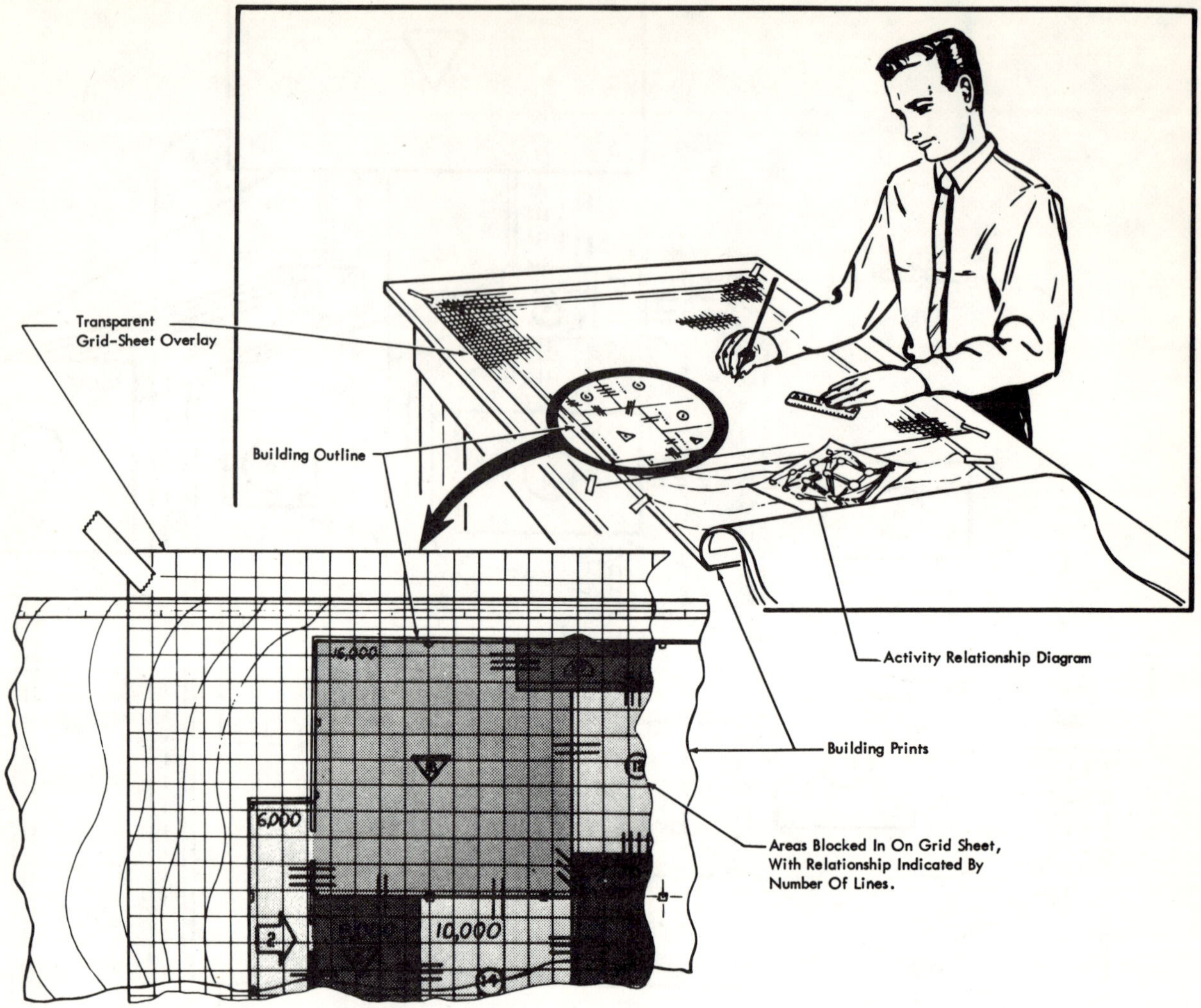


**Figure 8-4.** Space Relationship Diagram made on existing building plans. A clear grid-sheet is placed over a print of the existing building. Then the space for each activity is fitted to the building according to the activity relationship diagram.

space can be used to interest operating people, top management people, and others who are hesitant to look at drawings but willing to help physically manipulate "their" area.

### Drawing on Cross-Section Paper

The most common method of representing layout space is to use pencil on a piece of cross-section grid paper. The grid provides a scale and units of space and so saves measuring and squaring areas. The planner simply accounts for the number of squares and draws straight lines. Because it is known that the diagram must be adjusted later, each area can even be drawn with a unit of space hanging outside (or notched out inside) the normally rectangular shape of activity. Experienced planners have learned that they can save time and decrease errors if all areas are converted together to a number-of-squares count, and then each activity's number of squares accounted for when diagraming. Indeed the extra space in the first column of the Activities Area & Features Sheet is frequently used to record a number-of-squares count (although the example in Figure 7–9 uses that column for a letter identification of each activity).

True, it is necessary to select a logical scale for the cross-section sheet. More often than not, this scale has already been established on a site survey or drawings of existing buildings or floor areas. Where practical, the planner should make his scale accordingly. Later on, in Phase III detailing, use the generally accepted quarter-inch or one-eighth inch to a foot (1 to 50 or 1 to 100 metric). But at this overall planning stage, the layout project should, as a practical matter, integrate with other overall drawings available.

If the choice as to scale is open, pick one where the unit of area is a round number such as 1 inch equals 10 feet or 1 inch equals 50 feet (or such as 1 to 100 or 1 to 500 metric). A scale like one-sixteenth inch to the foot does not yield a round number for one marked-off square on the cross-section paper, and therefore, makes it more difficult to apportion and account for the various amounts of space. Where column spacings are regular, it is often practical to use the area in one bay as the unit of space.

When drawing the diagram, generally apply the following:

1. Retain the same identifying symbols, numbers, and other conventions as used in prior charting and diagrams.
2. Retain the same geographical arrangement as the activity relationship diagram—impose the space on top of each symbol. (You will have time later to join, rotate, take the reverse image, or otherwise adjust.)
3. Spread out the activities so the relationship lines can be drawn in. (You will find the diagram easier to read if you extend the relationship lines to run inside each block of space.)
4. Draw generally compact and rectangular blocks at first, except where shape is a requirement or an odd number-of-squares count makes it easier to distort the rectangle. (You will shape the areas as you fit them together later.)
5. Try to avoid letting relationship lines touch areas other than the ones they connect. This practice sometimes does lead to congestion of the lines, in which case, it is best to draw them straight through another area but as dotted lines.
6. Work in black and white, for both speed and fidelity in reproduction. When diagram is complete, make the areas stand out by coloring them (in total or by edge-of-area border), or by shading with a black-and-white shading code. SLP incorporates the use of the IMMS–approved standard for color coding and indexed with it for black-and-white shading the tincture code of heraldry. (See Figure 6–4, Appendix V, and the Capsule Summary on the back cover.)

For those unskilled in the advanced techniques of layout planning, pencils and cross-section paper are the most practical materials to use. Figure 8–5 shows a typical space relationship diagram.

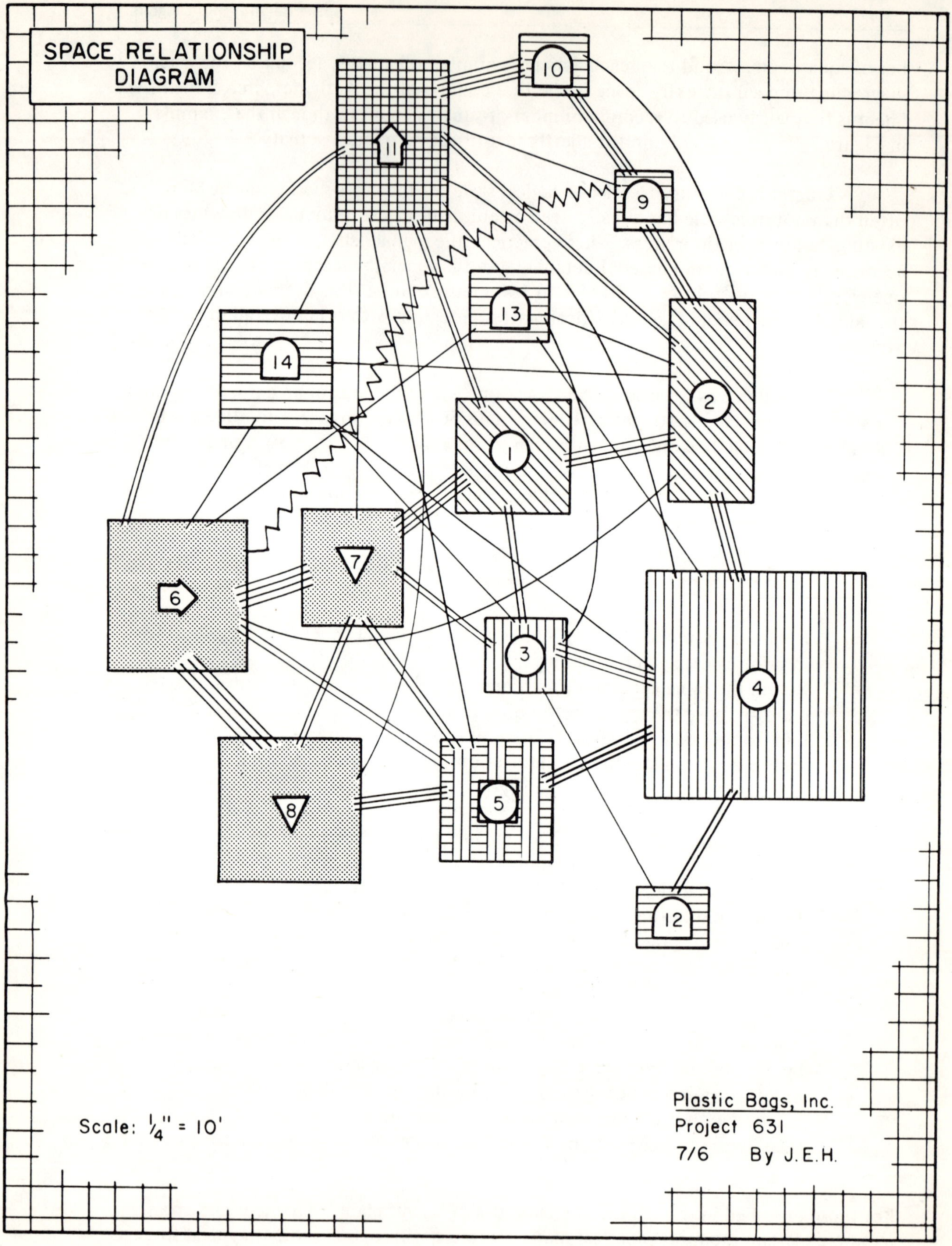
SPACE RELATIONSHIP DIAGRAM
10
11
9
13
14
2
1
7
6
3
4
8
5
12
Scale: 1/4" = 10'
Plastic Bags, Inc.
Project 631
7/6 By J.E.H.

## Unit-Area Blocks for Space Planning

The theory of *unit-area space planning* is based on the concept that any space or area can be shaped into almost any form if it is divided into small enough increments or units. Any major area, department, or plant space divided into very small units of space, can be rearranged in an infinite number of ways—just as any design of wall or house can be made from the standard-size brick or building block. See Figure 8–6.

Figure 8–7 shows how unit-area blocks are used in practice.

Unit-area blocks are similar to areas blocked on cross-section paper except that they are generally larger, are free to move, and are heavier than paper. In using them for space planning, the procedure is as follows:

1. Divide all areas required for various activities or departments being planned into a common basic unit of area.
2. Prepare blocks equal in number to the areas allowed for each activity and identify the blocks by marking on them.
3. Arrange these area units into fits and configurations based on the activity relationship diagram—or space diagrams roughed out on paper based on the relationship diagram.
4. Substitute large blocks for small and vice versa as needed or helpful, making sure not to change the space representing each activity by losing blocks or miscounting during substitutions.
5. Rearrange the unit-area blocks into as many layouts of the space involved as seems warranted. (Note: Much of this rearranging will be done as the modifying considerations and practical limitations, to be discussed in the next chapter, are woven into the space planning.)

Like cross-section sheets, unit-area blocks are made to *scale*. The base scale selected should be no smaller than one-half inch (or one centimeter metric) square, at least to start. Otherwise, the unit-area block is too small to be moved about readily and quickly, and, at the same time, is too easily disturbed when adjacent blocks are moved. The figures established earlier for space allowed are rounded off to the closest round number equal to the scale area of the smallest unit-area block.

Unit-area blocks should all be square and should be multiples of one another, usually 1, 2, 4 on the side length and 1, 4, and 16 in area to simplify counting of dis-

**Figure 8-5.** Typical Space Relationship Diagram for a plant producing plastic bags for airlines, tourist groups, schools and beach resorts. The space for each activity-area is drawn to scale—generally compact. Activity 2 has a minimum length so is drawn elongated. One square of grid equals a convenient 100 square feet. After the diagram was made, it was colored according to the type-of-space color convention. This makes the areas stand-out and helps identify further the type of function in each activity. Either colors or black-and-white shadings can be used to accentuate and further identify the general use of the space. The colors can be placed as an edge-of-area border or throughout the whole of each activity's space. Note that a combination of colors is used for a combined activity. Here, shadings are used. They permit photo reproduction. For ease of use and clarity, colors are more frequently used as shown inside back cover.

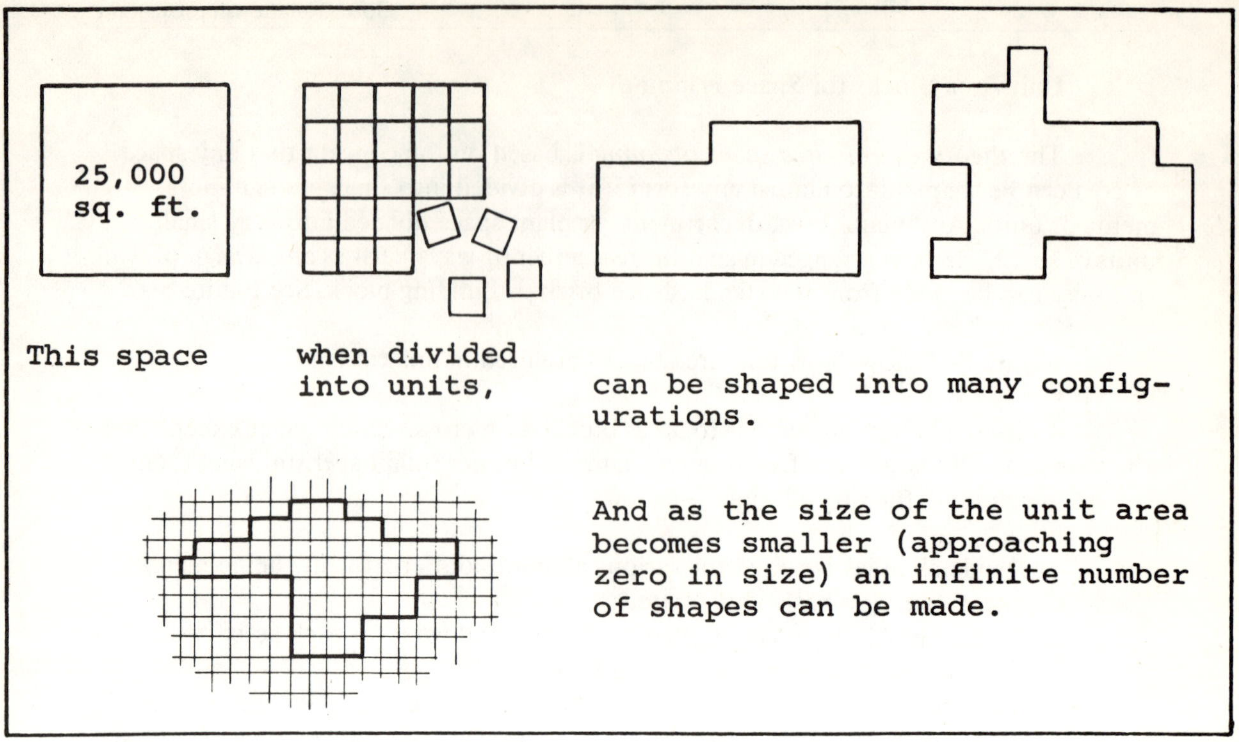


**Figure 8-6.** The theory of unit-area space planning. By dividing the space into units, it can be manipulated into many shapes or configurations.

There is a practical limit to the minimum size, however, for problems in identifying, counting, and manipulating units offset the greater accuracy. Besides, detailed accuracy of configuration is not needed in planning general overall layouts—where the unit-area technique is usable.

As in drawing, where cross-section grids smaller than one-tenth inch (or about two and one-half millimeters) are not practical, so in using unit-area blocks one quarter inch (or about five millimeters) is the minimum limit. In practice, units made of matte-surface plastic are best. They should be in layout-standard colors, with edges smooth and corners rounded for easy manipulation. They should be in modules of area such that blocks of different size are interchangeable at 4 to 1, 16 to 1, and 64 to 1.

---

**Figure 8-7.** Examples of Unit-Area Blocks in use. Figure "a" shows the planning of over 1,000,000 square feet of factory space. These blocks are made of wood. They are individually coded by identification symbol and by color. Note the way in which the planner is moving—with two straight-edge rules—several thousand square feet of plant area represented by unit-area blocks.

Figure "b" shows the use of colored plastic unit-area blocks in the rearrangement of a printing plant. Here the existing building was not to be altered in any way and therefore the planning is being done directly on a print of the floor plan of the plant. Note how the planner is converting his activity relationship diagram and his space required figures (shown on the two sheets under his left hand) into his space relationship diagram—and how, as he places blocks for his space relationship diagram, he makes his adjustments to form an alternative layout plan.

Figure "c" shows how unit-area blocks have been replaced by cellulose plastic, after the space has been formed with blocks into its desired configuration. This saves moving the many unit-area blocks around and keeping them lined up. In the upper portion of the photograph, for example, a press shop and sheet metal assembly shop are shown as block areas with the unit-area blocks on top. Below, an engineer is planning the layout of a second floor. The ground floor has already had its unit-area blocks replaced by the light-weight, easily cut cellulose plastic. A transparent grid sheet of plastic has been placed over this, and the unit-area blocks required for the second floor are being arranged in alternative patterns on the transparent plastic sheet.

a

b

c
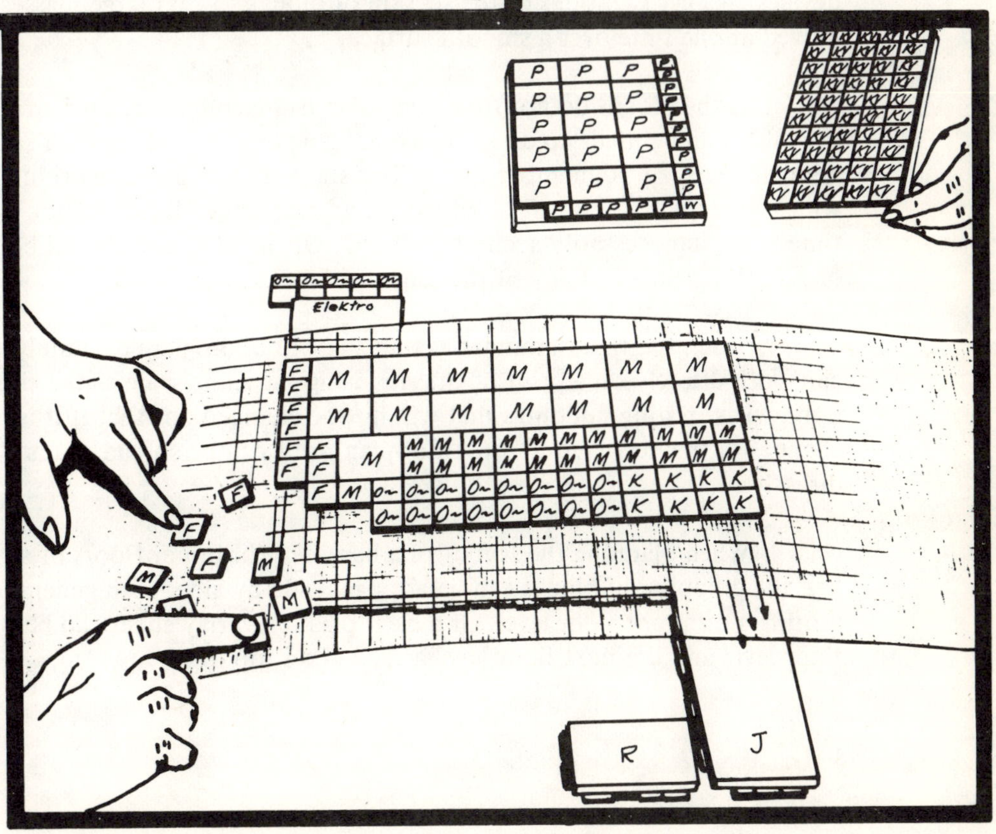

tance and calculating of areas, and so the edges of the areas line up. If a module or column spacing is already established, work the unit-area blocks to fit that spacing as closely as is practical.

In fine-area planning a one-quarter inch size may also be used. These small blocks can be substituted into the planning after the work has been refined to logical alternatives.

As pointed out in reference to drawings on cross-section paper, the unit-area block scale should match that of existing drawings when practical. In any event, the scale should be such as to allow the total area being planned to be included on *one* convenient surface. Again, use round numbers for the units of space, rather than a scale that requires working in 256 square feet or 1025 square feet.

Several *materials* can be used for unit-area blocks. Plastic, with minimum thickness of 1/32 inch, is best, although wood, tile, and sugar cubes have all been used. Paper and cardboard are too light. They blow away and do not stay at rest when other area blocks are fitted against them. Neither adhesive-backed plastic nor pinned-down cardboard is recommended for this technique. Frequently, of course, the choice of materials is a matter of using something that is readily available. The blocks should be colored and, as emphasized above, identified by numbers and symbols representing each activity or area (use the SLP conventions or the code standards of the plant involved).

*Arranging the blocks* is simple enough. Start with an arrangement that most closely resembles the activity-relationship diagram or the diagram of materials flow.

It is easiest to move the blocks by holding two rulers, or other firm-but-thin devices, against two sides of the section of blocks so that large masses of area can be moved at one time over a smooth surface.

If the blocks do not fit a particular arrangement or configuration of space, they probably can be made to do so by exchanging one unit of area for four blocks of the next smaller size. Remember that at this stage the planner is working in overall areas, which in all likelihood will be adjusted and rearranged later; do not waste too much time being unnecessarily accurate initially. Obviously, care should be taken not to lose or mix the identity of any of the required blocks.

When the fixed boundaries of a site or building are definitely established, one can work directly on a print of the site using the same scale as the site drawing. Similarly, with existing buildings that will not be changed—or that if it is not practical to change for the new layout—the floor-plan print can be spread out and worked on directly.

When planning the layout of a basement or upper floors of a multi-story building, the grouping of blocks can be arranged side by side, each general complex representing a different floor level. Or a clear plastic overlay sheet can be placed over one floor level and the next floor level arranged on top of it.

### Modifications

There are various modifications of these two basic systems—drawing on cross-section paper and unit-area blocks. For example, it may be practical to make replicas of certain machines—normally considered a Phase III technique—and fit them into a block template representing a battery or group of machines. This is generally referred to as a strip template.Or the planner may work with a block of area representing one bay of the building rather than one square unit of area. See Figure 8–8.

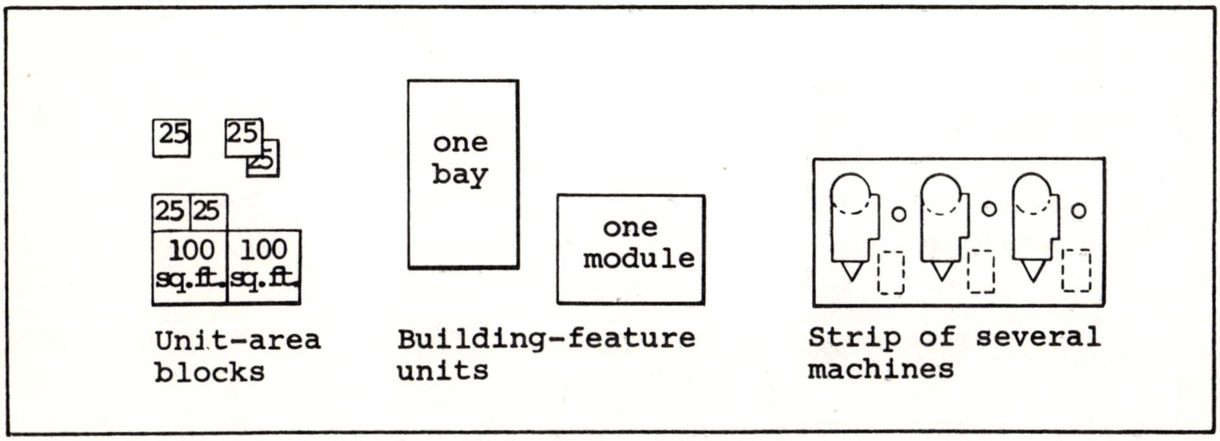


**Figure 8–8.** Block templates used in general-overall-layout space planning.

It is possible, of course, to use a combination of the two methods. And this combination we frequently recommend: First, rough out various arrangements on cross-section paper, discarding those fits and configurations which do not seem practical; then work out the alternatives that do seem practical with unit-area blocks. The resulting plan will undoubtedly have to be sketched for final analysis or presentation, which brings us back to the cross-section paper drawing again.

Unit-area blocks do not have the reproducing advantages of drawings or print-through, two-dimensional clear plastic templates. If the blocks are held down or fastened for reproduction, the advantage of manipulating them into various fits and configurations is lost. As a result, arrangement of blocks have to be redrawn or photographed.

Sometimes a large block template—usually cardboard, easily cut cellulose plastic, or styrofoam—is substituted for a large group of unit-areas, especially when the space configuration has been decided. Wherever cardboard is used, we recommend taking advantage of colored cardboard on which a grid sheet has been printed.

### Recording The Alternatives

The space relationship diagram serves as a bridge between Relationship and Space. Indeed it is often difficult for the planner, when preparing his diagram, to resist immediately making an Adjustment of it into a layout alternative. See Figure 8–9. By the time he does very much adjusting of his initial space relationship diagram,

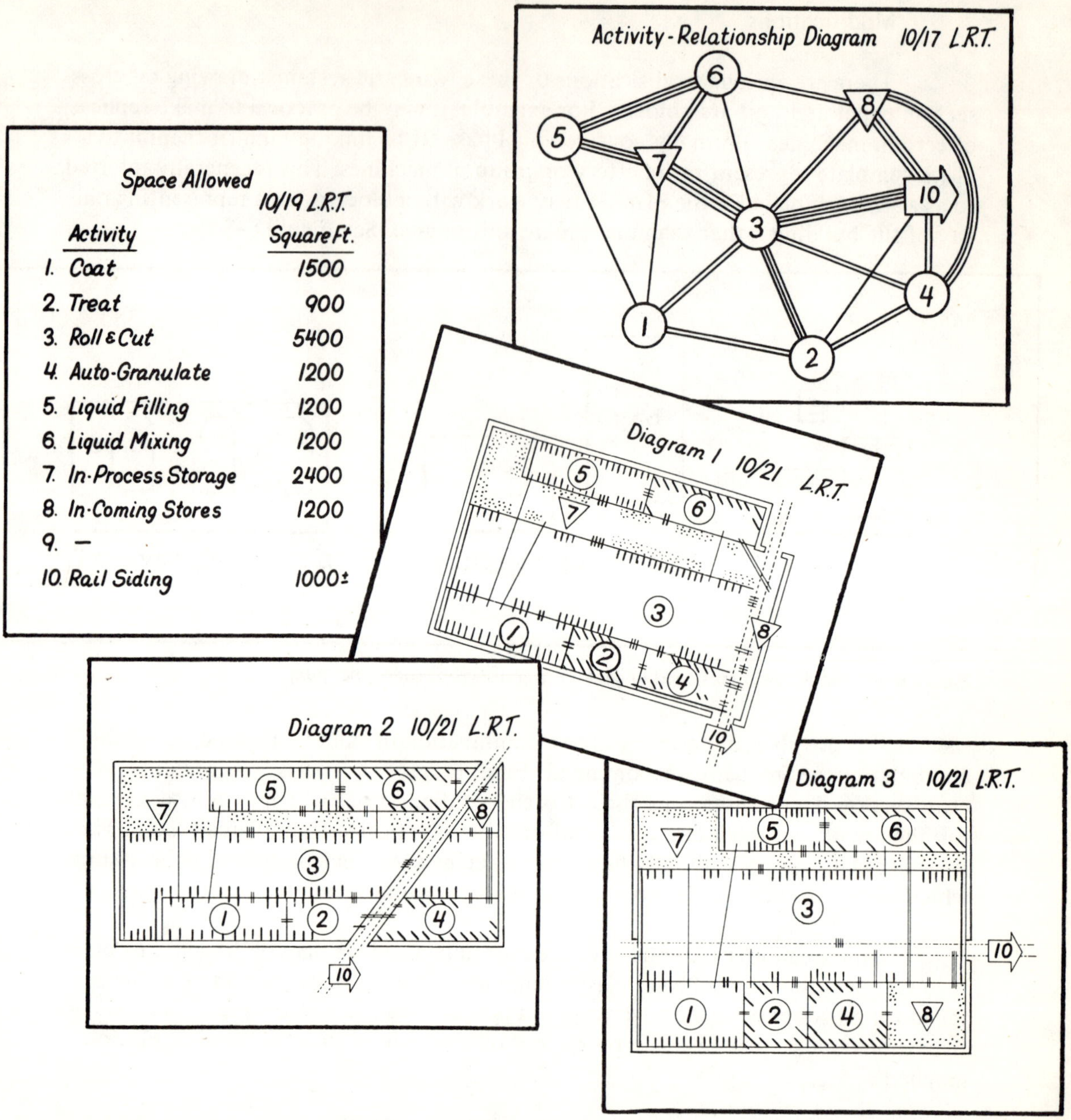


**Figure 8-9.** The Space Relationship Diagram translates relationships and space into adjusted arrangements that are almost layout plans. It thus serves to tie together the three basic fundamentals of every layout-planning project: Relationships, Space, and Adjustment.

Three different arrangements are being proposed here—each with a different orientation to the rail siding. As the space is diagramed, the number-of-lines coding is marked between the appropriate activity pair on the diagram. This ties the diagram to the charted relationships and reveals weaknesses in the arrangement of the activities when four-line or three-line relationships are lengthy compared with the length of one-line and two-line relationships.

Note the black-and-white shading code is used here instead of colors, so each arrangement can be reproduced through conventional duplicating equipment.

he is already involved with the many modifying considerations and practical limitations, to be discussed in the next chapter.

As the diagram is fitted and arranged, a record should be made of it. With drawings, this means identifying the arrangement adequately and dating it. With unit-area blocks this is most easily done by camera. A small paper label to identify the alternative should be placed with the blocks and photographed with them. A record of the various alternative layouts allows a methodical evaluation of each one later on. The photos or drawings of the several layout alternatives can be compared side by side.

As might be expected, comparing various plans frequently results in combining the best features of each alternative and working out a new combination layout. The procedure is one of refining and weighing alternatives—usually with the aid of the operating department heads, service people, and others logically involved.

# Chapter 9
# Adjusting the Diagram

Up to this point in the SLP pattern, efforts are largely directed to assembling data and following somewhat straightforward procedures. Now comes the really creative part of layout planning. As soon as the space relationship diagram is completed, the planner begins to adjust and manipulate it, creating a host of possible overall arrangements. Operating and service managers, therefore, must be brought back into the project here; now there is something they can visualize. Futhermore, much of the adjustment is made in response to the desires or practices of these people.

The space relationship diagram is practically a layout plan. Probably it is not a very good one, because it does not incorporate modifications and/or limitations. In terms of the pattern, as soon as Box 6 (Space Relationship Diagram) is put together, the bottom drops out of it. The planner finds himself adjusting, modifying, integrating, blending, and massaging the diagram to get an acceptable layout. The diagram is derived directly from the best *Relationships* and *Space*, and therefore represents a theoretically ideal arrangement. The theoretical ideal is seldom usable without *Adjustment*. This chapter deals with the many modifying considerations (Box 7) which stimulate adjustments of the diagram, and with the practical limitations (Box 8) which hold within bounds the many ideas which come to mind as possible modifications.

The techniques of visualizing or of physically making the diagram and its adjustments having been discussed, we can turn our attention to the conditions initiating the adjustments.

### Modifying Considerations

Remember that modifying considerations usually fall into one of these categories:

1. Handling methods
2. Storage facilities
3. Site conditions or surroundings
4. Personnel requirements
5. Building features
6. Utilities and auxiliaries
7. Procedures and controls
8. Shape of detailed activities' layouts

Many others could be added, but those listed seem to occur most frequently.

Each layout project has its own set of important modifying considerations.

Again, certain of these modifying considerations may be so important to a particular layout plan, that the planner devotes much of his "adjusting time" to analysis of those specific considerations. Other modifying considerations may be negligible. A good example of this is in warehouse layout, for example, the planner is highly concerned with the handling and the storage facilities: The type of handling system and storage facilities selected could cause almost complete rearrangement of the space relationship diagram. Whereas, in office layout, these same considerations are unimportant compared to those of say the personnel requirements or procedures and control.

Because of the many different modifying considerations and their varying importance from project to project, there are all manner of techniques for analyzing them. Each modifying consideration may be investigated in several different ways. A single consideration on some projects may actually require several different analyses. Such individual, special analysis—consideration by consideration—is probably best. But where detailed investigation is not justified or convenient, there are two techniques of analysis that are highly practical in analyzing the influence of specific considerations.

One of these is the *universal problem-solving procedure* illustrated in Figure 9-1. To be sure, it is hardly detailed enough for any complex problem, but it is a simple, basic method of solving problems. It can help deal with the many considerations in an organized way.

The second method is *sampling*. Its use is limited to existing installations. That is, records are spot checked based on existing ways of doing the work, or sample observations of the existing layout are made. The purpose is to gather information which can be projected into the conditions anticipated to prevail in the new layout being planned.

Several types of sampling sheets can be made up, depending upon the type of information needed. Where records from which to make necessary tallies are not available, actual observations—termed *work sampling* or *ratio delay*—are practical. (A summary of a work sampling study is shown in Figure 9-2.) Recognize the necessity of obtaining representative sample and understand the "booby traps" in recording and in taking work sampling observations. There are several texts available on the subject; those not familiar with the techniques should refer to them.

### Materials Handling Consideration

In many plant layout problems, the materials handling methods are a dominant consideration. For this reason, very early in Phase II is Box 1, Flow of Materials. Where materials flow is important to the layout, most of Box 3 (Flow and/or Relationship

**Figure 9-1.** The general, six-step procedure for solving problems. It provides an organized way of attacking the many modifying considerations bound to crop up in almost every project. The problem illustrated here is a typical one. The form guides the planners in using this generally recognized industrial engineering procedure.

## UNIVERSAL PROBLEM-SOLVING PROCEDURE

Problem *Incorporate in layout planning complaint from City concerning excessive dirt and odors* Plant *Solid Iron Works*

Date Started *12/1*

Area/Dept. *Core Room and Furnace* Analyst *P. Gast*

1. STATE THE PROBLEM

*Eliminate excessive dirt and odor caused by furnace and core room as part of revising layout.*

2. GET THE FACTS

*a. Prevailing winds (per Weather Bureau) are WSW.*
*b. Complaints originating from bakery at 12th Street – 2 per year.*
*c. Cost of filtering system $75,000.*
*d. Cost to move furnace and oven – $250,000 est. – out of question.*
*e. Furnace 20 years old and completely depreciated.*
*f. Furnace maintenance last year $20,000 – excessive and trend is up.*
*g. Used core sand is dumped in yard.*
*(over)*

3. RESTATE THE PROBLEM

*Reduce the amount of dirt and odor being created.*

4. ANALYZE AND DECIDE

*(a) Complaint is justified*
*(b) Prevailing wind and plant location with respect to bakery will always create problem.*
*(c) Replace furnace for better combustion and reduce maintenance expense.*
*(d) Change area where used core sand is discarded.*
*(e) Use filters.*

5. TAKE ACTION -- What, Who, When

*Obtain and evaluate bids for new furnace – Peters 12/15*
*Review layout to incorporate furnace – Kirby 12/31*
*Install new furnace and remove old furnace – Kirby 5/3*
*Revise layout for handling and discarding used core sand – Kirby 1/8*
*Obtain detail information concerning filter systems and suggested layout revisions to incorporate such systems – Peters & Kirby 1/31*

6. FOLLOW-UP

*1/3 Kirby presented plans for furnace installation for approval.*
*1/15 Walters prepared Press Release for approval on what we are doing to combat problem. Should F.U. further by 5/3*

WORK SAMPLING SUMMARY
Studies taken from April 1 to May 1

| Dept. | Average No. of Employees 4/1--5/1 | Activity | | | | | | | |
|---|---|---|---|---|---|---|---|---|---|
| | | Noon Use of Cafeteria | | Rest Room Usage | | | | Tool Room | |
| | | | | Male | | Female | | | |
| | | Readings | % Total | Readings | % Total | Readings | % Total | Readings | % Total |
| 40 | 82 | 750 | 20.5 | 2405 | 27.6 | 400 | 8.1 | 2600 | 23.8 |
| 41 | 68 | 1000 | 27.4 | 2005 | 23.0 | --- | --- | 2375 | 21.7 |
| 42 | 53 | 600 | 16.4 | --- | --- | 2120 | 43.0 | 600 | 5.5 |
| 44 | 105 | 500 | 13.6 | 2600 | 29.6 | 800 | 16.3 | 5205 | 48.0 |
| 46 | 42 | 605 | 16.5 | 1545 | 17.6 | --- | --- | 102 | 1.0 |
| 49 | 50 | 205 | 5.6 | 195 | 2.2 | 1600 | 32.6 | 38 | neg. |
| Total Readings | | 3660 | 100.0 | 8750 | 100.0 | 4920 | 100.0 | 10920 | 100.0 |

**Figure 9-2.** Work sampling essentially is a method of tallying by sample observations at random times. It can be used to measure machine and worker utilization, most prominent causes for idleness, density of material in process or in storage, operation times, and a host of other things. By adequate breakdown it can be used to pick up the direction and destination of movements and the degree of personal contact between office areas.

The work sampling summary here shows results that will be used in locating the service areas—cafeteria, rest rooms, and the tool room. It summarizes the number of people (or frequency) from each department using the various services.

Diagram) and Box 6 (Space Relationship Diagram) are based on the intensity of materials movement. Refer to Figure 5–4 as a guide to how much attention should be given this modifying consideration in any given layout project.

It is not possible here to present a full discussion of materials handling analysis. However, note that handling analysis too can be broken down into four phases, which include

I. Integration with external transportation
II. General overall handling plan
III. Detailed handling plans
IV. Installation

Note the similarity of the handling analysis to the four-phase framework of SLP. (See Figure 9–3.) These phases also are sequential and provide best results when they overlap.

First, make sure what external transportation facilities the internal handling plan must be coordinated with, noting any changes that may be practical. Then adjust the general overall space relationship diagram, building into it the general overall handling plan. That is, concentrate on Phase II techniques of handling analysis–interdepartment or interarea moves. These are less detailed and less specific than those used later in Phase III. To be sure, overlap into detail handling analysis as necessary. For example, if an overhead traveling bridge crane looks like the general handling method to be used, and the moves this equipment make dominate the layout, then its capacity, space, speed, supporting-column spacing, and the like must be considered the overall handling plan, and, therefore, the general overall layout, is determined.

As soon as the space relationship diagram is adjusted to a layout–even though it is but one possible arrangement and still unrefined–there are distances. That is, the layout establishes actual configurations and distances between the various activities. Therefore, a handling analysis based on the intensities of materials moved and distances

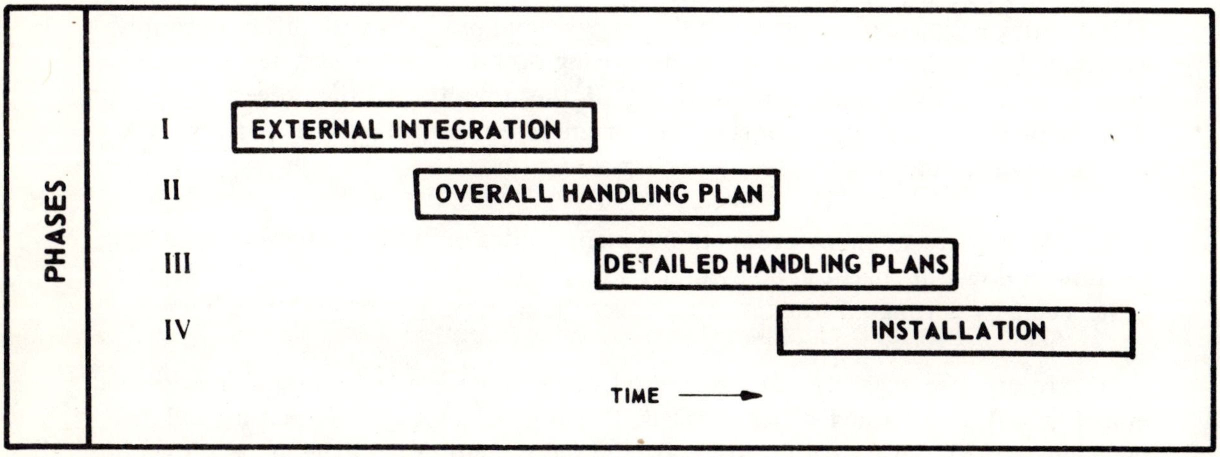


**Figure 9–3.** The four phases of materials handling analysis.

involved for each route can be developed. With those known, the planner can calculate transport work and can generally figure operating times for given methods of handling, and thereby estimate costs.

In discussing the flow diagram and/or space relationship diagram, we assumed that materials could always move from origin to destination over the shortest possible path. Now we must modify that assumption: the shortest distance is not always the most practical. Aisle locations, floor-load capacities, walls and partitions, all affect the actual distances. Based on an analysis of materials and moves, the planner must decide on an overall method(s) of moving each material between each pair of activities. *Materials handling methods* consist of (1) a system or pattern of moves; (2) the equipment or actual transport vehicle; and (3) the transport unit (box, pallet, bundle, drum).

By system is meant the general way in which the different movements are tied together, both as to geography and physical movement. Basically, materials can move between different places in three different systems:

1. Direct system
2. Channel (or Kanal) system
3. Central system

A description and use guide for each system is shown in Figure 9–4.

The planner must select the system or combination of systems he plans to use and fit the equipment and container (or transport unit) to it. The whole should integrate into the space relationship diagram which he is adjusting into an acceptable layout.

Just as materials flow is analyzed with different techniques depending on the position on the P-Q curve, so materials handling can be analyzed in a number of ways.

For one or a few products (or product groups), use the flow or transport process chart (see Figure 11–2), which traces the material all the way through the process. This charting is time-consuming when there are several products with different routes. Therefore, when there are many products moving over different routes, use the route chart or route-analysis chart. See Figure 9–5. Rather than tracing the process, the route chart helps analyze the movements on each *route*—from A to B on one chart, from A to C on another, and so on.

When there are many products and many routes of modest intensity in each, the flow-in-flow-out chart can be used. See Figure 9–6. With this method, the planner gathers data by getting information about everything moving in or out of each *area.*

In any case, regardless of the method, the planner is now interested in both the intensity of flow and the distance traveled. To aid in visualizing this intensity and distance, mark the intensity patterns right on the space relationship diagram, as shown in Figure 9–7.

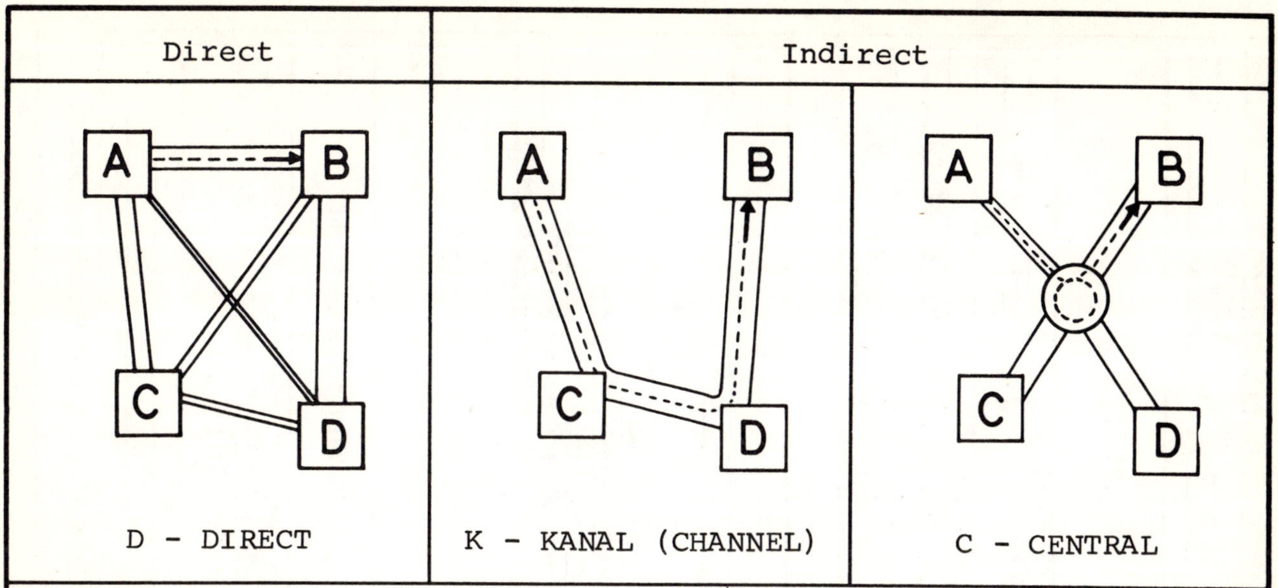


Direct System

Materials move from origin to destination over the shortest practical path.

When the intensity of flow is high and the distance short or moderate, this system is generally most economical to use--especially if the materials are special in some way(s) and time is urgent.

Kanal (Channel) System

Materials move in a pre-established route and flow to destination together with other materials moving to or from other areas.

When the intensity of flow is moderate or low and the distance moderate or long, this system is economical--especially if the layout is irregular or spread out.

Central System

Materials move in a pre-established route from origin to a centralized sorting and/or dispatching area and then on to destination.

When the intensity of flow is low and the distances moderate or long, this system quite often is economical--especially if the plant area is basically square and control is important.

**Figure 9-4.** The basic systems for movement of materials between departments or areas. (Layout planners should be aware that there is a companion technique to SLP which was specifically developed to help analyze internal transport problems. Called Systematic Handling Analysis (SHA), it is available in book form. A condensed recap of S H A is included here as Appendix XIII of this book).

## ROUTE CHART

From *Packing Material Storage* To *Packaging Department*
Distance *300 Feet*

Plant *B. Hansen Co.* Project *C3-12*
By *K.M. Higgins* With *R.N. Abel*
Date *3-10* Sheet *3* of *14*

| | PRODUCT-MATERIAL DESCRIPTION (ITEM OR ITEM GROUP) | P-M CLASS | QUANTITY PER *Week* UNIT | AV. | MIN. | MAX. | NOTES | |
|---|---|---|---|---|---|---|---|---|
| 1 | Carton, Knocked-down, size C | a | Bundle | 450 | 0 | 680 | Size of bundle = 5'x4'x3½' | For product 12-R |
| 2 | Paper sheet | b | Pallet | 17 | 0 | 26 | From supplier on pallets | |
| 3 | Cushion material | b | " | 34 | 0 | 51 | " " " | |
| 4 | Carton, Knocked-down, size E | a | Bundle | 63 | 63 | 63 | Size of bundle = 4'x4'x3' | For product 12-D |
| 5 | Cushion material | b | Pallet | 8 | 8 | 8 | From supplier on pallets | |
| 6 | Wood spacer | b | " | 19 | 19 | 19 | " " " | |
| 7 | Strapping material | c | Carton | 35 | 21 | 48 | | To both packaging lines. |
| 8 | Miscellaneous | c | " | 10 | - | - | | |
| 9 | | | | | | | | |
| 10 | | | | | | | | |
| 11 | | | | | | | | |
| 12 | | | | | | | | |
| 13 | | | | | | | | |
| 14 | | | | | | | | |
| 15 | | | | | | | | |
| 16 | | | | | | | Conversions — | |
| 17 | | | | | | | Bundle, Carton C: 1 bundle per pallet | |
| 18 | | | | | | | " " E: 1 " " " | |
| 19 | | | | | | | Carton, Strapping: 4 cartons per pallet | |
| 20 | | | | | | | " Miscellaneous: 5 cartons (av.) per pallet | |
| 21 | | | | | | | | |
| 22 | | | | | | | | |
| 23 | | | | | | | | |
| 24 | | | | | | | | |
| 25 | | | | | | | | |

**TOTAL MOVEMENT**

| PRODUCT-MATERIAL CLASS DESCRIPTION | P-M CLASS | INTENSITY OF FLOW UNIT | AV. | PLAN | INTENSITY x DISTANCE UNIT | AV. @ | PLAN @ | COMMENTS OR SUGGESTIONS |
|---|---|---|---|---|---|---|---|---|
| Knocked-down Cartons | a | pallet | 513 | 650 | pal.-ft. | 150 M | 195 M | When packaging stations are reoriented, |
| Paper, cushion materials, spacer | b | " | 78 | 93 | " | 23.4M | 27.9M | it is possible to move K.D. cartons |
| Strapping, miscellaneous | c | " | 11 | 14 | " | 3.3M | 4.2M | on Conveyors. |
| | | pallet | 602 | 757 | pal.-ft. | 176.7 M | 227.1M | |
| | | | | | | | | @M = 1000 (that is, 000 is replaced by M) |

RICHARD MUTHER & ASSOCIATES - 240

For each particular handling system (direct, channel, or central), determine the best handling method—the type of equipment and container (transport unit), best suited to that system. Establish the time and cost for that best method and the anticipated investment required. In the same way, the planner also considers combinations of the three systems and dual systems, for splits of product-material classes or different item groups. Then, each handling method chosen is compared with the others in terms of time, cost, and investment, so that the most suitable system and its accompanying method can be selected.

For the general type of equipment to be used, there is a host of possibilities. Appendix III gives a classification of handling equipment and containers. But for aid in selecting the general type of equipment, the equipment evaluation form shown in Figure 9–8 provides the planner with a method of screening and selecting the general type of equipment. In all likelihood, this selection will be followed by or supported with more detailed and specific analysis.

At this stage, the general method of handling for moves between areas or departments is being determined. In Phase III, Detail Layout Planning, the same kind of handling analyses is used, though more detailed and limited to smaller areas. (See Appendix XIII for a synopsis of Systematic Handling Analysis (SHA).)

### Other Modifying Considerations

Arranging *storage facilities* and areas is the heart of all warehouse layout projects. It is important to a greater or lesser extent in all industrial layouts, service-areas, offices, and retail store layouts as well. In fact, the storage of waiting materials, supplies, stationery, and other items is probably more common to every layout project than any other modifying consideration—though these may be detail layout problems, to be sure, in many instances.

The methods of stacking, racking, or supporting are numerous. Perhaps the planner got a pretty good idea of how to store materials when he established the storage space requirements. In any case, now he should go back to re-examine his idea in the light of the space relationship diagram and the other modifying considerations with which these storage facilities must be integrated. In doing so, remember that storage is the area most easily and most commonly squeezed for space in planning new layouts.

**Figure 9–5.** The Route Chart is used for analyzing the flow of materials between two areas. The different products, materials, or item-groups are listed; their product-or-material class is noted; and the necessary information is written down. In this example, the items are divided into three product-material classes each consisting of items which are similar. In the lower part of the chart the total movements for each product-material class is summarized. The intensity of flow is measured in pallets and the intensity times distance in pallet-feet. Even though the knocked-down cartons are the main item in the example, the other items (cushion materials, wooden spacers, etc.) cannot be overlooked; all movements have to be considered to find a practical solution.

*Site conditions or surroundings*—the "larger" constraints imposed by location—can certainly influence how a layout is arranged. Thus, if the location has not yet been selected, then the overall layout very likely will "underlap" the Phase I decision. Such factors as slope of the land, prevailing wind, southern exposure, or North light are natural conditions which can influence the arrangement of the space. Manmade site conditions include dirt or fumes from adjacent plants, rail or waterway access, and such public regulations interests as passing highways, and the effect of the company's own noise, hazards, and traffic on neighboring plants or homes. Within the plant, such considerations take such forms as the spread of contaminating materials, winter drafts from doors, glare from welding arcs, and vibrations from rapid or heavy-impact equipment. Modifying considerations like these must be identified and provisions to handle them worked into the space adjustment.

Closely akin to site conditions are the *building features* themselves: ceiling height, floor load, column spacing, types of windows, walls, and doors. Where an existing structure is to be used, these features may already be determined. Even so, certain of them may be moved or changed. Competent architectural advice can be very important here, especially if prints of the original structure are not available.

We do not presume to indicate the type of building features that any particular layout project should have. Figure 9–9 does, however, offer some general guides which should be of help for industrial and service plants.

In working out the building features, one significant consideration is modular arrangement of areas. Going onto a new site or into a new structure, this is especially important. Modular patterns offer many potential savings in real estate, construction costs, and the like. Among the advantages of modular patterns are

1. Straightness of roads—of aisles
2. Regularity of building faces or column lines—of departmental areas, walls, or aisles
3. Length and installation cost of utility main lines—of distribution lines
4. Ease of layout planning, construction, installation
5. Ease of logical expansion and later rearrangement
6. Savings in maintenance and repair costs

While use of a module may mean a little additional initial investment, its advantages, from the standpoint of construction and maintenance, of installing and servicing the utilities and auxiliary lines and equipment, and of integrating buildings or departmental areas with each other, usually outweigh the disadvantages.

**Figure 9-6.** The Flow-in Flow-out Chart shows the movements in and out of one area. On the left side are listed all items which go into the area. On the right side, all items leaving the area. The chart shows the quantity and condition of material moving in and out also. Further, it indicates the change that takes place in the condition of each item (notice how dry enamel ware changes to enamel ware). Filling out such a chart helps avoid overlooking the insignificant movements such as burning tools. If not provided for, these seemingly unimportant materials can cause a poor layout.

## FLOW-IN FLOW-OUT CHART

Area: *Finishing Department, Furnace Area*

Plant *Metal Utensil Company* Project *Finishing Expansion*
By *L. L. Latimore* With *E. McGee*
Date *7-3* Sheet *5* of *7*

| FLOW IN → | | | | | OPERATION OR AREA | FLOW OUT → | | | | |
|---|---|---|---|---|---|---|---|---|---|---|
| PRODUCT-MATERIAL DESCRIPTION (ITEM OR ITEM GROUP) | QUANTITY PER *Hour* | | | FROM | | TO | QUANTITY PER *Hour* | | | PRODUCT-MATERIAL DESCRIPTION (ITEM OR ITEM GROUP) |
| | UNIT | AV. | MAX. | | | | UNIT | AV. | MAX. | |
| 1 Standard Dry Enamel Ware | Piece | 1050 | 1200 | Dryer Conveyor | | Coat Dip Conv. | Piece | 1050 | 1200 | Standard Enamel Ware |
| 2 Special " " " ⓐ | Racks | 10 | 12 | Dryer | Burning | Coat Dip | Crate | 8 | 11 | Special Enamel Ware ⓒ |
| 3 Empty Crates | Crate | 8 | 11 | Coat Dip | Furnaces | Dryer | Racks | 10 | 12 | Empty Racks With Boards |
| 4 Burning Tools ⓑ | Set | 1 | | Tool Storage | No. I-II-III | Tool Storage | Set | 1 | | Burning Tools, Used |
| 5 Heating Gas | | | | Gas Producer | | Stack | | | | Exhaust Gas |
| 6 | | | | | | | | | | |
| 7 | | | | | | | | | | |
| 8 | | | | | | | | | | |
| 9 | | | | | | | | | | |
| 10 | | | | | | | | | | |
| 11 | | | | | | | | | | |
| 12 | | | | | | | | | | |
| 13 | | | | | | | | | | |
| 14 | | | | | | | | | | |
| 15 | | | | | | | | | | |
| 16 | | | | | | | | | | |
| 17 | | | | | | | | | | |
| 18 | | | | | | | | | | |
| 19 | | | | | | | | | | |
| 20 | | | | | | | | | | |
| 21 | | | | | | | | | | |
| 22 | | | | | | | | | | |
| 23 | | | | | | | | | | |
| 24 | | | | | | | | | | |
| 25 | | | | | | | | | | |
| 26 | | | | | | | | | | |
| 27 | | | | | | | | | | |
| 28 | | | | | | | | | | |
| 29 | | | | | | | | | | |
| 30 | | | | | | | | | | |

NOTES: ⓐ Special Dry Enamel Ware: 4-12 pieces per board and 15 boards in one rack.
ⓑ Burning Tools: One set changed each hour.
ⓒ Special Enamel Ware: 90-115 pieces in each crate.

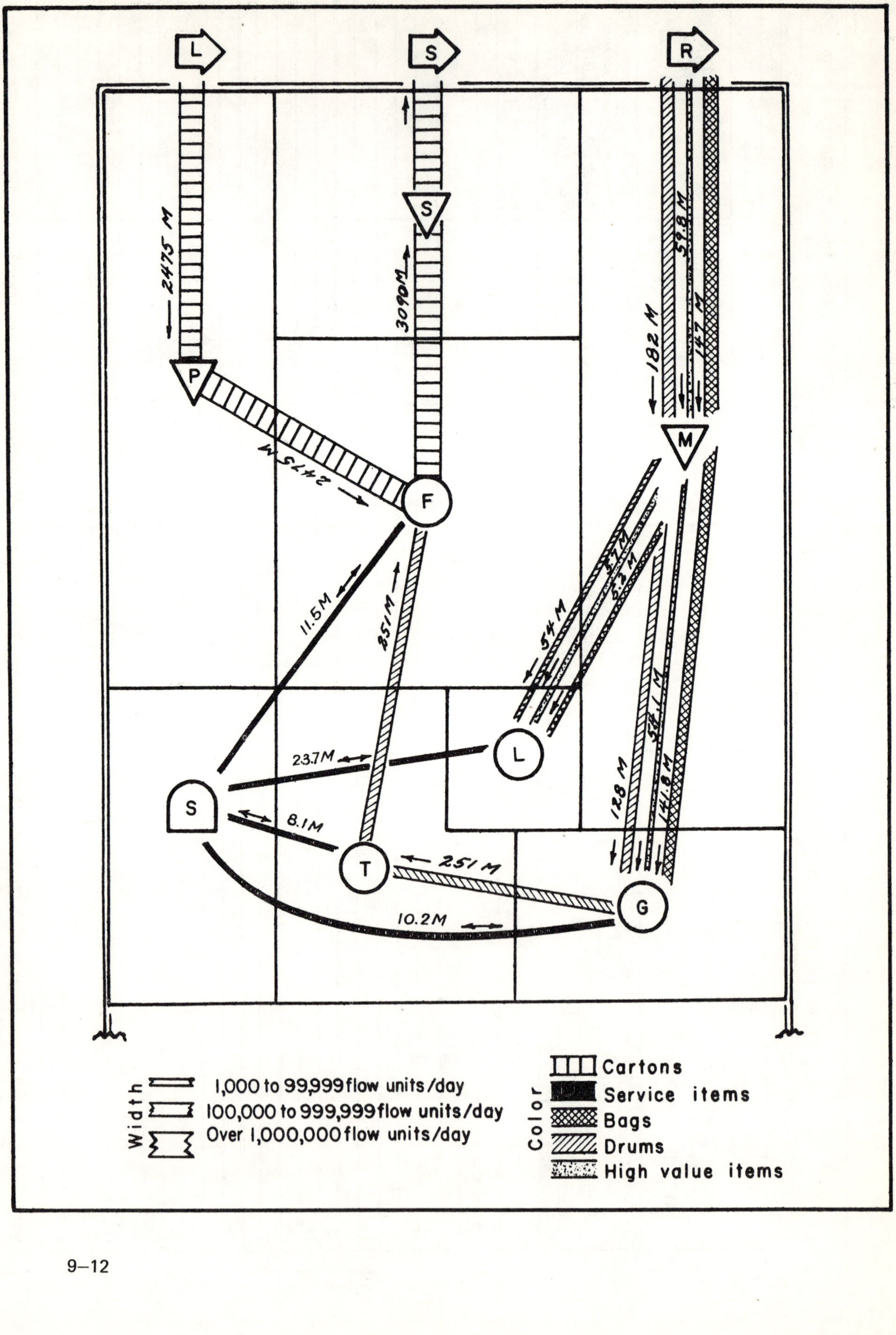
L
S
R
2475 M
3090 M
59.8 M
182 M
147 M
P
S
M
2475 M
F
11.5 M
851 M
54 M
23.7 M
L
S
8.1 M
T
251 M
G
10.2 M
128 M
141.8 M
Width
1,000 to 99,999 flow units/day
100,000 to 999,999 flow units/day
Over 1,000,000 flow units/day
Color
Cartons
Service items
Bags
Drums
High value items

In any event, building features must be integrated with the layout plan as the planner adjusts the space. Certainly, for new construction, the architect-engineer should be available for consultation as the general overall plan takes shape, for his preliminary drawings should be integrated with and approved at the same time as the overall layout.

Along with buildings, the distribution of *utilities* and *maintenance* must be considered. These are becoming more and more important as industry generally continues its trend toward increased use of automatic equipment. Adequate space must be provided for the maintenance or repair department itself. Additionally, employees of the maintenance department need access, for themselves and their equipment, through the areas and around the building structures involved. In detail layout planning, access around individual pieces of machinery and equipment must be provided.

*Personnel* requirements include questions of centralized versus decentralized plant entrances, of locker rooms, food service, and the like. Whether to put restrooms on the ground floor, on a mezzanine, or in a basement, must be decided. *Safety* considerations, convenience of employee access, and questions of communicating systems may all be involved in the project and should be considered as the general overall layout takes shape. Even housekeeping and appearance considerations can modify layout plans.

An exhaustive analysis (in connection with the relationship chart) may have been made to determine for example, the closeness of the credit union to all of the other activities. Or it may have been considered too minor an activity and so kept off the relationship chart, to be picked up as a modifying consideration. In either case, when the planner adjusts the space relationship diagram, he may fit in the space and features required for the credit union with all of the other activities involved. That is, the relationship between all major activities is determined in making the relationship chart. Less important activities are picked up as modifying considerations. The less important activities or features therefore adjust, rather than determine, the space relationship diagram. This is always true when the relationships are based on flow of materials only.

Another area of major consideration is operating *procedures* and *controls.* Any arrangement of space can have its advantages completely offset if practical operating procedures cannot be followed. The production planning and control system, the inventory paperwork, the scheduling system, the time-keeping routine, the methods of controlling quality, the procedure for count, the manner of keeping track of rework or scrap, and a host of other operating procedures and their supporting paperwork are important to the performance which the installed layout actually can give. For example, the layout shown in Figure 3–4 can only be made to work a certain way. The slow

**Figure 9–7.** Flow diagramed to show the intensities of movement by class of material directly on the space relationship diagram. This analysis was made by a firm producing ethical drugs to analyze its overall handling system, equipment, and containers. Materials handled in pipelines are not included.

## M. H. EQUIPMENT EVALUATION

Plant *National Pacific* Project *Enlarge Fab. Shop*
By *R. M. Powers* With *Wirth and Martin*
Date *5-7* Sheet *1* of
Handling Situation *Move plates from Cutting thru Fab. to Clean*

Alternative A: *Over-riding Bridge Crane with local jibs.*
Alternative B: *Under-hung (overhead-supported) Cranes*
Alternative C: *Traveling Wall Cranes*
Alternative D:
Alternative E:

| OBJECTIVE – Does it-- | WHAT IS WANTED FROM ANY HANDLING EQUIPMENT? Enter the requirements (what is required of this specific equipment?)—That is, The importance to the project of each Objective and Sub-Objective, using numerical values. Rate each alternative by rating letter; convert, extend, total. | REQMTS. Importance To the Project O | S-O | ALTERNATIVE A | | B | | C | | D | | E | |
|---|---|---|---|---|---|---|---|---|---|---|---|---|---|
| MOVEMENT --move materials 1. | a. freely? | 8 | 3 | A | 12 | A | 12 | A | 12 | | | | |
| | b. to the right place? | | 5 | E | 15 | E | 15 | O | 5 | | | | |
| | c. | | | | | | | | | | | | |
| DIRECTNESS --move materials 2. | a. without transfers? | 7 | – | | | | | | | | | | |
| | b. directly to point of use? | | 7 | A | 28 | O | 7 | U | 0 | | | | |
| | c. without unnecessary delays? | | – | | | | | | | | | | |
| | d. | | | | | | | | | | | | |
| CONVENIENCE --allow minimum 3. | a. loading time? | 5 | – | | | | | | | | | | |
| | b. unloading time? | | – | | | | | | | | | | |
| | c. rehandling time? | | 3 | A | 12 | O | 3 | A | 12 | | | | |
| | d. *synchronization time* | | 2 | E | 6 | I | 4 | O | 2 | | | | |
| SAFETY AND QUALITY 4. | a. keep materials safe from breakage or damage? | 5 | – | | | | | | | | | | |
| | b. keep material free from contamination, deterioration? | | – | | | | | | | | | | |
| | c. avoid hazard to workers or facilities? | | 5 | I | 10 | U | 0 | O | 5 | | | | |
| | d. | | | | | | | | | | | | |
| SPACE --accomplish 5. more | a. without consuming much floor-space? | 3 | 2 | O | 2 | A | 8 | O | 2 | | | | |
| | b. without obstructing workers, machines, storage services? | | 1 | O | 1 | O | 1 | U | 0 | | | | |
| | c. | | | | | | | | | | | | |
| COMBINED ACTIVITY --allow equipt. to be used as 6. | a. a worktable or work-holding device? | 0 | ✓ | | | | | | | | | | |
| | b. a storage device? | | | | | | | | | | | | |
| | c. an inspection or checking device? | | | | | | | | | | | | |
| | d. a pace setter? | | | | | | | | | | | | |
| | e. | | | | | | | | | | | | |
| FLEXIBILITY 7. | a. handle several different materials, products, containers? | 10 | 5 | A | 20 | E | 15 | E | 15 | | | | |
| | b. have adaptability to change, removal, or relocation? | | – | | | | | | | | | | |
| | c. have adaptability for increased volume, weight, size? | | 5 | E | 15 | E | 15 | I | 10 | | | | |
| | d. | | | | | | | | | | | | |
| PRODUCTIVE AID --keep materials 8. | a. in sequence or easily rearrangeable? | 5 | – | | | | | | | | | | |
| | b. from getting lost? | | – | | | | | | | | | | |
| | c. timed, scheduled, or synchronized? | | – | | | | | | | | | | |
| | d. free of workers' watching or attention? | | 3 | I | 6 | U | 0 | U | 0 | | | | |
| | e. easy to count, check, oversee? | | – | | | | | | | | | | |
| | f. readily available to operators? | | 2 | U | 0 | I | 4 | O | 2 | | | | |
| | g. | | | | | | | | | | | | |
| OPERATING ECONOMY --operate with 9. optimum | a. operating man-hours? | 15 | 5 | E | 15 | O | 5 | O | 5 | | | | |
| | b. maintenance and repair cost? | | 10 | O | 10 | A- | 35 | E | 30 | | | | |
| | c. power or fuel cost? | | – | | | | | | | | | | |
| | d. | | | | | | | | | | | | |
| DEPRECIATION 10. INVESTMENT | a. carry optimum rate of depreciation? | 8 | 4 | O | 4 | A | 16 | E | 12 | | | | |
| | b. require reasonable outlay of funds? | | 4 | O | 4 | A | 16 | E | 12 | | | | |
| | c. | | | | | | | | | | | | |
| OTHERS 11. | | ✓ | ✓ | | | | | | | | | | |
| | | | TOTAL | | 160 | | 156 | | 124 | | | | |

NOTES/ADDITIONAL SOLUTION SUGGESTED:

RICHARD MUTHER & ASSOCIATES - 202

movers—made to order daily—are collected first, and then the instock items picked from their finished-goods shelving. In order to make this layout work, a revised system of releasing orders to the plant also had to be worked out. Furthermore, in order to make an effective incentive system for the operators producing fast-mover items, a system of counting and controlling the work of each operator had to be planned into the layout.

It is easy for staff analysts, and especially for engineers, to overlook these practical operating procedures. They are a most important part of layout planning. The planner should lean over backwards not to neglect them.

Actually, many overall modifications may involve details particular to one subactivity. The planner will have delved into the *detail layout* of that area before deciding on its configuration and therefore the overall layout. Overlapping into Phase III is a significant part of adjusting the Phase II layout.

### Practical Limitations

Considerations which open for development, design, or decision, we call modifying considerations; those which impose constraints on our planning are called *practical limitations*. The latter include such restrictions as may be built into an existing building, existing handling methods, or a not-to-be-changed production-control-and-dispatching system. Company policy, building codes, labor union contract, and community regulations on waste disposal can all affect the layout; and, as noted earlier, the physical characteristics of the location—a Phase I decision—always exert limitations on the layout. Even a need for straight aisles of adequate width imposes requirements, for main aisles are part of any overall layout plan. Certainly one of the most important limitations is the question of cost savings and available investment money.

As the planner works with each of these modifying considerations, various ideas for rearranging the space come to mind. Yet each of these ideas is faced with its own set of practical limitations. For example, in a certain layout project, it may be desirable to have a highly synchronized, fully automatic conveying system. But practical limitations counter the advantages of such an installation. Return on investment, placing too great a dependence on one piece of equipment, obstruction to cross-traffic which the conveyor may cause, and similar limitations present themselves.

**Figure 9-8.** Evaluation form for selecting the general type of materials handling equipment. This example shows the evaluation of three alternatives for handling materials for the enlargement of the fabrication shop of a yard making small ships. The breakdown into objectives includes a list of features we would want from any piece of handling equipment on any project. Indicated in the requirements column is the importance of each objective, and subobjective, to the particular project in mind. The alternatives are rated for each subobjective using the SLP vowel-letter rating. When the rating is complete, the letters are converted to numbers (see Figure 10-4) and multiplied by the S-O requirements weight. Totals indicate a weighted and rated value for each alternative.

## INDUSTRIAL BUILDING FEATURES GUIDE

Use general-purpose or multipurpose building when the following are important or predominant:

- Initial cost
- Speed of getting the layout into production
- Probability of selling the building later for:
  - Profit
  - A better location
  - Foreclosure
- Frequency of changes in:
  - Products or materials
  - Processes or methods
  - Machinery and equipment
  - Volume or output

Use single-story construction, possibly including balconies and/or basement, when the following conditions exist:

- Product is large or heavy
- Weight of equipment causes heavy floor loads
- Large, more-or-less unobstructed space is needed
- Land value is low
- Land is available for expansion
- Product is not adapted to handling by gravity
- Erection time is limited
- Frequent changes in layout are anticipated

Use a relatively square building where there are:

- Frequent changes in product design
- Frequent improvements in process
- Frequent rearrangements of layout
- Restrictions on building material availability or substantial savings desired in amount of materials used

Use other shapes or separate buildings when there are:

- Limitations in physiography of the land
- Property lines at awkward angles
- Operations that cause dirt, odors, noise, vibration
- Operations not part of production
- Operations susceptible to fire, explosion, contamination

Use a basement if these features can be obtained:

- Ample headroom
- Ample lighting
- Good ventilation
- Waterproofed walls
- Sound foundations
- Freedom from seepage or flooding

Use balconies for these typical situations:

- Light subassembly above final assembly on ground level
- Assembly operations with heavier forming machinery below
- Light-machine operations with heavier machines below
- Treating operations with forming operations and assembly of bulky units on ground level
- Supporting activities that can be kept off the production floor--storage, wash and locker rooms, production offices, packing, auxiliary equipment, and the like
- Operating or servicing upper parts of tall, high machinery
- Liquid or bulk material storage and preparation areas, involving mixing, aging, blending, and the like

Use _no windows_ generally in cases when:
- Plant is underground
- Work is affected by changes in temperature, humidity, light
- Work is subject to dust, dirt, contamination
- Workers or work are affected by external noise
- Artificial light and power is inexpensive
- Seeing things outside is not necessary
- Windows would get dirty fast

Use these desired _floor characteristics_ where practical:
- Floors of various buildings at same level
- Strong enough to carry machines and equipment
- Made from inexpensive materials
- Not too expensive to install
- Ready for use quickly after laying down
- Easily and quickly repaired, removed, and replaced
- Resistant to shock, abrasion, heat, vibration
- Not slippery under any condition
- Noiseless and sound absorbing
- Attractive to eye and with numerous colors available
- Unaffected by changes in temperature and humidity, or by oils, acids, alkalies, salts, solvents, or water
- Odorless and sanitary
- Resilient enough to seem soft underfoot and to minimize damage to articles dropped on it
- Easy to fasten machines and equipment to
- Will dissipate static electricity and not cause sparks
- Easily kept clean

Use these _roof and ceiling features_ when applicable:
- Overhead space and height clearance for:
  - Production machines
  - Process equipment--treating vats, drying ovens, etc.
  - Handling equipment--cranes, conveyors, etc.
  - Elevated traffic ways
  - Sprinkler (with 24 inches clear underneath)
  - Electrical distribution
  - Heating and ventilating systems
  - Air circulation
  - Washroom and toilets, service and storage areas built off the production floor
- Strength for underside (or above) support of:
  - Machinery and process equipment
  - Handling equipment for material or machinery
  - Heating and ventilating systems
  - Elevated traffic ways, storage or service areas
- Light:
  - Roof lighting independent of walls or expansion plans
- Heat conductor for:
  - Heat losses in winter
  - Effect on personnel in summer
- False ceiling:
  - Dust accumulation and dust drop-off
  - Appearance

**Figure 9-9.** General guide for industrial building features.

For each good idea, there is a set of practical limitations which must be weighed against it. As the pro's and con's of each consideration are weighed, the obviously poor or weak possibilities are abandoned and the seemingly worthwhile ideas retained. Worthwhile ideas are incorporated into the various adjustments of the space relationship diagram—each to help the planner develop a more satisfactory layout.

Essentially, the process is making compromises, the planner striving for an arrangement of activities which will give the most practical overall compromise of all considerations.

As you work through the many considerations involved, keep a record by making a drawing or a photograph of the various alternative plans. Finally, there are relatively few alternatives. Very seldom is there only one clear-cut obvious arrangement. Rather, the alternatives sometimes comprise as many as six or eight possibilities. If the alternatives haven't boiled down at least to this few, then the planner has not finished exploring the modifying considerations and their practical limitations. That is, he is not ready for the next step in the SLP pattern.

By integrating the modifying considerations with the space relationship diagram, and discarding all impractical ideas, the planner should end up with generally two to five alternative plans—Plan X, Plan Y, and Plan Z, in SLP pattern. Any one of them can be made to work. The next problem is to decide which layout alternative to adopt. In the next chapter we discuss how to evaluate these alternatives.

Chapter 10

# Selecting the Layout

We are now at a point where we have a relatively few alternative layouts. The SLP pattern refers to them as Plans X, Y, and Z. Any one of these plans will do the job. Any of them can be made to work satisfactorily. However, each has its own peculiar set of advantages and disadvantages. The problem is to decide which of the alternative plans to select. Of the number of ways by which this selection can be made, there are three basic methods:

1. Balancing advantages against disadvantages
2. Factor analysis rating
3. Cost comparison and justification

## Clear Presentation

Before attempting any selection of one plan over another, each plan should be clearly represented. Usually this involves "dressing up" the alternative layouts. Get a clear replica of each plan. And, recognize that others whom we may invite to participate in this selection process are not familiar with the codes, symbols, and activity designations used to do the planning. The replicas or reproductions should leave no doubt about how various arrangements of features and space are meant to work. The planner can well afford to color layout reproductions, to print in the familiar company names, and to designate clearly main aisleways, walls, craneways, and the like, eliminating or at least holding into the background, the cross-section or unit-area features of earlier space manipulations.

Two examples of how unit-area blocks can be replaced with more readily understood replicas are shown in Figure 10–1. But, regardless of the materials actually used, the main requirement is that to have an easily understood replica of the competing alternatives.

Experience has shown that each plan should be clearly labeled. We prefer to use letters as identification, to avoid any unintentional implication of preference which numbers tend to give. Additionally, a title or brief description of each plan, small enough to be written on the plan and on the evaluation sheet, helps eliminate any mix-ups and reduces reference conversation both during and following the evaluation.

Also, be careful to make each layout exactly as it is intended to be. Prepare replicas of the actual layout to be evaluated. Evaluate the *actual* plan, not what could be done with a few changes. To evaluate plans other than those actually before us, is to reduce the accuracy of the evaluation. The time involved becomes significantly longer.

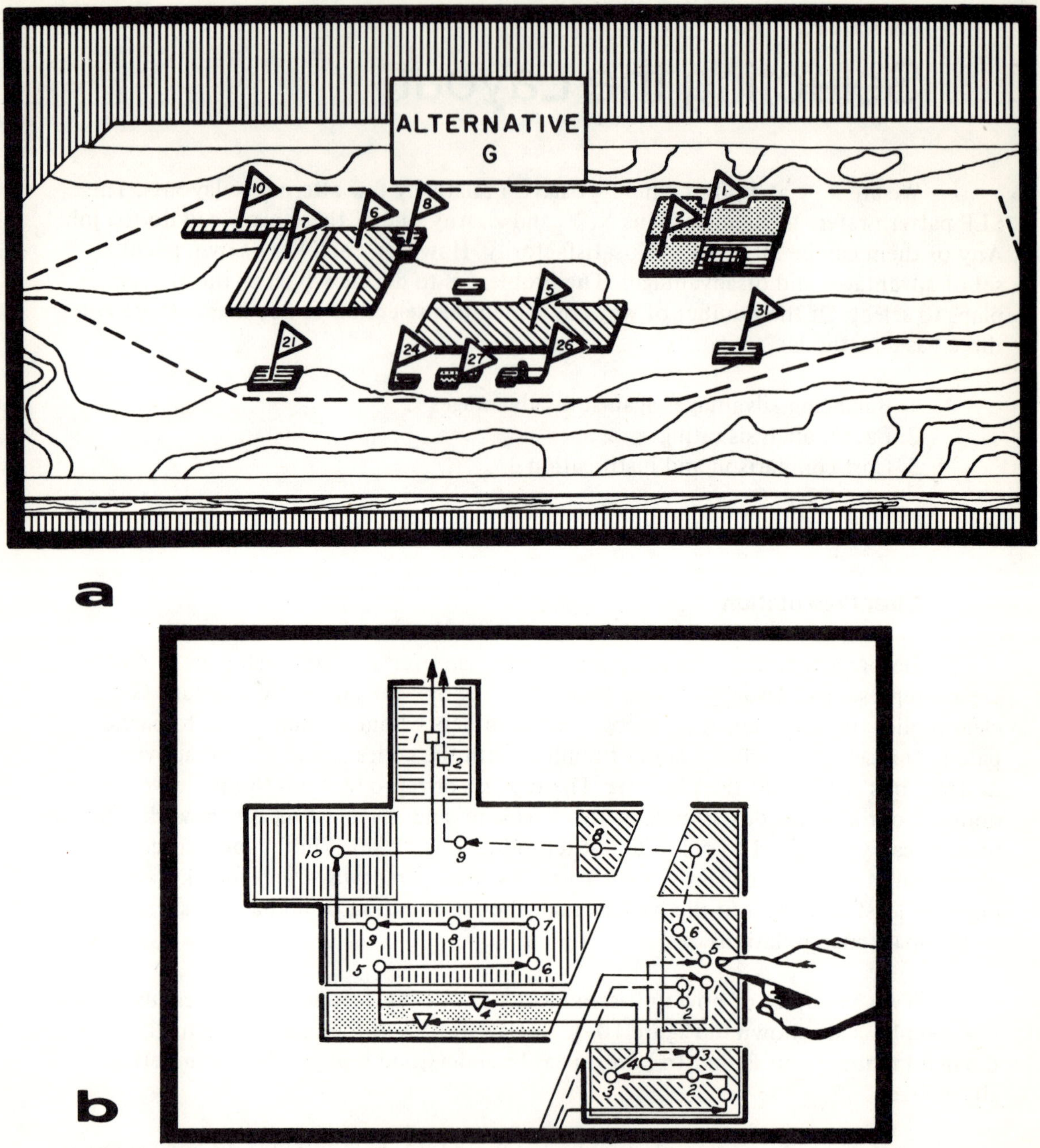


**Figure 10-1.** Before evaluating alternative layouts, have replicas of the actual plans on hand. In "a", unit-area blocks used in the early space planning have been replaced by cut-out sheets of cellulose plastic for the final evaluation of five alternatives. This offers a clear visualization of what is planned. Plans involving approximately a million and half square feet under roof, were set up right on the site survey (itself color marked for increased clarity) and photographed. The evaluation was made from photos of various alternatives, with the actual replicas available at any time for more extensive study. Below, "b" shows one of three alternative overall department layouts planned in blocks of sub-department areas—in this case, cardboard sheets on a drawing of the building outline, with flow of major materials shown thereon.

This is not to say that changes in the plans cannot be made. Actually, the very process of evaluating layouts frequently brings new ideas to light. As a result, the end product is often a new combination of two alternatives or a further modification of one of the alternatives. But the point here is that if and when there are changes, the layout should be redrawn or the replica prepared anew so the new layout plan can be evaluated. In other words, always have before you as you make the selection, the true replica of each layout being evaluated.

In this connection, and for convenient reference, it is well for the planner to clean up and organize his work sheets and diagrams. Clearly understood documentation of relationships and space provide others with a quick appreciation of what has been planned.

Probably the best results are gained when the choice is between alternative plans, each of them sound, but each significantly different. Plans which are all classically the same offer little choice or possibility for further improvement. This doesn't make the evaluation process fun; rather, it makes it a necessary step in planning (Box 9 of the SLP pattern).

### Advantages Versus Disadvantages

Probably the easiest of the three evaluation methods mentioned is that of listing *advantages* and *disadvantages*—the pro's and con's system. It is also the least accurate. Therefore, it is used more for preliminary screening of rough alternatives, or in Phases I and II where data is not so specific and not so readily available.

The pro's-and-con's system is merely listing in columns or on adjacent sheets all the advantages of each alternative. Below them are listed the disadvantages. This simple comparison is surprisingly effective and certainly not a time-consuming procedure. An example of the method's use is shown in Figure 10–2.

The same method of weighing pro's and con's can be made more accurate by rating the significance of each advantage and seriousness of each disadvantage, as shown in Figure 10–3. The same vowel-letter rating used throughout the SLP procedures now is given a numerical value scale, as explained in Figure 10–4.

### Weighted Factor Analysis

Every layout plan has intangible costs which for several practical reasons cannot be measured in terms of dollars and cents. Moreover, a comparative cost analysis of alternatives sometimes doesn't aid in the decision-making, no one plan having a clear-cut financial advantage over the other. As a result, perhaps the most effective general method of evaluating layout alternatives is that termed *factor analysis.*

The factor analysis method follows the engineering concept of breaking down

Building Expansion

Alternative C Add one Furnace

Building Expansion

Alternative B Add two Furnaces

(2 Shift Operation)

Step

Adva

Disa

*Lay

Building Expansion

Alternative A Add one Furnace

(2 Shift Dipping and Burning)

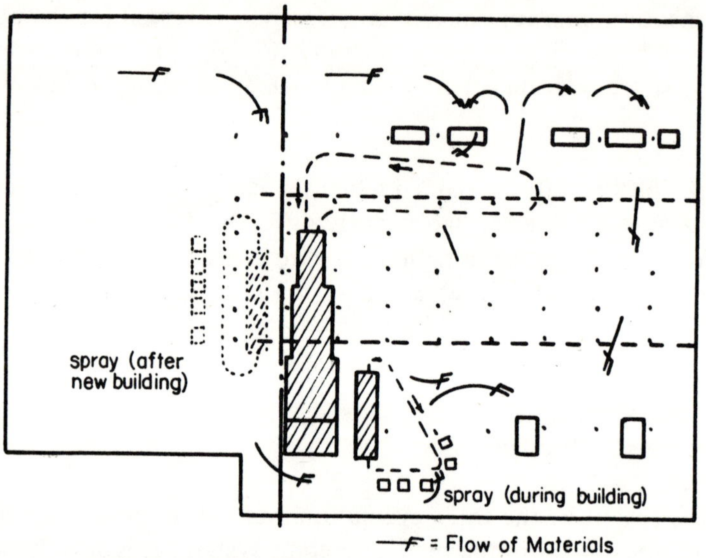


Step 1. Tear out one section of balcony wall and floor
" 2. Tear out furnace no. 4 and build continuous furnace
" 3. Move one set of dryers (3)
" 4. Begin 2-shift operation and continuous furnace
*" 5. Install spray dryer between old and contin. furnace
" 6. Rebuild old section of building
" 7. Move spray department to new building; extend chain and dipping area; and use under-balcony-area for wrapping

Advantages of this Layout:

1. Have opportunity to make use of new building in most effective way--including use of this furnace
2. Economy in burning tools and chain (short length)
3. No blocking of changes or dipping expansion in future
4. Could reconvert to one shift easily if forced to
5. Have ample room for storage of sprayed items before burning in case you cannot burn them with catalogue-ware
6. A minimum of cross-traffic

Disadvantages:

1. Must tear out 1 section of balcony (Est. $2000)
2. Must tear out furnace no. 4
3. Must relocate spray department after new building is complete (Est. $5000)
4. Must move one section of vertical dryers (Est. $2500)

*Layout shown--step 5

the problem into its elements and analyzing each one. This makes it more objective. Essentially, the procedure is as follows:

1. List all of the factors which are considered important or significant to deciding which layout to select.
2. Weigh the relative importance of each of these factors to each other.
3. Rate the alternative plans against one factor at a time.
4. Extend the weighted, rated values, and compare the total value of the various plans.

The procedure is set forth in Figures 10–5 and 10–6. An example of it is shown in Figure 10–7. The factor analysis method is highly flexible, yet it is precise, even though its accuracy is based on a series of judgments or estimates of probability.

The overall objectives of the layout planning are broken down into so-called factors or considerations—the things that are important for the layout to achieve.

The point is, the objectives or criteria, should be established, preferably by one person, after discussions with those who will approve the layout. The planner should list and define them briefly. Subsequently, before or during the time the weight or relative importance of each factor is determined, the list should be scrutinized, modified as necessary, and increased by any factors overlooked.

In listing the factors, be certain they are clearly defined, easily clearly understood. Fuzziness or duplication can be as serious as omissions. To list "flow of materials," "progressive routing," "smooth feed-in of materials" all as separate factors only causes confusion.

A list of the factors or considerations most commonly involved follows—not in the order of importance. (See Appendix VI for explanations of these factors.)

1. Ease of future expansion
2. Adaptability and versatility
3. Flexibility of layout
4. Flow of materials effectiveness
5. Materials handling effectiveness
6. Storage effectiveness
7. Space utilization
8. Effectiveness of supporting-service integration
9. Safety and housekeeping
10. Working conditions and employee satisfaction
11. Ease of supervision and control
12. Appearance, promotional value, public or community relations

**Figure 10–2.** Evaluating alternative layouts by listing advantages and disadvantages is a simple and often very effective way of selecting the best of several plans—especially in rough or early screening stages of evaluating.

ADVANTAGES

| Plan A--Receive at Rear; Flow up and through. | | Plan B--Receive at Side; Figure-eight flow. | |
|---|---|---|---|
| 1. Can utilize basement for raw material storage | A- | 1. Tool and Die Shop near Engineering Office | E |
| 2. Can receive by rail | I | 2. Majority of workers near parking lot | I- |
| 3. Can use present Shop Office without disruption | O | 3. Expansion for new process will be | I |

**Figure 10-3.** A rating scale can be used in conjunction with the list of advantages and disadvantages to give a value to each one. Here the simple SLP vowel-letter rating was used. Disadvantages are rated with a negative vowel—and converted into a negative value. A- and I- are ratings of half a degree less than A and I respectively.

RATING CODE AND VALUES

| Vowel Coding | Description of Rate | Numerical Value |
|---|---|---|
| A | Almost Perfect--(Excellent) | 4 |
| E | Especially Good--(Very Good) | 3 |
| I | Important Results Obtained--(Good) | 2 |
| O | Ordinary Results Provided--(Fair) | 1 |
| U | Unimportant Results--(Poor) | 0 |
| X | Not Acceptable--(Not Satisfactory) | ? |

**Figure 10-4.** The simple vowel-letter coding used throughout the SLP procedure is used as above for assigning values to ratings. "A" is still equivalent to four—just as it was in the number-of-lines scale when we diagramed relationships. This gives a 25% spread between letters, or 12½% when minus ratings (or half degrees) are used. Although letters are readily converted to numbers, evaluating in letters is preferred because of the unintentional implied accuracy of numerical ratings and because it is too easy, and tempting, to add up numbers before the rating is done and thus see how the various plans are coming out.

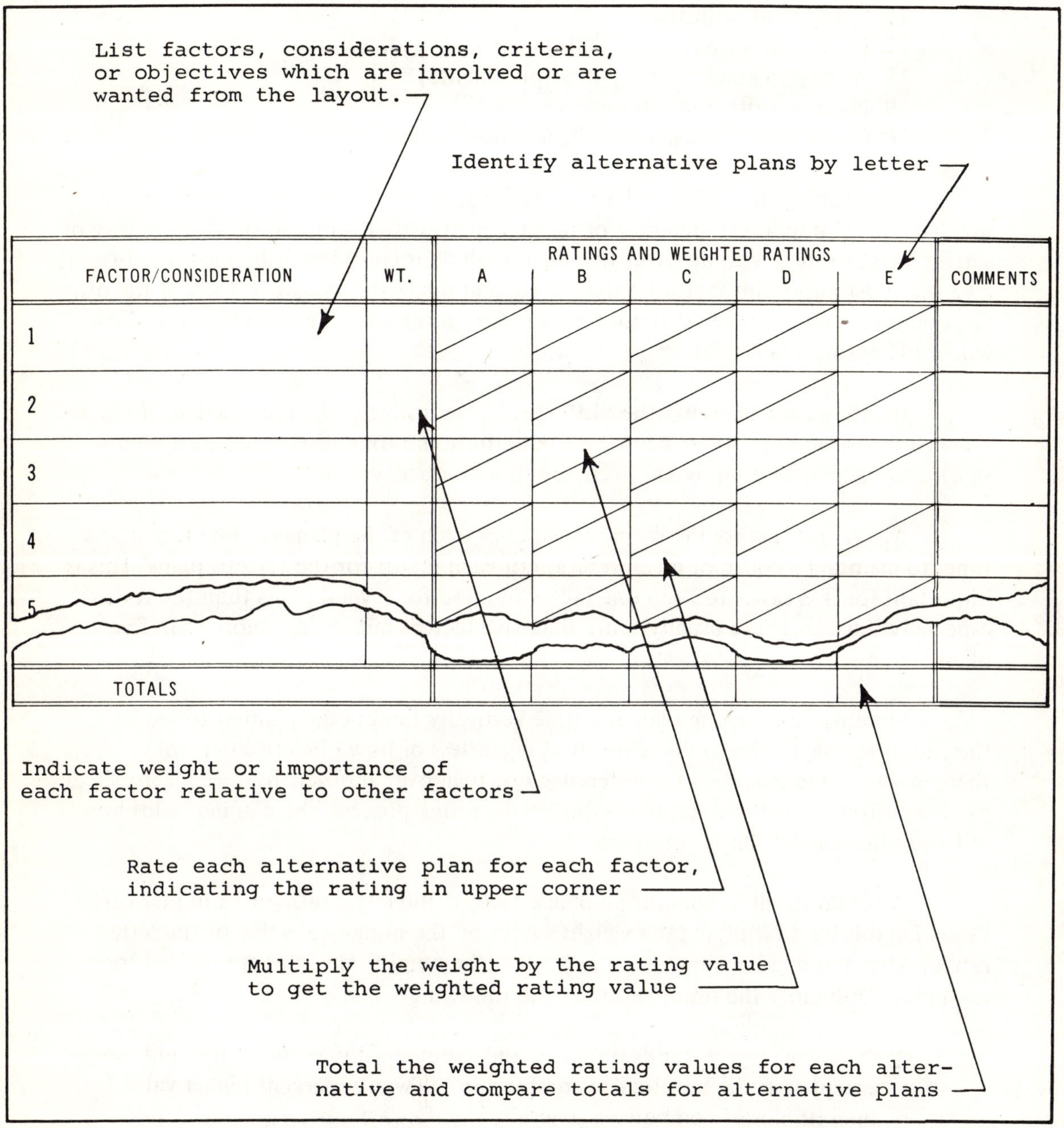


**Figure 10-5.** Mechanics of factor analysis method of evaluation. SLP recommends a range of 1 to 10 for weights and the vowel-letter code with its equivalent values for the ratings.

13. Quality of product
14. Maintenance problems
15. Fit with company organization structure
16. Equipment utilization
17. Utilization of natural conditions or surroundings
18. Ability to meet capacity or requirements
19. Plant security and pilferage
20. Compatibility with long-range company plans

Establishing the *weight values* for each factor is usually a joint effort, often involving fairly important members of management. Perhaps the most effective way of setting weight values is to pick out that factor which is considered the most important; give this most important factor a value of 10; and relate the weight of each of the other factors to 10. Get approval of these values before going ahead, especially from those who will approve the layout.

In rating each plan, use the SLP vowel-letter rating code, modified as in Figure 10–4. It is easy to remember, significant in definition, refinable (by adding a minus sign), and yet does not imply the accuracy of a number system.

Always *rate across* the form, considering each of the plans for one factor at a time, to maintain a constant interpretation of each factor for the various plans. This is important for it is easier to keep one factor in mind for several plans than the reverse, especially since there are usually more than five factors but seldom more than five plans.

Moreover, rating one plan at a time vertically tempts the planner to see how the results are beginning to shape up. And regardless of how objective one tries to be, there is always the tendency to preference one plan over another. By rating across and by using letters rather than numbers during the rating process, the planner helps himself and others avoid this temptation.

After rating all factors for all plans, convert the letter rating to a numerical value. Do this by multiplying the weight factor by the numerical value of the letter rating. After the numbers have been extended, the numerical values are totaled for each plan. Ordinarily the result is one of the following:

1. One plan clearly stands out head and shoulders above the others and can be accepted as the best logical compromise. (Twenty percent higher value total than its closest competitor generally signifies a winner.)
2. Two plans come out very close. In this case, reevaluation of the two plans, involving more factors, closer scrutiny of the weighting and rating, or inviting more people to share in the weighting and rating process should be undertaken.
3. The planner (and/or other "evaluators") will see possible improvements in one or more of the alternatives. For example, he may see where the two or

**Figure 10-6.** Procedure for factor analysis method of evaluation.

## PROCEDURE FOR EVALUATING ALTERNATIVES BY FACTOR ANALYSIS

1. <u>Identify the Plans to be Evaluated</u>
    a. Select the layout plans that are to be evaluated.
    b. Have a visual plan or sketch of each layout in front of each rater, and clearly understood by him, during the evaluation process.
    c. Identify each visual plan by letter--A, B, C, etc. Also give it a brief three-to-five word description.

2. <u>Establish the Factors or Considerations</u>
    a. Establish what factors, considerations, criteria, or objectives are involved or are wanted from the layout.
    b. Define the factors so they are clearly understood. Avoid duplication between terms and confusion as to meaning.

3. <u>Arrange a Rating Sheet</u>
    a. List the factors or considerations vertically on lines of one sheet of paper.
    b. Array the identification letter for each alternative plan horizontally in columns across the same sheet.
    c. Leave room for adequate reference notes.

4. <u>Determine the Relative Importance of Each Factor</u>
    a. Determine a weight or importance value of each factor relative to the other factors.
    b. Record by whom the weight values were determined.

5. <u>Rate Each Factor for Alternative Plans</u>
    a. Establish a rating code or system.
    b. Rate each alternative on the extent to which it achieves or affords the ends represented by the factor in question--rating each layout exactly as it is planned.
    c. Rate all plans for one factor; then take next factor.
    d. Enter rating symbol above slant line on rating sheet.
    e. Record by whom the rating(s) was (were) made.

6. <u>Calculate Weighted Values and Total</u>
    a. Translate the rating symbols into numerical values and multiply the weight value by each rating number.
    b. Total the weighted rating values for each alternative plan by adding the respective columns.
    c. Record by whom the extension and tally were made.
    d. Take action as appropriate, based on the totals.

## EVALUATING ALTERNATIVES

Plant/Area *Plant #3 Manufacturing* Project *#60121* Date *6-2*

Description of Alternatives: A. *Split Storage Areas*

B. *Cleared Areas (partitions removed)* C. *Cluster Service*

D. *Semi-Clear, Semi-Cluster* E.

Weight set by *JDW & RM* Ratings by *JDW & RM* Tally by *JDW*

| | FACTOR/CONSIDERATION | WT. | A | B | C | D | E | COMMENTS |
|---|---|---|---|---|---|---|---|---|
| | | | RATINGS AND WEIGHTED RATINGS | | | | | |
| 1 | *Convenience of Service* | *10* | *U* / *0* | *I* / *20* | *I* / *20* | *E* / *30* | | |
| 2 | *Ease of Supervision* | *6* | *O* / *6* | *A* / *24* | *E-* / *15* | *E-* / *15* | | |
| 3 | *Expansion Possibilities Use of Basement and 2nd Floor* | *5* | *O-* / *3* | *I* / *10* | *O* / *5* | *O* / *5* | | |
| 4 | *Flow of Materials and Handling Economy* | *10* | *O* / *10* | *U* / *0* | *I* / *20* | *E* / *30* | | |
| 5 | *Flexibility* | *9* | *O* / *9* | *E* / *27* | *I* / *18* | *E* / *27* | | |
| 6 | *Investment Cost (Building alterations)* | *8* | *O-* / *4* | *O* / *8* | *I* / *16* | *E* / *24* | | |
| 7 | *Ability to provide additional fabrication capacity (Woodshop)* | *5* | *A* / *20* | *O* / *5* | *U* / *0* | *O-* / *3* | | |
| 8 | | | | | | | | |
| 9 | | | | | | | | |
| 10 | | | | | | | | |
| 11 | | | | | | | | |
| 12 | | | | | | | | |
| 13 | | | | | | | | |
| 14 | | | | | | | | |
| | TOTALS | | *52* | *94* | *94* | *134* | | |

NOTES *A - Almost perfect (4), E - Especially Good (3), I - Important Results Obtained (2), O - OK, Ordinary Results (1), U - Unimportant Results Obtained (0)*

three best alternatives rated lowest. The U's and O's as a red flag. By concentrating on these particular weaknesses of the respective plans, the planner may be able to improve each layout still more.

4. During the rating process, it is discovered that a combination of two or more of the plans can be worked out. A replica of that combination layout must be made. By adding another column to the form (or another sheet), the planner rates the new combination plan on the same basis as the others.

Note that at the top of the evaluation form, provision is made for several important records. Not only should the plan be at hand during the evaluation, but a brief description of each alternative plan should be at the top of the form. Additionally, those who establish the weighting should be listed; those who do the rating should be listed; and the person who actually enters the figures, makes the extensions, and totals the columns should be recorded in the "tally by" space.

The rating itself can be done by the layout analyst alone, or it can be done *in conjunction with others*. When the people most interested are asked to participate in the evaluating, several advantages are gained. If general agreement is reached by those participating, the layout has, in essence, sold itself; there is no problem of going back and convincing or trying to convince these people as to the advantages and disadvantages of one layout over another. Participation helps focus the minds of several people on the layout, balancing their own personal preferences for any one plan. Participation at this stage, like participation at the stages of establishing relationships and determining space, brings in for consultation and discussion the people who have to make the layout work.

When the joint-participation method is used for evaluating alternatives, ratings can be done in two ways: (a) by individuals, then compared, or (b) by joint discussion. It is generally better to do the former. Comparisons of individual ratings usually show better than half of the ratings the same, and discussion can be limited to only those areas where differences occur.

The factor analysis method is one that makes a systematic evaluation out of many otherwise subjective views, and it is, therefore, particularly adaptable where investment costs or savings between plans are not accurately measurable or significant. The procedure is especially suitable, too, for projects where the degree of opinion is high in relation to measurable economic considerations. The procedure is especially appropriate for general overall layouts, service areas, and office layouts—less so for detailed layouts of production machinery and equipment where the factors involved are more tangible or specific. Recognize, however, that investment required and savings or return can be entered on the evaluation sheet themselves.

The planner should not overlook the genuine psychological benefits of this

**Figure 10-7.** Example of evaluating alternative layouts by the factor analysis method. Vowel-letter rating and values are those shown in Figure 10-4. Here, one alternative stands out as best. More often, the decision is not so clear and further appraisal is necessary.

technique. It affords a convenient, organized way to involve both those who must make the layout work (in the rating) and those who must approve the expenditure of funds (in setting the factors and/or the weight values). Most important, it provides a device for allowing the planner to get his relative sense of values coordinated with the thinking of management and operating people before he makes his recommendation on a specific plan.

Actually, the factor analysis system of selecting alternative layouts is practically the same as that in most job evaluation plans. But making the layout evaluation system more elaborate has not proven practical, largely because the factors and weight values change from company to company and from project to project within the same company.

### Cost Comparisons

The method of evaluating layouts which has the most substance is some form of cost comparison or financial analysis. In most cases, if cost analysis is not the chief basis for decision, it is used to supplement other evaluation methods.

There are two entirely different reasons for making a cost analysis, although the data used in both cases may be identical. In the first case, the purpose will be to *justify* a particular project—to find out whether it is economically feasible. In the second case, the problem is to *compare* alternative proposals with each other, and/or to an existing operation, assuming each to be adequately justified. The layout planner is usually concerned with the second case, though frequently with the justification as well.

There are basically two approaches to preparing a cost analysis. Either consider the *total costs* involved or consider *only those costs that will be affected by the project* under consideration. If alternative proposals for an entirely new layout are being compared, total costs must be used. With a relayout, it is often simpler, and just as effective, to deal only with the changes in costs reflected in the various proposals under consideration.

The size of the project also must be considered in deciding what kind of comparison to make. If the project is extensive and complex, it should be evaluated for its effect on the total operation of the company. A relatively small relayout project, on the other hand, can be evaluated and alternatives compared simply by calculating the change in contribution to profit.

Before making any computations, the layout planner should be thoroughly familiar with the accounting and financial policies of his company. There is usually competition for available funds among different areas of the company, and proposals must be subject to a rule of measure common to all. Additionally, it is most important that the layout planner know what decision management is planning to make on the basis of the data he furnishes them. Company policy on handling depreciation, tax considerations, cost distribution procedures, and the assignment of costs to specific accounts are areas for potential error unless clearly understood.

There are also several methods of figuring the practicality of an investment in a new layout. Return on investment, return on capital employed, and payout period are all methods that can be used, and there is no uniform agreement as to which is best. Our rule is to *use that method which your own financial officer recommends or will accept.*

Before starting to collect data for an analysis, the layout planner, or analyst, must recognize that there are two forms of expenditures: *capital expenditures* and *operating expenses.* Improper or inconsistent classification of these expenditures can lead to serious errors. Questions of proper classification should be resolved with the accountants before the analysis is begun.

With these considerations in mind, the planner needs a systematic way of classifying cost elements and accumulating cost figures in such a manner that any of the methods of analysis can be used to measure the performance expected. We recommend the following procedure:

1. Prepare a work sheet(s) that picks up *investment* requirements for each alternative.
2. Prepare a work sheet that establishes *operating* cost estimates.
3. Make *calculations* to compare or justify expenditures for alternative layout plans.

Figure 10–8a, the Estimate of Investment Requirements, is a form that provides adequate space for recording the various kinds of *investment expenditures*, and relates those expenditures to the period in which they occur. Most of the items or classes of expenditures necessary are listed down the left side of the sheet, with several blank spaces provided for special items. We have included space for the expense items that occur prior to the start of operations, because they are often overlooked. Remember that any items listed in this category are carried forward to an expense account and are not to be capitalized. However, some of the items listed as expense, such as site preparation and engineering services, may well be handled as capital expenditures depending upon the circumstances under which they occur.

A section of the form is provided for working capital—again because it is frequently omitted from calculations, where it belongs. The vertical columns are identified at the top by the year in which a given expenditure is to be made. The time value of money is important, and it is easy to forget that certain funds may be spent well in advance of the beginning of operations. For example, if $500,000 is paid for land three years before operations start rather than one year before, the difference in total cost at a simple interest rate of 6% per year for two years is $60,000. This kind of expenditure is easy to miss unless space for "before" expenditures is provided on the form being used.

Sometimes after operations begin, additional capital investments will be made. For example, a terminal warehouse may be added to a production facility some time

a

**ESTIMATE OF INVESTMENT REQUIREMENTS**

Company *Mity - Fine Products, Inc.*
Plant *Milford* Project *60103*
Project Description *Modernization of Material Handling System and Production Equipment* Estimate by *JDW* With *KH*
Date *12-1* Sheet *1* of *1*
Reason for Project (check one): ☒ Cost Reduction, ☐ Support new product or line, ☐ Other____

| CLASS OF EXPENDITURE | BEFORE ZERO POINT | | | AFTER ZERO POINT | | | | TOTAL BY CLASS |
|---|---|---|---|---|---|---|---|---|
| | 19__ -3 | 19__ -2 | 19*71* -1 | 19*72* +1 | 19__ +2 | 19__ +3 | 19__ +4 | |
| **Capital** | $ | $ | $ | $ | $ | $ | $ | $ |
| 1. Land | | | | | | | | |
| 2. Buildings | | | | | | | | |
| 3. Production Equipment | | | 250,000 | | | | | 250,000 |
| 4. Mobile Equipment | | | | | | | | |
| 5. Auxiliary Equipment | | | | | | | | |
| 6. *Mtl. Handling Equip* | | | 100,000 | 50,000 | | | | 150,000 |
| 7.___ | | | | | | | | |
| 8. Sub-Total | $ | $ | $ 350,000 | $ 50,000 | $ | $ | $ | $ 400,000 |
| **Expense** | $ | $ | $ | $ | $ | $ | $ | $ |
| 9. Site Preparation | | | | | | | | |
| 10. Moving costs | | | | | | | | |
| 11. Engineering Services | | | | | | | | |
| 12. *Start-Up (one-time)* | | | 40,000 | | | | | 40,000 |
| 13. *Obsolescence* | | | (10,000) | | | | | (10,000) |
| 14. Sub-Total | $ | $ | $ 40,000 | $ | $ | $ | $ | $ 40,000 |
| 15. Total requested or Authorized (Lines 8 + Line 14) | $ | $ | $ 390,000 | $ 50,000 | $ | $ | $ | $ 440,000 |
| **Working Capital** | $ | $ | $ | $ | $ | $ | $ | $ |
| 16. Min. Cash Balance | | | | | | | | |
| 17. Accounts Receivable | | | | | | | | |
| 18. Fin. Goods Inventory | | | | | | | | |
| 19. WIP Inventory | | | | | | | | |
| 20. Raw Material Inventory | | | | | | | | |
| 21.___ | | | | | | | | |
| 22. Total | $ | $ | $ | $ | $ | $ | $ | $ |
| 23. GRAND TOTAL (Line 15 + Line 22) | $ | $ | $ 390,000 | $ 50,000 | $ | $ | $ | $ 440,000 |

NOTE: Zero Point is the beginning of the year in which operations are begun, or the year in which revenues are first generated, whichever is the later.
Use "Expense" Section to list non-capitalized expenditures prior to Zero Point.

b

**ESTIMATE OF OPERATING COSTS**

Company *Mity - Fine Products, Inc.*
Plant *Milford* Project *60103*
Projection Description *Modernization of Material Handling System and Production Equipment* Estimate by *JDW* With *KH*
Date *12-1* Sheet *1* of *1*
Reason for Project (check one): ☒ Cost Reduction, ☐ Support new product or line, ☐ Other____

| COST ELEMENT | Period___ Alt. *Present* Diff.___ | Period *72+* Alt. *A* Diff.___ | Period *72+* Alt.___ Diff. ✓ | Period___ Alt.___ Diff.___ | Period___ Alt.___ Diff.___ | Period___ Alt.___ Diff.___ | Period___ Alt.___ Diff.___ | Period___ Alt.___ Diff.___ |
|---|---|---|---|---|---|---|---|---|
| **Material** | $ | $ | $ | $ | $ | $ | $ | $ |
| 1. Direct Mtl. | 500,000 | 470,000 | 30,000 | | | | | |
| 2. Scrap or waste | 50,000 | 20,000 | 30,000 | | | | | |
| 3. Supplies and Packing | | | | | | | | |
| 4. Maintenance Parts | 10,000 | 5,000 | 5,000 | | | | | |
| 5.___ | | | | | | | | |
| 6.___ | | | | | | | | |
| 7. Sub-Total | $ 560,000 | $ 495,000 | $ 65,000 | $ | $ | $ | $ | $ |
| **Labor** | $ | $ | $ | $ | $ | $ | $ | $ |
| 8. Direct Labor | 75,000 | 10,000 | 65,000 | | | | | |
| 9. Overtime-shift premium | | | | | | | | |
| 10. Idle time-downtime | | | | | | | | |
| 11. Maintenance* | 10,000 | 5,000 | 5,000 | | | | | |
| 12. Inspection* | | | | | | | | |
| 13. Handling and stores* | 40,000 | 6,000 | 34,000 | | | | | |
| 14. Supervision* | | | | | | | | |
| 15. Engineering* | | | | | | | | |
| 16. Other service or indirect* | 13,000 | 3,000 | 10,000 | | | | | |
| 17. Fringe benefits | 25,000 | 3,000 | 22,000 | | | | | |
| 18.___ | | | | | | | | |
| 19. Sub-Total | $ 163,000 | $ 27,000 | $ 136,000 | $ | $ | $ | $ | $ |
| **Burden-Overhead** | $ | $ | $ | $ | $ | $ | $ | $ |
| 20. Interest on Investment | 1,000 | 16,000 | + (15,000) | | | | | |
| 21. Floor space, rentals | | | | | | | | |
| 22. Fuel, power | 5,000 | 8,000 | + ( 3,000) | | | | | |
| 23. Taxes and Insurance | | | | | | | | |
| 24. Depreciation | 2,000 | 40,000 | + (38,000) | | | | | |
| 25. *Misc.* | 7,000 | 2,000 | 5,000 | | | | | |
| 26. Sub-Total | $ 15,000 | $ 66,000 | $ + (51,000) | $ | $ | $ | $ | $ |
| 27. TOTAL (Lines 7, 19, 26) | $ 738,000 | $ 588,000 | $ 150,000 | $ | $ | $ | $ | $ |

*Denotes items pulled out of overhead for greater accuracy.

c

**ESTIMATES OF PAYOUT PERIOD**

Company *Mity - Fine Products, Inc.*
Plant *Milford* Project *60103*
Projection Description *Modernization of Material Handling System and Production Equipment* Estimate by *JDW* With *KH*
Date *12-1* Sheet *1* of *1*
Reason for Project (check one): ☒ Cost Reduction, ☐ Support new product or line, ☐ Other____

| ITEM* | Present Costs* | Costs For Alt. *A* | Costs For Alt.___ | Costs For Alt.___ | Costs For Alt.___ | Costs For Alt.___ |
|---|---|---|---|---|---|---|
| 1. Total Operating Costs (Line 27, RMA 721) | $ 738,000 | $ 588,000 | $ | $ | $ | $ |
| | | Diff: From *Present* | Diff: From ___ | Diff: From ___ | Diff: From ___ | Diff: From ___ |
| 2. Difference in Costs (Present vs. Alternative or Alternative vs. Alternative) | | $ @ 150,000 | $ | $ | $ | $ |
| 3. Estimated Income Tax | | @ 75,000 | | | | |
| 4. Added Profit after Tax (Item 2 minus Item 3) | | @ 75,000 | | | | |
| 5. Difference in Charges for Depreciation (Line 24, RMA 721) | | 38,000 | | | | |
| 6. Amount Recovered in Period (Savings, or Profit Increase on a cash-flow basis) | | 113,000 | | | | |
| 7. Total Funds to be Recovered (Line 15 minus *50* % of Line 14, RMA 720) | | 420,000 | | | | |
| 8. PAYOUT PERIOD (Item 7 ÷ Item 6) | | 3.7 Yrs. | Yrs. | Yrs. | Yrs. | Yrs. |

Explanatory Notes:
- * All figures as based on properly comparable annual or period costs.
- Item 2. This figure is equivalent to added profit before income tax.
- Item 3. & 5. Apportion to period if other than annual figures are being used.
- Item 5. To convert calculations to a cash-flow basis, since depreciation charges do not require cash outlays. Normally added to Item 4.
- Item 7. Include capital investment and expenditures for expense items but not working capital additions. Figures must be from column headed "Total by Class" on RMA 720. Fill in percent appropriate to individual tax structure in blank space provided. For example, with a corporate tax structure and a profit situation, use 50%.
- Item 8. Convert to years if costs are based on a period other than year.

NOTES: *@ Before one-time charge for Obsolescence and Expense (after tax) of $30,000*

after full production is reached. The form provides a reminder that such things are to be ferreted out and included.[1] All entries on this form should be supported by appropriate engineering estimates or financial forecasts. One sheet of this form is generally prepared for each alternative plan.

The same general format is used on the worksheet for recording *operating cost* data. A sample worksheet, the Estimate of Operating Costs, is shown in Figure 10–8b. Again the cost elements are listed down the left side of the sheet. The vertical columns may be used for recording the cost amounts either by period or by alternative, or in combination. The form can therefore be used to compute differences between alterna-

1. The data on this worksheet is the basis for requests for capital appropriations to be included in the Capital Expenditure Budget (Chapter 15).

**Figure 10-8.** Examples of cost comparisons. Figure "a" shows Estimate of Investment Requirements for the modernization of the material handling system and production equipment. This form provides space for classifying and listing the funds needed to get the project underway. In the example, a third of the material handling equipment will not be purchased until next year. The periods before and after zero point can be any increment of time that is convenient to use; whether we use calendar years, fiscal years, quarters, or months does not matter.

The expense columns can be used after the zero point, but only if they are expenses other than normal operating expenses and are incurred as part of getting the project fully underway. If there is a one-time charge for obsolescense, as there is in this example, it should be entered in the expense section, but the amount should be circled and not included in the totals.

Any entry in the Working Capital Section should show *added* working capital required to support this project. When comparing alternate proposals involving differences in investment, a separate sheet of this form should be used for each alternative proposal.

Figure "b' shows an Estimate of Operating Costs. Here comparative operating cost figures for both the existing method and a proposed method (Alternative A), and the differences between the two, have been entered. In the example shown, we have compared the average costs of the present method with the average costs of Alternative A, and have calculated the difference in the costs for each element and for the total amounts.

In the column of figures farthest to the right, the entry on Line 27 represents an average reduction in annual costs, or added profits before tax, of $150,000 that would be the result of making the expenditures listed in Figure "a".

Figure "c" shows an Estimate of Payout Period. This was used to compute the payout period using dollar amounts from Figures "a" and "b". Any working capital requirement is excluded from payout period calculations, for we assume that the entire amount of added working capital will be recovered when the installation is terminated or abandoned. We should remember that the total funds requested, the total capital investment to be authorized, and the total funds to be recovered during the payout period will frequently be different amounts. For convenience in computing the payout period, any one-time charge for obsolescense and/or start-up expense is ordinarily excluded from the cost figures, unless it is appreciably different for given alternatives.

tives, total period costs, or combinations of both. All entries should be supported by detailed cost analysis sheets.

The breakdown of cost elements may seem unusually detailed, but it is deliberately so. It is far too easy to miss some of the less common elements; and this listing is helpful as a check to ensure that all applicable costs are included in the calculations.

Both forms can be used on any project, regardless of complexity, by making entries only where appropriate.

With these two forms filled out for a given project, the planner has the data needed for calculating a solution. Now he must decide what kind of analysis of the data accumulated to make. Often, the analyst is restricted by company policy to a given method of evaluation; even so, he should be familiar with the theory and the inherent assumptions on which that method is based.

Space does not permit a detailed discussion of the various methods of analysis. A wealth of information is available in current literature on the subject, although all authorities do not agree on which is actually the best method to use.

We suggest use of the discounted-cash-flow method or profitability-index method for justification studies and/or comparison between alternatives for the following cases:

1. A major relayout project
2. Any layout project that involves increments of investment in successive years
3. Any layout project that involves a number of assets with large differences in useful lives
4. Any layout project in which the pattern of revenues and/or costs varies extensively as time progresses
5. When the project under consideration has to compete with projects from other divisions or areas for a limited amount of funds

Most layout projects do not require such extensive analysis, and a less complex method is adequate. Many layouts do not involve large investments, but are rearrangements of equipment already on hand. In such cases, the major portion of the costs to be recovered are those incurred in planning and moving. Extensive analysis is also usually unnecessary for relayout projects where proposed alternatives can be compared with existing operations; for projects that affect only a small part of a total operation; or for "necessity" projects that must be done to keep a larger operation functioning properly. In such cases, the calculation and comparison of payout periods is adequate for a proper decision. But a word of warning: Do not use payout period as a crutch to avoid using a more detailed and accurate method when it would be more appropriate.

Inasmuch as all methods of analysis depend on essentially the same expenditure and cost data, it is often helpful to evaluate by a simple method, such as payout period or average return, and then go on to more precise calculations if a clear advantage is not apparent.

In Figure 10–8c, the data from the figures above have been used to calculate the payout period. Here, with changes in costs between an existing method and a proposed method, the savings in cost (profit increase on a cash-flow basis) are compared with the total funds to be recovered in order to determine the payout period.

This example of payout is included because it is a popular method of cost justification and comparison; but, we have warned of its limitations. Especially, it is not advisable to depend on payout period calculations if the useful life of the project is considerably longer than the payout period.

Costs are not the only element to be evaluated in choosing among proposed layouts. The intangibles evaluated by the factor analysis method are frequently more significant than cost justification and comparison. In any event, the planner can well afford to apply both methods to many if not most projects.

### Request for Approval

Although the method of getting the layout approved depends upon the nature of the project, the company, and what kind of approval is required, generally speaking, it involves the following:

1. As a check on your layout, review it with the other people most involved—including operating and service personnel. Others gain a feeling of participation, and the review helps to presell the layout. Others understand better what is planned. They help catch errors which may have been made. All this allows presentation to higher authorities with the support of those who are involved.
2. Prepare a clear and accurate "picture" of what is proposed. Clear visualization helps others understand quickly what they are being asked to approve. Use present-and-proposed illustrations, with explanatory drawings or models, and clear plans showing recommended and alternative solutions.
3. Provide a recap or synopsis of how the recommended plan was developed. The SLP methodology serves here beautifully, for its pure logic and plain orderliness command respect. And, if both the relationships and space have been signed off as approved earlier, the submission for final approval can be highly convincing.
4. Prepare a presentation for approval, either oral or in written report form. The specific request should be just that—a specific request for approval and allocation of funds. It should include a summary statement indicating the investment funds required; when the money will be spent; what the returns will be, and when; a breakdown of how the money will be spent; as well as intangible advantages and disadvantages. And, it should answer the questions which people most often have in mind when asked to give their approval, namely:
   a. What do we stand to *gain* from this layout?
   b. What do we *risk* with this layout?
   c. How does this layout *affect me* personally and the group I represent?

### Recognizing Approval

When approval is given to the overall or block layout plan, the SLP pattern of procedures is complete. Having reached this point, the planner can move aggressively into the detailed planning phase of the project. It is important, therefore, to draw up, duplicate, or otherwise reproduce the approved plan so that all further planning will be integrated with what has generally been approved. This helps ensure coordination and avoids some groups' misdirecting their efforts by working to nonapproved plans.

### Overlapping Considerations

Planning the general overall layout is discussed, in the eight chapters of Part Two, as one straightforward sequence. Keep in mind, however, that Phase II is overlapped by Phase I and itself overlaps Phase III. Frequently important Phase III details of critical subareas are considered in the work on the general overall layout. Moreover, it often means making a preliminary general overall plan. This approved Phase II is then submitted to others for study, comment, and rework. The home office, architectural engineers, layout consultants, and others are frequently involved here, if they have not already participated during integration of the modifying considerations or evaluating of alternatives.

Tentative plans on the Phase III details can now move forward without much likelihood that major changes in the general overall layout will cancel any detail planning. Still, it is wasteful of manhours and engineering talent to go too far into Phase II details without having general agreement on the major physical features (column spacing, walls, main aisleways, utility distribution, and the like) that are part of, or approved with, the overall layout.

# Part Three

# EXAMPLE OF SYSTEMATIC LAYOUT PLANNING, PHASE II

Inputs: P,Q,R,S,T & Activities

**1**

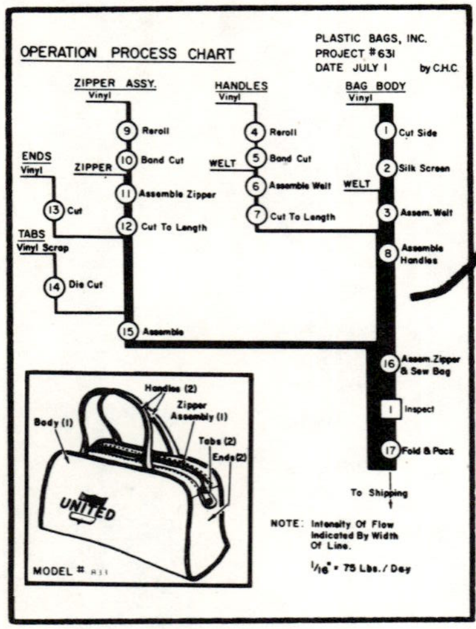


Flow of Materials

**1 & 2**

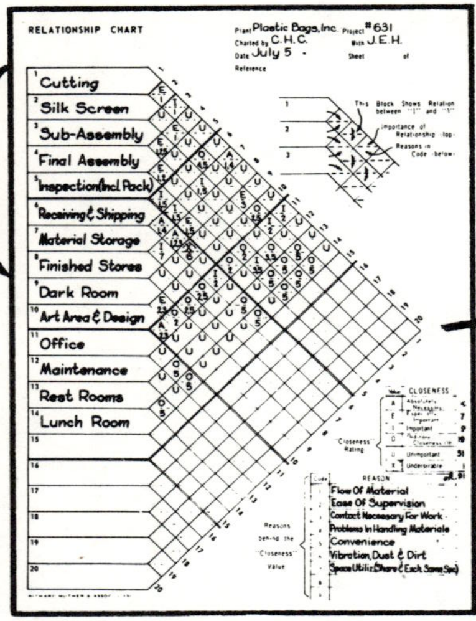


Relationship Chart of Activity-Areas (based on flow and services)

**3**

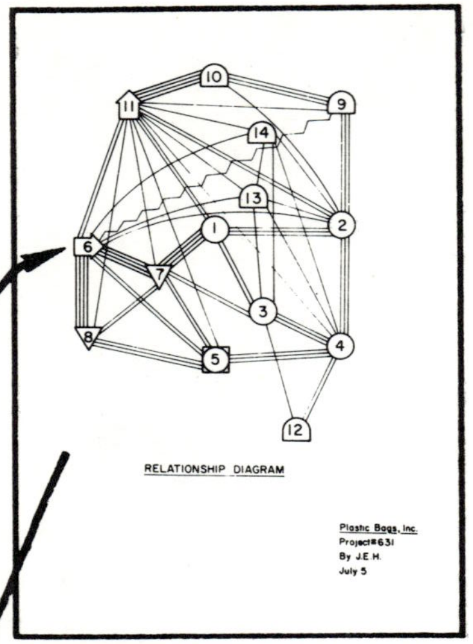


Relationship Diagram

**4 & 5**

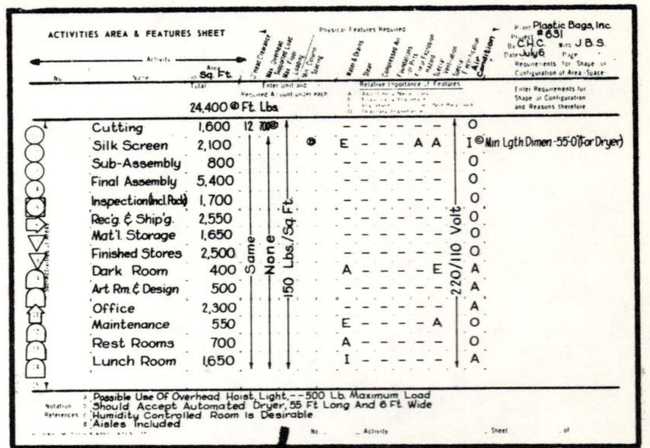


Space Requirements

**6**

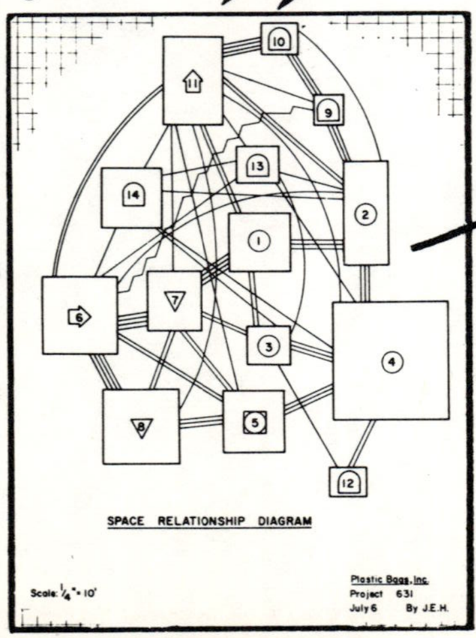


Space Relationship Diagram

**7 & 8**

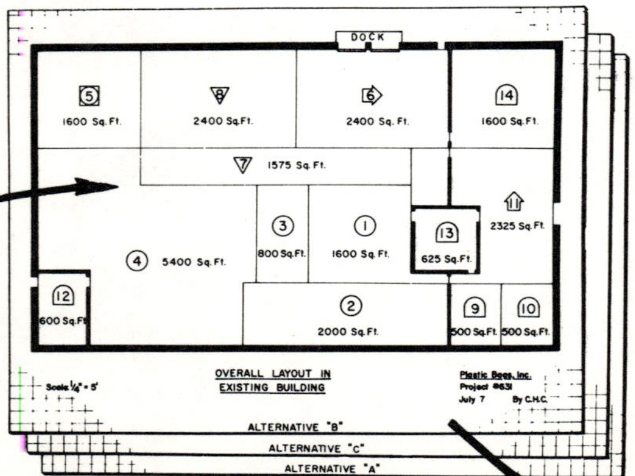


Alternative based on Modifying Considerations and Practical Limitations

**9**

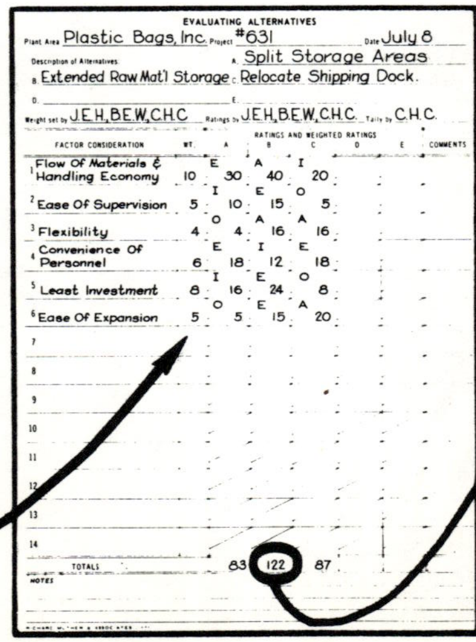


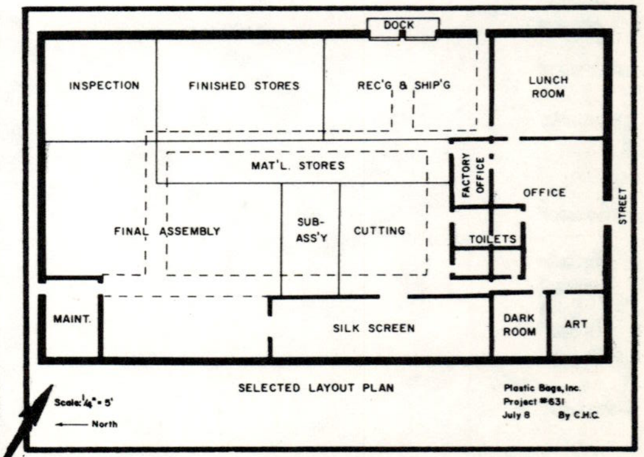


Selected Layout Plan Phase II — Overall Block Layout

# Introduction to Part Three

*Part Three covers planning detailed layouts, Phase III of the framework. It involves arranging the location of each individual machine, piece of equipment, employee working space, storage rack of material, and supporting features.*

*As we have pointed out, the phases are sequential and, for best results, should overlap. Also, we have stated that the SLP pattern of procedures is essentially the same for both Phases II and III. The pattern merely repeats itself in more detail for each of the subareas to be laid out.*

*Perhaps Part Two is best summarized by an example. The* Figure *accompanying this introduction illustrates each step of the Phase II sequence. It represents the planning for a new factory, of about 25,000 square feet, to produce plastic bags for airlines, beach bags, tennis racket covers, and the like.*

*Each step is numbered above its illustration. The complete sequence represents the pattern as followed for this particular example.*

*An example of detail layout planning, Phase III, for Activity Number 2 (Silk Screen Department) is shown later on as Figure 11-6. It follows the pattern of procedures, but, of course, puts different emphasis on different boxes.*

## Chapter 11

# Detail Layout Planning

Detail layout planning, locating each individual machine and piece of equipment, requires more specific and more detailed data, dimensions, and techniques of analysis. (To refresh views on this, refer to Figure 1–6.) Still, when laying out each area in detail, the planner follows basically the same pattern set forth for Phase II, only in greater detail.

Another significant difference is that detail planning is done within more specific confines. Space Available, for example, is pretty well established; the limits of area and configuration are set when the General Overall Layout is approved; column spacing, floor loading, and other building features are also fixed; the interdepartmental handling system is agreed upon; scheduling, inventory, and timekeeping policies and procedures are now set.

It is logical and conserving of time to get the overall (generally interarea) questions resolved before tackling the details, for detail layout planning requires considerable time. This is not to say detail planning is more important or requires more skill and experience. On the contrary, Phase II is where major and costly layout decisions are made. But Phase III requires more manhours—and it has to be done to save money in installation, readjustments, and operating downtime. As a result, it is usually possible, especially on large projects, to delegate detail layout planning to those in the company at an echelon of responsibility one level below those who have been intimately concerned with planning the overall layout.

Still one further difference between Phase II and Phase III is the level of those who will participate and assist in the layout planning. In overall planning, the participation and approval of division officers or departmental managers is obtained. In detail layout planning, the opinions of those who will be directly responsible for the operation of each department or subarea being planned are solicited. This involves foremen, supervisors, and supporting-service group or section leaders; they cannot be overlooked. Failure to include in the planning the people who are actually involved and vitally interested is a serious error, one that can lead to a completely impractical layout or an installation that does not function effectively.

### Degrees of Facilities Planning

Because detail planning does take considerable time, and every layout project has its practical time limit, it is well to set a schedule for each area to be laid out. Coordinate the times for adjacent areas so their effect on each other can be worked out. The schedule stimulates improved performance on the part of those planning the layouts. And, in general, the schedule better coordinates the overall project.

Discriminative scheduling of each area is in order. By this we mean: Recognize and designate ahead of time differences in degree, kind, or emphasis of the layout planning to be done. One area may involve entirely new equipment; a second area will be moved "as is," with all but the barest necessary detail planning postponed until later.

To accomplish this planning in various degrees, again use the vowel-rating conventions. See Figure 11-1.

This A-to-U rating, when assigned to each area, designates to subordinates and operating personnel where the planning time is to be spent and which areas are to remain "about the same." Areas designated A are those involving new processes completely unfamiliar to the company. U-designated areas are to be left essentially as is; or the department supervisor involved will arrange his own area. In either event, no layout planning time is scheduled.

This system of rating layout planning effort or attention is termed *Degree of Facilities Planning.* Generally, degree of facilities planning indicates, in addition to the degree of layout planning, the extent of financial investment as well. It is logical, therefore, that this rating method should be applied broadly and not limited to scheduling layout-planning manhours.

### The SLP Pattern Repeated

Apply the SLP pattern to the layout of each detail area. Establish what products or materials are involved and how many of each for each area. Analyze the other input data for each area, reorganize them into a finer breakdown, and establish the activities for the area (actually, the subactivities within the earlier designated activity).

```
A - Absolutely necessary to do extensive layout planning at
    this time.

E - Especially important to do extensive layout planning at
    this time.

I - Important to do extensive layout planning at this time.

O - Ordinary layout planning is OK.

U - Unimportant--don't spend any time on layout planning now.
```

**Figure 11-1.** Rating designations for Degree of Facilities Planning. This is a systematic way of indicating the degree of layout attention each area can justify in the light of the total project schedule.

To analyze *flow of materials* for the overall layout, the technique of analysis varied with the relative quantity. The same holds true for detail layout planning. Use essentially the same type of analyses—adjusting each to serve best the particular project at hand. For example, in Phase II, for charting flow of one or a few products or materials, an *operation* process chart is used, whereas here fuller details are demanded; therefore use a *flow* process chart. See Figure 11-2. Similarly, when planning the details of one assembly line, a form to show the flow-in from both sides can be more definitive. See Figure 11-3. Any project in which rearrangements and realigning of layouts are a frequent necessity may require its own analysis form.

Individual *relationship charts* may be made for the activities within each area. In Phase III, an entire chart may be made for what, in the overall layout phase, was just one activity on the chart. In the detail relationship chart, such activities as the scrap barrel, the time clock, and the drinking fountain may be included, while in the overall relationship chart, comparable activities would have been the salvage department, the timekeeping office, and the cafeteria. Of course, it is not necessary to build a relationship chart for each line on the overall relationship chart, but where the area is a sizeable one, or where services are important or complex, time is saved in the long run by going through the entire relationship chart procedure for that area.

In *diagraming* the flow or relationships within the area, follow the same procedures as in the overall layout. Some adjustments and modifications are to be expected, but generally stick with the symbols and codes of the SLP conventions. They are meaningful, clear, and consistent. See Figure 11-4. (Also, refer to Figure 6-5. The office area diagramed there could well have been one department of a large office, in which case, the diagram shown would be considered Phase III.)

In figuring *space required*, the planner deals with the space for each individual machine, man, set-down space, and the like. This means working with actual dimensions of each space-occupying element to be included. It is likely that overall space figures in Phase II did not include dimensions of each detailed element. In Phase III, the planner must determine the actual space for each element.

The overall *space available* is already established in amount and condition with the general approval of the overall layout. However, it is still possible to exchange space between departments and in most cases to adjust somewhat the general overall layout within its major fixed features. The project after all, is still "on paper." But, if Phase II has properly overlapped Phase III, there should be a minimum of interarea adjustments affecting the overall layout.

In converting the activity relationship diagram to a space relationship diagram, the planner uses actual replicas of the space required for each machine or piece of equipment. This leads us to the point of discussing equipment replicas: drawings, templates, and models. However, we leave this discussion to Chapter 12, where adequate attention is given to methods of visualizing and duplicating the layout. Just as in Phase II, however, to build a space relationship diagram, the planner must pick a scale and work out the space by means of visual analysis.

## FLOW PROCESS CHART

Process charted: *End Bracket No.70-B*

☐ Man or ☒ Material

Starting point *In-Process Storage*

Ending point *Assembly Line*

| Conversions for Charted Unit to End Unit | | |
|---|---|---|
| Charted Unit | ~~Size~~/Weight | Quantity/End Unit |
| *Box* | *56 lbs (filled)* | *1/11* |
| *Tray* | *25 lbs.* | *1/5* |
| *Piece (Brkt.)* | *5 lbs.* | *1* |

Plant *LANTING COMPANY* Project *259F*

By *K.M. Higgins* With

Date *6/5* Sheet *3* of *4*

Quantity of End Unit (unit of end item) per (time) *235 Brack't/Day*

☒ Present ☐ Proposed (Alternative # )

Description of Alternative:

| | CHARTED UNIT (Unit of Product or Material Charted) | UNITS PER LOAD | Operation / Handling / Transport / Inspection / Delay / Storage | DESCRIPTION OF ACTION | Weight or Size of Load in *Pounds* | Number of trips per *Day* | Distance in *Feet* | Time in *Min.* per *Day* | Cost in *$* per *Day* *(Labor)* | NOTES Verify: Product-Quantity-Route-Support-Time Analyse: Why-What-Where-When-Who-How Eliminate-Combine-Rearrange |
|---|---|---|---|---|---|---|---|---|---|---|
| 1 | *Box* | | Storage | *In-process storage (shelves)* | | | | | | |
| 2 | *Box* | *1* | Handling (1) | *Placed on 4-wheel handtruck* | *56* | | | | | *Why not use fork lift truck* |
| 3 | *Box* | *4* | Transport (1) | *To assembly department* | *224* | *5.3* | *310* | *46* (rows 2–4) | *1.53* | *235 pcs.÷11 pcs/box = 21.4 box/day* |
| 4 | *Box* | *1* | Handling (2) | *Placed on inspection table* | *56* | | | | | *21.4÷4 = 5.3 loads/day* |
| 5 | *Box* | | Delay (1) | *On inspection table* | | | | | | *Max. 8 boxes* |
| 6 | *Piece* | *1* | Handling (3) | *Removed from box* | *5* | | | | | |
| 7 | *Piece* | | Inspection (1) | *Checked for quality* | | | | | | |
| 8 | *Piece* | *2* | Handling (4) | *Placed in tray* | *10* | | | *370* (rows 6–10) | *12.35* | |
| 9 | *Tray* | *1* | Transport (2) | *Carried to production line* | *25* | *47* | *40* | | | *Revise layout* |
| 10 | *Tray* | *1* | Handling (5) | *Loaded in roller rack* | *25* | | | | | |
| 11 | *Tray* | | Delay (2) | *In roller rack* | | | | | | *Max. 7 trays in rack* |
| 12 | *Piece* | *1* | Handling (6) | *Removed by assembler* | *5* | | | | | |
| 13 | *Piece* | | Operation (1) | *Bolted to main unit* | | | | *510* (rows 12–13) | *17.00* | |
| 14 | | | | | | | | | | |
| 15 | | | | | | | | | | |
| 16 | | | | | | | | | | |
| 17 | | | | | | | | | | |
| 18 | | | | | | | | | | |
| 19 | | | | | | | | | | |
| 20 | | | | | | | | | | |
| 21 | | | | | | | | | | |
| 22 | | | | | | | | | | |
| 23 | | | | | | | | | | |
| 24 | | | | | | | | | | |
| 25 | | | | | | | | | | |
| 26 | | | | | | | | | | |
| 27 | | | | | | | | | | |
| 28 | | | | | | | | | | |
| 29 | | | | | | | | | | |
| 30 | | | | | | | | | | |
| | Totals | | *1 / 6 / 2 / 1 / 2 / 1* | | | Totals | *350* | *926* | *30.88* | |

The space diagram leads almost immediately to a rough arrangement of equipment replicas. Next incorporate the *modifying considerations* and weigh their *practical limitations*. Here all manner of analytical techniques, too numerous to mention here, are applied. For an explanation of these techniques, refer to any of the good handbooks on procedures or industrial engineering.[1]

### The Scale

There is no great advantage in having the scale of the detail layout plan a multiple of the general overall layout. Pick the scale that best suits the purpose. While the scale for Phase II can have almost any dimensions, the scale detail layout planning recommended is ¼ inch equals one foot (or two centimeters equal one meter metric).[2] Anything much bigger is too space-consuming. Anything smaller begins to lose accuracy.

---

1. *Industrial Engineering Handbook*, Edited by Harold B. Maynard (New York: McGraw-Hill Book Co., Inc., 1971)
*Handbook of Industrial Engineering & Management*, Edited by W. Grant Ireson & Eugene L. Grant (Englewood Cliffs: Prentice-Hall, Inc., 1970)
*Production Handbook*, Edited by Gordon B. Carson (New York: The Ronald Press Co., 1972)
*Plant Engineering Handbook*, William Staniar, Editor-in-Chief (New York: McGraw-Hill Book Co., 1959)
*Materials Handling Handbook*, Edited by Harold A. Bolz (New York: The Ronald Press co., 1958)
*Systems and Procedures, Handbook for Business and Industry*, Edited by Victor Lazzaro (Englewood Cliffs: Prentice–Hall, Inc., 1968)

2. Two centimeters equal one meter is a scale of 1 to 50; ¼ inch equals one foot is 1 to 48. One centimeter equals one meter is a scale of 1 to 100; 1/8 inch to the foot is 1 to 96.

---

**Figure 11-2.** Flow Process Chart traces the movement of material through each detailed step of the process. It employs all the process chart symbols. We use it in Phase III, just as we used the operation process chart in Phase II. This doesn't limit flow process charting to one departmental area, however. It can be used whenever the full detailed steps of the sequence of flow is wanted, especially when analyzing for material handling improvements, such as when modifying the space relationship diagram for materials handling considerations.

The handling symbol in this chart is made by combining two halves of the operation and transport symbols. By segregating the handling activities, it is easier to distinguish the true operations and true transports, and thereby to analyze more clearly the present or proposed process method. "Handling" occurs when an object is arranged, prepared, transferred, or positioned prior to another action. This definition then limits the "operation" symbol to when an object is intentionally changed in any of its physical or chemical characteristics, or is assembled to or disassembled from another object. It also limits the "transport" symbol to when an object is moved from one place to another.

At the top of the form is a place to make conversions between the various units charted. Even though only one product or material (or product-material group) is recorded, conversions are necessary to develop the information on time and cost.

Note that the time column can be used in a number of ways. It can show the time used *per day* (as in this example) or per *end unit*. It can also show time per charted unit, although then the column can not be totaled, for the frequency varies from one charted unit to another. The same applies to the cost column. In the heading of the cost column is room to note whether we are dealing with labor cost, operating cost, total cost, or some other figures.

**LINE-FEEDING FLOW CHART**

Line or Product *Tooth brush assort. 12-pack*
Starting at *Empty belt*

Plant *Milwaukee*
By *O. R. Davis*
Date *10/10*
Present Method ☐

Project *Assem. Dept. Relayout*
With
Sheet *1* of *4*
Proposed Method ☐

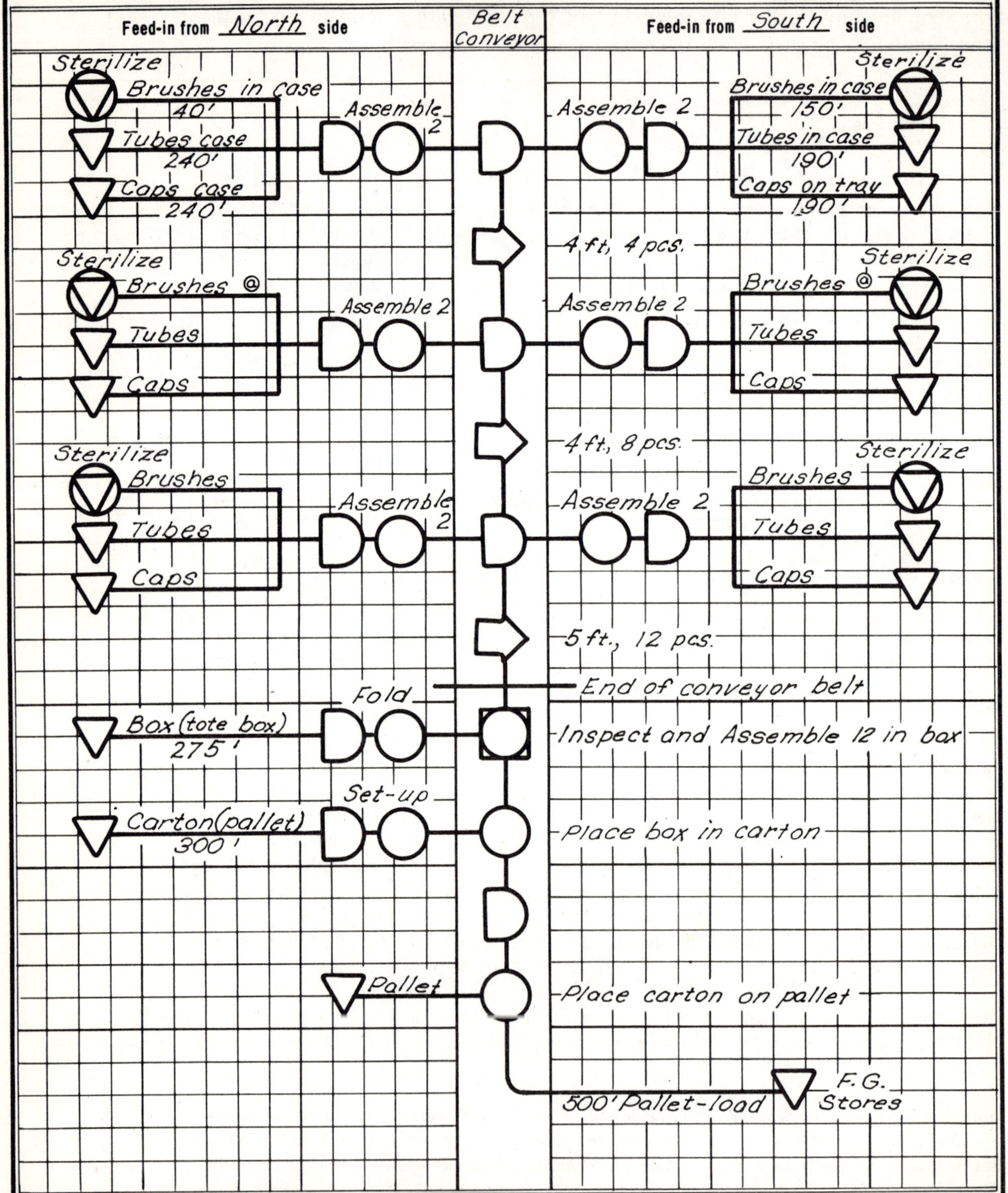


Reference Notes: *@ Brushes, Tubes and Caps each come from same source. Distances and containers are same for assembly stations two and three as for first assembly station.*

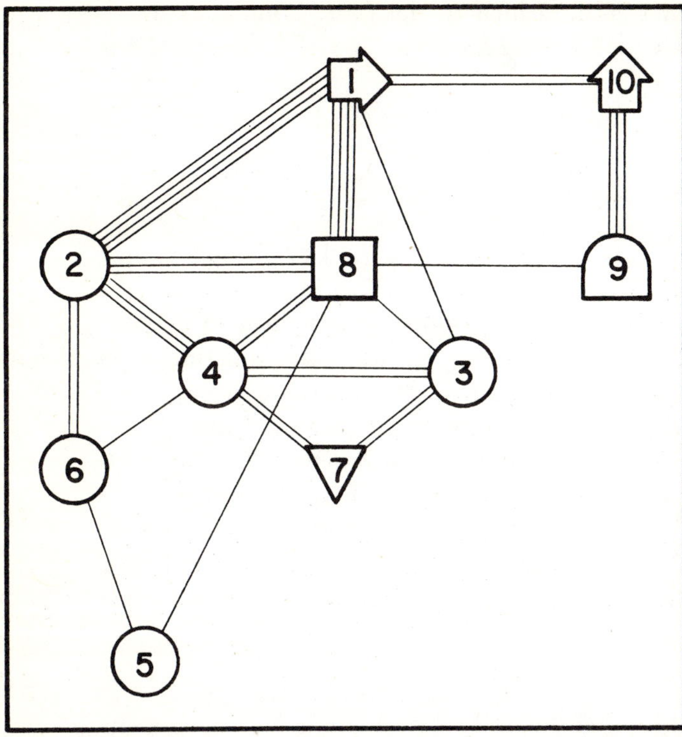


1. In-Out Material Staging Area
2. Power Shear #53
3. Federal 30-ton Punch Press #55
4. Johnson 55-ton Punch Press #54
5. Federal 30-ton Punch Press #56
6. Power Brake #58
7. Die Rack
8. Inspection Bench
9. Time Clock and Job Assignment Rack
10. Foreman's Desk

**Figure 11-4.** Activity Relationship Diagram for detail layout of a press department. This Phase III diagram uses the codes and symbols of the SLP conventions just as in Phase II. Here they represent individual machines, racks, or other detail activities, whereas in Phase II they represented entire departments or larger activities. This diagram will be improved, of course, by shortening the four-line and three-line relationships between activities 1, 2, and 8. Compare this diagram with the one in Figure 6-3 to note the similarity in diagraming for Phases II and III.

A scale of 1/8 inch equals one foot (or one centimeter equals one meter metric) can be used under the following situations:

1. There is a good deal of space to plan (approximately 100,000 square feet or more).
2. The production materials are relatively large (like airplanes or motor trucks).
3. The operations involve very large equipment (like large paint booths or steel-mill rolling equipment).
4. Prints of existing structures involved are already drawn at that scale.

**Figure 11-3.** Planning the detail flow of material into, through, and away from a production line by means of the line-feeding flow chart. This illustration shows the assembly of an assortment of twelve toothbrushes (in tubes and capped) over a belt conveyor. Companies producing automobiles, appliances, or any other products produced on a line fed from both sides can use this type of chart for their detail layout planning. Several lengths of this sheet can be fastened end-to-end for longer lines. Similarly, the grid can be used for line spacing—although this simple example does not do so.

Otherwise, best practice is to stick with ¼ inch to the foot. Commercially available materials are more readily available at this scale. The A.S.M.E. standard for templates and models recommends the ¼-inch scale.

### Adjustments

As the detailed layout is worked out, the planner may find he can get even better arrangements by changing or shifting the space and configuration of each area already allocated in the overall layout. Within practical limits, it is logical to make these adjustments in the overall layout. And it is also logical to make adjustments in the detailed layouts of other adjacent areas being planned. Moving main aisles, relocating departmental area limits, and altering the position of proposed utility trenches are examples of such adjustments. As a result, it is best not to to fix too finally any single detailed area until all detailed areas are planned, or at least until all areas surrounding the one in question are planned. Time will not always permit this; one may have to move ahead with a schedule of installation right behind detail planning. Recognize, however, that in so doing one is foregoing certain opportunities for making improvements.

### Checks

Checking the layout before it is finalized is sound practice. Generally checks are made in one or more of four ways:

1. Challenging the layout planned with check questions
2. Asking others to review it
3. Making further refinements, such as a three-dimensional model
4. Making a computer run for any other alternative

*Check Questions* involve such challenges as, "What would happen if we oriented the layout 180 degrees? . . . if we flopped it over the other way? . . . if we placed the second floor layout on the ground level or vice versa?" If his plan cannot be justified in the face of these questions, the planner is not yet ready with a recommendation. Other check questions are

Will it increase production?
Will it reduce costs?
Will it improve housekeeping?
Will it eliminate accidents?
Will it produce a better product?
Will it increase floor space?
Will it improve morale?
Will it reduce waste?
Will it reduce scrap?
Will it improve other working conditions?
Will it increase or decrease maintenance?

Will it provide insurance benefits?
Will it provide tax benefits?
Will it improve sanitation?

*Reviews* can be made individually or in joint meetings. Either can be practical. First having individual and then joint meetings on the same project is probably best. These reviews are over and above any solicitation of ideas from these same people or participation by them during the actual planning—which have been emphasized as extremely important, from both a practical and psychological view. Reviews with others include:

Heads of operating departments involved
Heads of supporting-service departments involved
Other staff planning engineers responsible for industrial engineering, materials handling, methods, plant engineering and/or maintenance, office manager, and the like, as practical
Safety director or safety engineer
Personnel director
Your direct supervisor

Making a *three-dimensional model* is both a form of checking and a refinement of planning. There is nothing like a model to help others visualize a plan. When others review the planned layout in scale model, they feel the impact of possible improvements. And by going through the extra refinement himself, the planner finds still additional improvements he can make in the layout.

Moreover, clearances can be checked by moving a model of the material, container, or transport equipment right through its route on the layout model.

A *computer run* on one of four programs currently available sometimes reveals an overlooked alternative. See Appendix XIV for a discussion of how a computer program may be used as an aid to supplement the mental/manual SLP methodology.

Recognize, of course, that the computer is even more valuable in processing input data. And be sure to appreciate that any layout printed out from computer analysis still requires the supplying of relationship and space data as inputs to the computer, and still requires adjustment of the print-out.

### Approval

In Chapter 10, we discussed getting approval for the general overall layout from higher management since at that point long-range or major commitments are agreed upon. Where there is ample time and where the project involves new construction or a major capital investment, it is good practice to have a topside reappraisal of the layout details before the final authorization of funds.

On the other hand, in rearranging existing layouts or where the project does not involve a major investment, approval of the detail layout plans logically rests "down the line." As a general rule, this final approval lies with the person responsible for making the general overall layout function. He usually goes along with detail layout plans which do not violently readjust the already-approved overall layout, and that are agreed to by his subordinates in their areas. Approval for detail layouts is sought from:

1. The people responsible for making each detail area operate
2. The people in charge of the services which support the detail operating areas
3. Safety personnel and interested engineering people
4. The group(s) responsible for making the installation
5. The person responsible for the operation of the area covered by the overall layout

Regardless of whom you ask to appraise or approve the layout, retain the consistency of the SLP conventions. It is helpful to others when the planner always shows the conveyors and storage equipment in orange or yellow, or when four lines always mean the greatest, most significant, or best. Indeed, the SLP conventions can serve as an added means of communications, so long as the planner provides adequate information on the plans.

Because of the many people involved, it is good practice to have each approving person sign the detail layout plan or an approval record similar to Figure 11-5. This is not done with the intent of "holding their feet to the fire" later on, when they change their minds or when they think up a further improvement. The sign-off does encourage critical examination and makes it more difficult for people to change their minds. The layout would never be installed if everyone involved were given a free hand with changes. In any case, if new or unforeseen possibilities crop up later on, or if the layout planner himself has made a serious error, no sign-off signature is going to hold back making a change.

In fact, changes during installation are bound to occur, and should be expected. Planning costs would be astronomical if the planner tried to avoid every possible change. No; what the planner wants is:

1. To have each person involved take time from his otherwise busy day to study the proposed layout
2. To have each person understand how the layout is planned to operate so he will get full value from the layout later on
3. To have each person agree that the layout is satisfactory and workable so far as he can tell
4. To have an organized way of checking off those who have reviewed the layout plan
5. To force the layout man actually to check with others to be sure all good ideas are incorporated in his plans
6. To create feelings on the part of those responsible for the functioning of

## REVIEW AND APPROVAL RECORD

Plant *Overall Clothes* Project *256-93*
By *H.M. Kirk* With *C.W. Rhule*
Date *10/2* Sheet *1* of *1*

Project Identification Name and Description:

Move Trim Department to area now occupied by Sewing Room and relocate portion of Sewing Room in East wing.

According to good practice and established procedure all persons having responsibility for subsequent performance of any installation should review and approve the plans beforehand.

I, the undersigned, have carefully reviewed the plans for the above mentioned installation. It is complete and workable as far as I can tell and I will fully support having this project installed as planned and function as designed.

| Title or Department / Name | Approval Signature | Date | Drawing/Print Number and Revision or Model Identification |
|---|---|---|---|
| 1. Supr. Trim Dept. / Al Payne | *A. Payne* | ~~10/20~~ 10/21 | O.C. 36952 ~~Rev 1~~ Rev. 2 A. Payne |
| 2. Supr. Sewing / R. Ewald | *R. Ewald* | 10/21 | O.C. 36952 Rev 2 |
| 3. Production Scheduling / D. Sherman | *D. Sherman* | 10/25 | O C 36952 Rev. 2 |
| 4. Maintenance Supr. / L. Cantilozzi | *L. Cantilozzi* | 10/25 | O.C. 36952 Rev. 2 |
| 5. Factory Manager / J. L. Loesch | *J. L. L.* | 10/25 | ditto |
| 6. | | | |
| 7. | | | |
| 8. | | | |
| 9. | | | |
| 10. | | | |
| 11. | | | |
| 12. | | | |
| 13. | | | |
| 14. | | | |
| 15. | | | |
| 16. | | | |
| 17. | | | |
| 18. | | | |

RICHARD MUTHER & ASSOCIATES – 305

**Figure 11-5.** Review and Approval Record is basically a sign-off sheet signifying that the individual concerned has reviewed, and is in agreement with, the proposed layout.

the layout that they have been consulted and that they are not being "pushed around"

7. To show a person with final responsibility for approval that his subordinates are all in agreement

The sign-off procedure ensures that these advantages are obtained. And, if it is made a formal procedure, it saves the layout planner time in getting his check-review and approval. When "the procedure"—not the "individual"—calls for sign-off approvals, relations are on a businesslike basis rather than a personal one. This is an important distinction when subordinate layout men must ask those in superior positions for some written indication of their approvals.

### Detail Planning Example

In the introduction to Part Three, the sequence of SLP planning for an overall layout is illustrated. Figure 11-6 shows the sequence of steps for planning the detail layout of one department, Activity No. 2—Silk Screen.

Note that each step of the pattern of procedures is considered, even though not shown in the example. In other departmental areas, different emphasis is placed on certain steps and different kinds of worksheets are used.

### Simplified Systematic Layout Planning

A short form of SLP is often useful in Phase III planning. For those who have mastered the full SLP methodology, it offers no particular advantage so far as actual planning is concerned. But, for new planners, or for small departments where the supervisor wishes to lay out his own area, it can be of real value.

Simplified SLP is described briefly in Appendix XVI. Essentially, it condenses Phases II and III into a six-step pattern, as shown in Figure 11-7. Note that Simplified SLP uses only one symbol, does not employ color or shading codes, and provides no quantitative means of determining flow of materials. It should not be used for large projects or projects having a major flow of materials. Rather, it lends itself especially well to layout planning of small areas like offices, laboratories, service shops, tool rooms, and similar "nonproduction" activities.

The experienced planner should be cognizant of the very practical use he can make of Simplified SLP to subcontract the layout planning of certain areas to those

---

**Figure 11-6.** An illustrative example of the application of SLP in Phase III—Detailed Layout Planning. This is Activity Area 2 (Silk Screening Department) of the overall block layout, the planning sequence of which was illustrated in the Introduction to Part Three.

# PLANNING A DETAIL LAYOUT

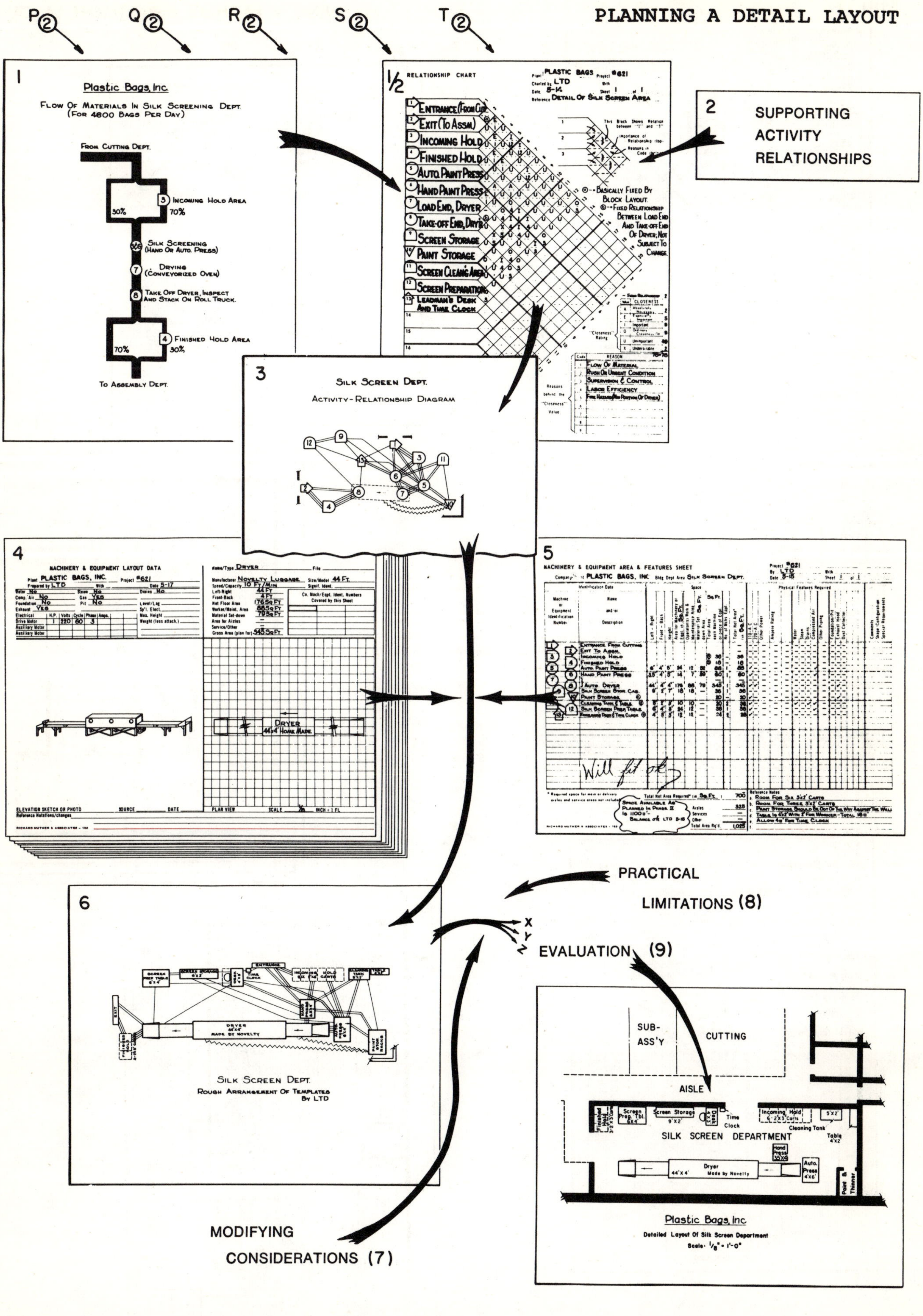

# PLANNING A FACTORY DEPARTMENT LAYOUT

1

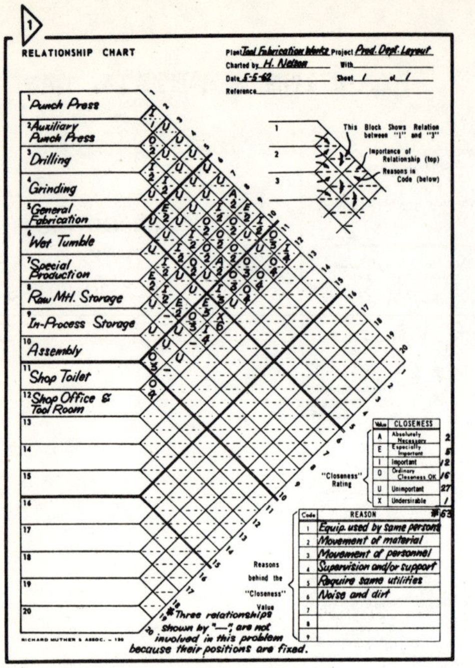


2

ACTIVITIES AREA & FEATURES SHEET

Plant Tool Fab. Works Project Prod. Dept. Layout By H. Nelson With Date 5-5-62 Page 1 of 1

| No. | Activity Name | Area in Sq. Ft. | Physical Features Required (Enter Unit and Required Amount under each) | Relative Importance of Features (A, E, I, O, —) | Requirements for Shape or Configuration of Area (Space) |
|---|---|---|---|---|---|
| | Production Department | Total: 2585 | | | |
| 1 | Punch Press | 550 | Existing clearance OK; Not applicable; 250 pounds per sq. ft.; No columns required | – – I – – – – | |
| 2 | Auxiliary Punch Press | 200 | | – – I – – – – | |
| 3 | Drilling | 250 | | – – E – – – – | |
| 4 | Grinding | 250 | | E – E – – – – | |
| 5 | General Fabrication | 400 | | – – O – – – – | |
| 6 | Wet Tumble | 100 | | A – – – – – – | Min. 10 x 6 feet |
| 7 | Special Production | 360 | | I – E (b) – – – | |
| 8 | Raw Mtl. Storage (a) | 0 | | | |
| 9 | In-Process Storage | 400 | | – – – – – – – | Provide setdown space 4 feet wide; plus aisle (c) |
| 10 | Assembly (a) | 0 | | | |
| 11 | Shop Toilet | 75 | | A – – – – – – | |
| 12 | Shop Office & Tool Room (a) | 0 | | | |

Key: A – Absolutely Necessary; E – Especially Important; I – Important; O – Ordinary Importance; — Not Required

Notation References:
a. Not part of the area involved in the layout—no space requirement.
b. Machine #17 requires special foundation.
c. Permits stacking two rows of 2'x2' or one row of 3'x3' containers with access to each.

RICHARD MUTHER & ASSOCIATES

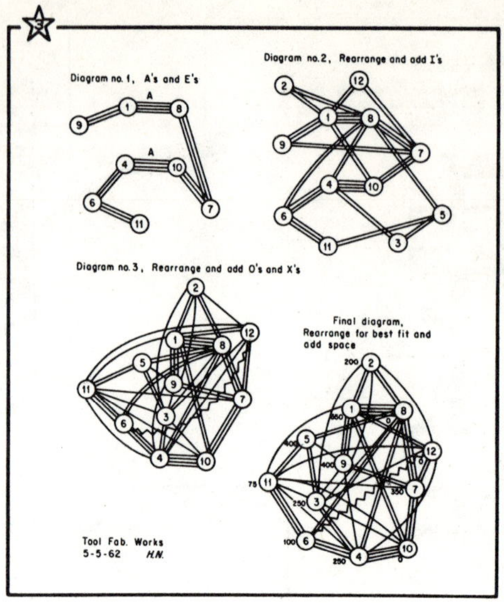


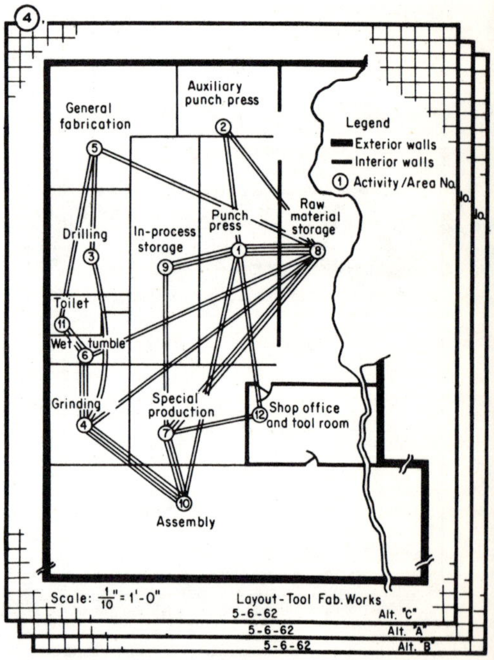


5

EVALUATING ALTERNATIVES

Plant/Area Tool Fabrication Works Project Production Dept. Layout Date 5-7-62

Description of Alternatives:
A. Punch Presses at Top; Gen'l. Fab. Right
B. Punch Presses at Left; Gen'l. Fab. Right
C. Punch Presses at Right; Gen'l. Fab Left

Weight set by H. Nelson & R.W.L Ratings by H. Nelson & R.W.L Tally by H. Nelson

| | Factor/Consideration | Wt. | A | B | C | D | E | Comments |
|---|---|---|---|---|---|---|---|---|
| 1 | Convenience of Service | 6 | I 12 | E 18 | E 18 | | | |
| 2 | Ease of Supervision | 5 | O 5 | A 20 | I 10 | | | |
| 3 | Flow of Material & Handling Economy | 10 | I 20 | E 30 | A 40 | | | |
| 4 | Flexibility | 8 | O 8 | O 8 | I 16 | | | |
| 5 | Least Investment | 8 | E 24 | I 16 | A 32 | | | |
| 6 | Appearance & Ease of Housekeeping | 3 | A 12 | E 9 | I 6 | | | |
| 7 | | | | | | | | |
| 8 | | | | | | | | |
| 9 | | | | | | | | |
| 10 | | | | | | | | |
| 11 | | | | | | | | |
| 12 | | | | | | | | |
| 13 | | | | | | | | |
| 14 | | | | | | | | |
| | Totals | | 81 | 101 | 122 | | | |

Notes: Rating Values: A (Almost Perfect) = 4; E (Especially Good) = 3; I (Important Results) = 2; O (Ordinary Results) = 1

RICHARD MUTHER & ASSOCIATES

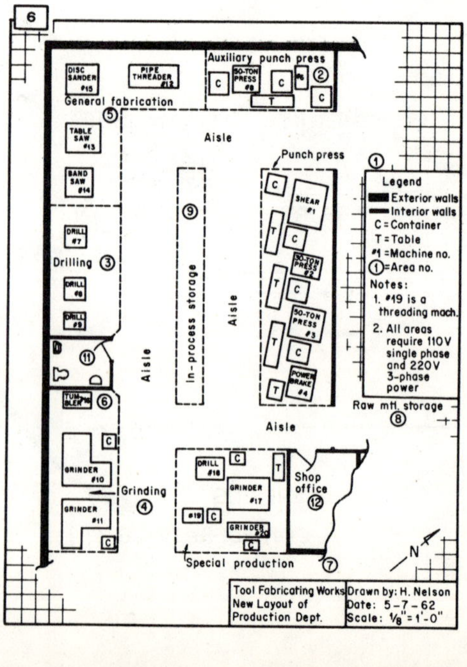

responsible for operating that area. This does not mean releasing one of his proprietary rights; on the contrary, he can expand his scope by indirectly and temporarily getting others to work for him. Others trained to use Simplified SLP (the training can be done for a group of supervisors in one day) can apply its simple organized procedures. The planner assists each trainee as needed, and audits the layouts submitted. Thus he retains his position of responsibility for layout planning but demonstrates his capability of getting results from others. Most important, he saves himself time to devote to the critical areas that need his planning expertise.

**Figure 11-7.** Example of the application of Simplified SLP to planning a factory department layout.

# Chapter 12
# Drawings, Templates and Models

In this chapter we describe further how to plan and visualize detail layout plans and how to reproduce them.

First of all, there is no single way of representing the planned layout that is best for all projects. What is good in one situation is time-consuming or expensive in another. As a result, the SLP practice includes use of all three major methods of visualization and sets up a guide as to which to use under what circumstances.

### Visualization

There are three basic methods of visually representing layouts: (1) drawings, (2) templates, and (3) models.

Regardless of which is used, if color and a standard code of identification symbols are used throughout planning, a sizeable additional advantage is gained.

### Drawings

Probably most frequently used—and the most universally usable—are *drawings*. This is true even in companies completely and elaborately outfitted with template or model facilities. In other words, one can always use a drawing. The planner needs them to supplement other means of visualization and falls back on drawings whenever it is not practical to use one of the other methods of layout visualization.

Away from the layout-planning desk, without layout prints, the planner finds himself sketching ideas on paper as notes or explanations to others. Moreover, the planner commonly roughs out sketches or drawings of possible arrangements to help think through possibilities before arranging or changing templates or models.

(Although we are now discussing Phase III—Detail Layout Planning, we must recognize that a drawing or print is still generally the most practical way to make and distribute the overall layout. In fact, prints or copies of the general overall layout definitely must be available for reference as work on the layout details progress.)
details progress.)

We are speaking here of *planning* or sketches, as compared to installation drawings or artistic renderings of buildings. And for planning, most experienced layout planners no longer depend on the techniques and skills of the draftsman. Instead, the planner uses paper preprinted with a grid of cross-section lines. This way, the

T-square and measuring rule are superceded by a straight-edge and the counting spaces. Cross-section paper is readily available; no one doing layout work should be without a simple pad or sheets of it.

Cross-section grids are also preprinted on tracing paper and translucent plastic sheets, which allows drawings to be made right over prints of existing layouts or building plans.

In making drawings, the use of a mechanical multicolored pencil is also recommended. Usually these have four slots. If the planner loads the mechanical pencil with red, orange-yellow, blue, and green and carries another, more comfortable mechanical or wood pencil for normal writing of letters or reports, he has a range of five colors. This combination is the standard color code, even though it uses gray/black in place of brown. Additionally, the planner can easily take advantage of the other conventions with the diecut symbol rule (provided in the Working Forms section). Similar template-cut rules with outlines of frequently used equipment like office furniture also aid drawing. Figure 12–1 shows two uses of drawings that offer advantages not available with templates or models. Figure 12–2 illustrates a handy short-cut for planning with drawings.

### Templates

The most flexible and adjustable method of visualizing layout plans is to use *templates*. Templates (or templets) are essentially predrawn or preprinted replicas of the equipment involved in the layout. They can be made in several degrees of detail. See Figure 12–3.

In preparing templates, best practice is to follow the lines recommended in the ASME Standard (see Figure 12–4) and to include the details called out in Figure 12–5 as fits one's needs. Templates may be prepared in any number of materials. However, of materials available today, only plastic or cardboard is recommended.

*Cardboard* has the advantage of being a material that everyone can work with; it requires no photoprinting or reproducing machine; and it is readily available

---

**Figure 12–1.** Examples of effective use of drawings in layout visualization. The top illustration (a) shows a simple overall layout drawing for an existing plant. Beside it, using the same existing building outline, is a proposed expansion for the plant. The areas assigned to the different functions are shown by letters and shading.

The bottom illustration (b) shows a section of a detail layout drawing. These views show a plan and evaluation sketch of flaking rolls on second floor and bagging machines below. This shows alignment and orientation of the equipment and the proper placement of it and the supporting handling equipment.

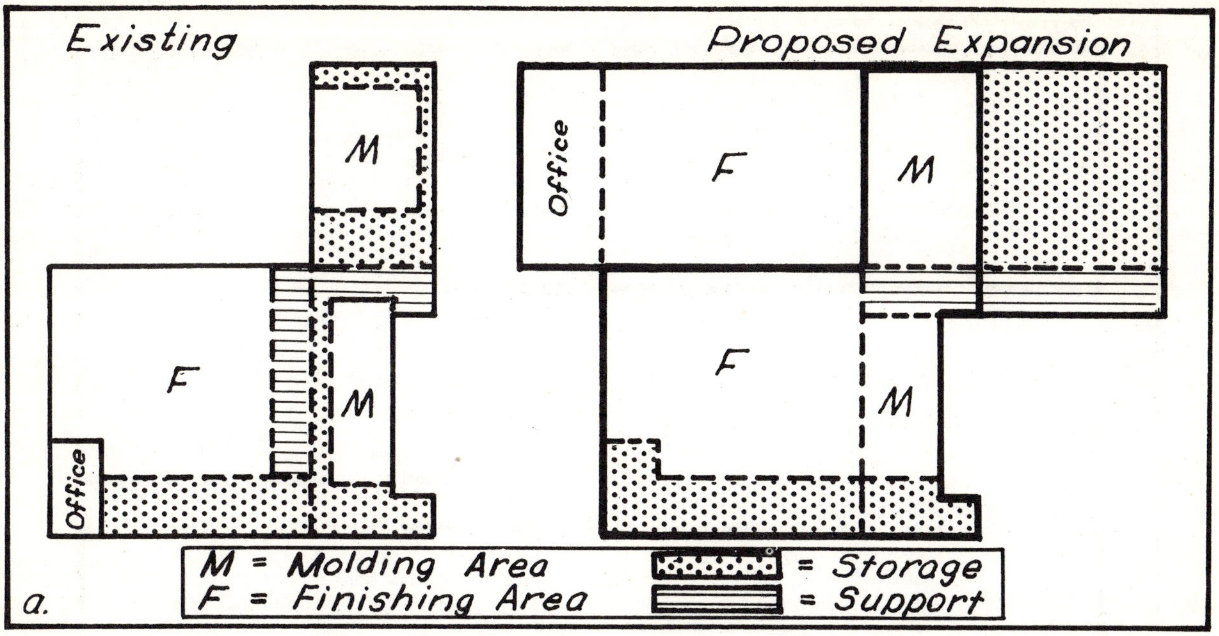
Existing
Proposed Expansion
M
F
Office
M = Molding Area
F = Finishing Area
= Storage
= Support
a.

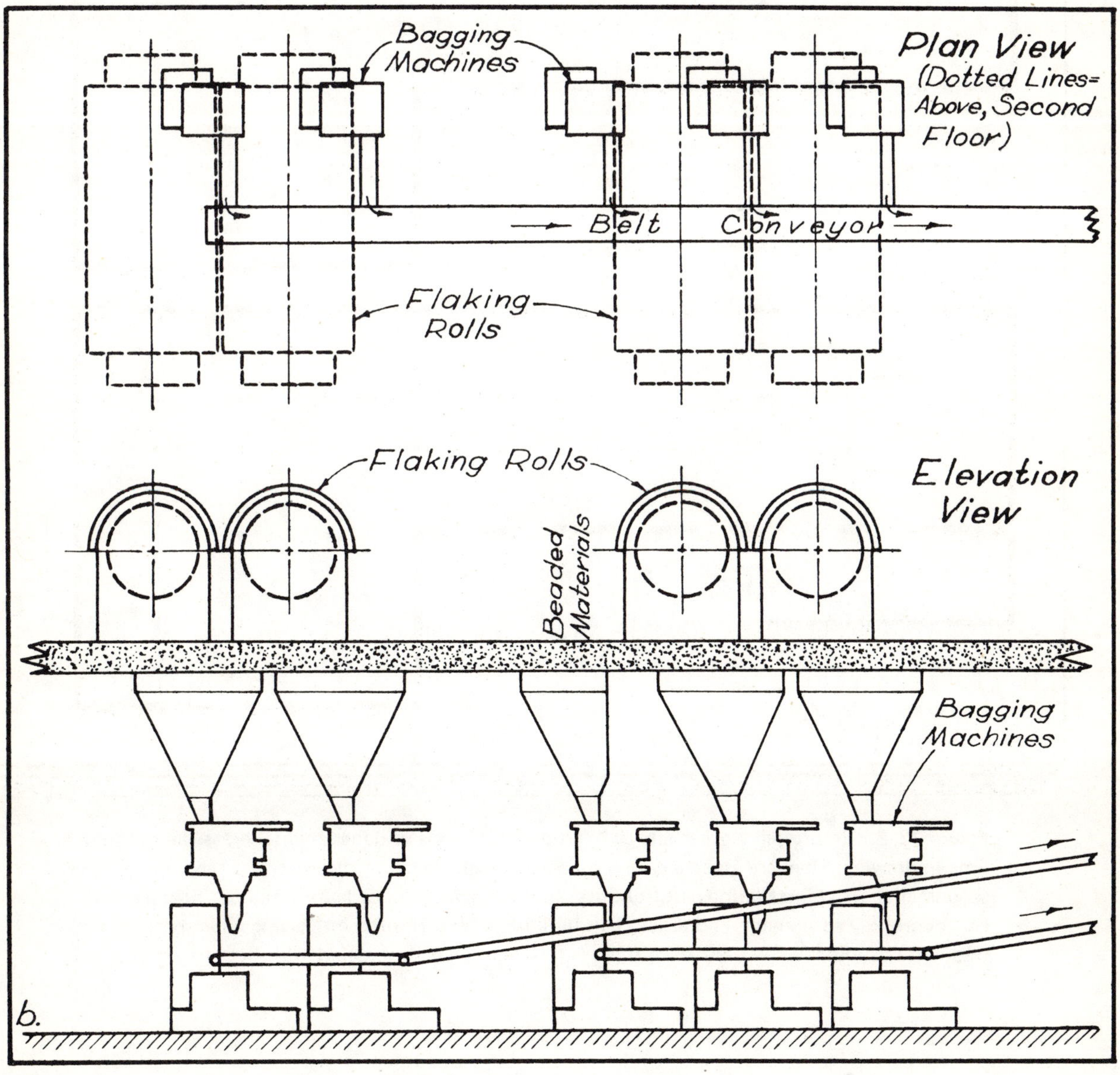
Bagging Machines
Plan View (Dotted Lines= Above, Second Floor)
Belt Conveyor
Flaking Rolls
Flaking Rolls
Elevation View
Beaded Materials
Bagging Machines
b.

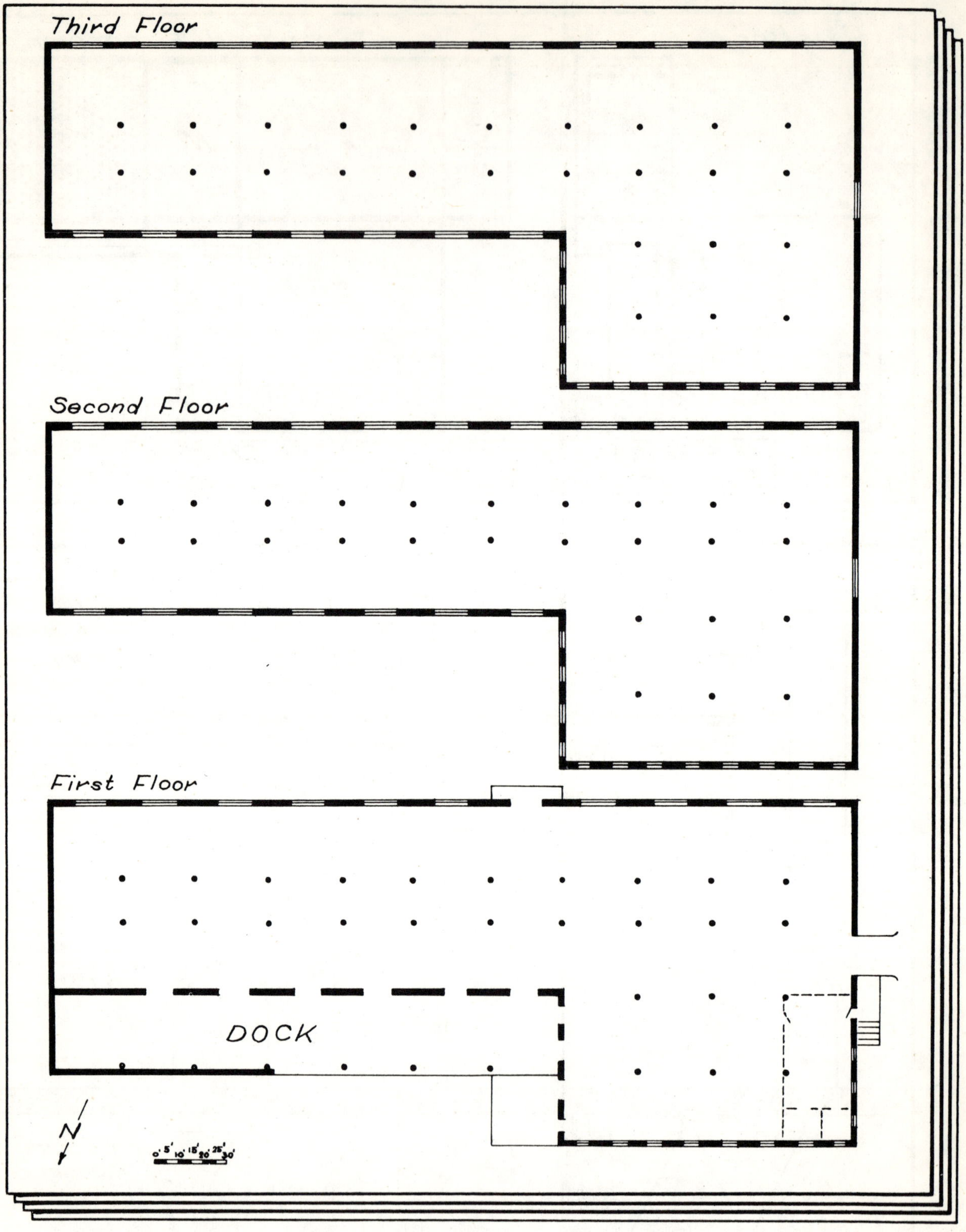


**Figure 12-2.** Preprinted worksheets of building floor levels or departmental areas can save much time in drawing. They are especially helpful when rough sketching alternative arrangements of area assignments, fixed installations, utility lines, and major features of layout like the flow pattern, conveyors, key equipment, and aisles. This building shell is printed on tracing paper because of frequent relayouts. When used, each sheet is fastened down to a scaled grid sheet to aid in drawing.

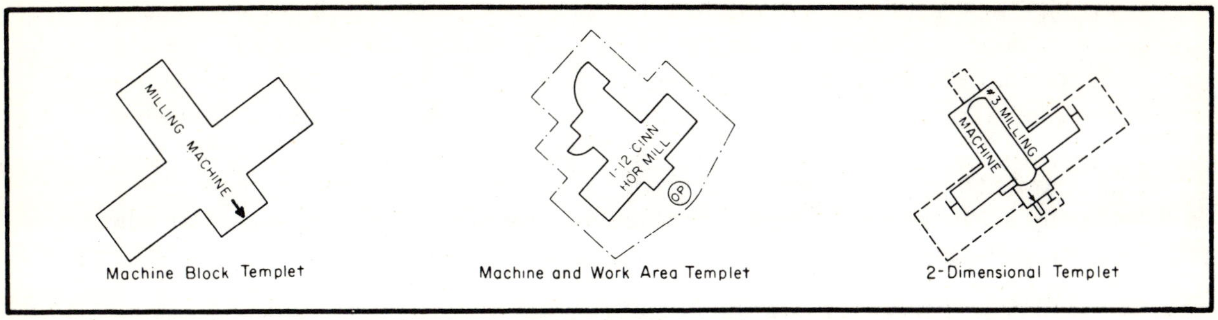


**Figure 12-3.** Examples of various kinds of templates for detail layout planning.

| | |
|---|---|
| THICK | *Fixed outline of machine tool or equipment from floor level to 7′-0″ elevation above floor level.* |
| THIN | *Detail parts and substructures.* |
| THICK DASH | *Clearances for moving parts of machine tools or equipment.* |
| MEDIUM DOT AND DASH | *Overhead or underground elements – foundations, pits, service clearances and other elements important to the template.* |
| THIN CENTER LINE | *Center lines* |

**Figure 12-4.** Recommended lines for preparing templates. From A.S.M.E. Standard: Plant-Layout Templates and Models.

| | |
|---|---|
| 1. Type of machine or equipment | 6. Left-right, front-back, height dimensions |
| 2. Name of manufacturer | 7. M - Position of motor |
| 3. Style, model, size, or capacity | 8. C - Position of control panel, switch, or adjustment |
| 4. Owner's identifying (tag) number | 9. A or E - Position of Elect. or Air Connection |
| 5. Operator's normal position (by arrow) | 10. * - Point of maximum height |

**Figure 12-5.** Data to be included on templates. The first three items are mandatory. Four through ten are optional depending on the circumstances.

in colors. In situations where the final layout plan must be traced anyway—because back-up board or undersheet can not be run through a print machine or otherwise reproduced from flat condition—it may be faster simply to draw machine and equipment replicas on cardboard and cut them out as templates. This usually occurs when a company has only an occasional move or rearrangement to plan and therefore is not knowledgeable about more time-saving ways to reproduce layout plans. This typically exists where a company produces products not highly engineered or where for other reasons draftsmen and a print machine are not at hand.

Cardboard should be used in colors and with cross-section grids already printed lightly on the sheets. The grid lines save many hours in preparing templates; the color aids understanding, makes selecting and sorting of templates easier, and offers greater eye appeal. We recommend 110-pound stock or heavier. Markings in black ink are best, though pencil is faster. Both can be sprayed with clear lacquer or fixative to keep them clean.

On the other hand, the use of *plastic* film (or foil) templates probably offers more practical advantages than any other system of layout visualization today. By fastening with adhesive, the planner needs no maptacks or staples—usually the most practical fasteners with cardboard. And by combining plastic templates with a plastic undersheet (usually having cross-section grid preprinted thereon) and adhesive backed tapes, he gets a pliable layout plan that can be run through a standard print machine and produce copies directly. Actually, it is the combination of materials used and the reproduction method(s) available that make this method effective for a given company.

There are all manner of variations in the plastic materials that can be used.

Templates
- a. Polyester or acetate
- b. Matte (frosted) or clear (glossy) surface
- c. Black or brown line—or dark with white line
- d. Translucent color tints or plain

Undersheet grid
- a. Polyester or acetate
- b. Matte (frosted) or clear (glossy) surface
- c. Grid line on sheet or clear sheet with grid on separate sheet below
- d. Reproducible or non-reproducible grid lines

Fastening material
- a. Ready-fixed, pressure-sensitive adhesive
- b. Double-faced, pressure-sensitive plastic tape
- c. Magnetic attraction to metal back-up
- d. Liquid adhesive

All of these have their advantages and limitations—and there are even other variations. Different systems have their staunch supporters—depending on when each adopted his system and what kind of reproducing means he has available. A system like that shown in Figure 12-6 will generally do the job well.

**Figure 12-6.** Plastic templates and plastic grid sheet. View shows the positioning of a template on a panel of sheet plastic on which a scaled grid is preprinted. Templates and colored tapes, made of plastic and fastened by pressure-sensitive adhesive, are used to signify the details of the layout.

All materials necessary to get set up with modern plastic layout methods are available commercially. See list of suppliers in Appendix VII. This raises the question of whether to purchase templates or make them. Generally speaking, a commercial template producer can save time and cost, especially at the start of any system. He can give a more clean-cut template for the same, or less, cost. His adhesive is already applied and protected. He can offer other features like colors or filing aids. Dealing with a template supplier forces a company to get its own template records and specifications in good order. However, being wed to one supplier has its limitations. Rush jobs requiring new templates can not always wait for mail; flexibility in design or nature of templates is more limited; the commercial source may have a fire or otherwise delay or stop deliveries. As a general rule, therefore, the layout engineer should be able to produce his own templates when and if he wants to in a way that will integrate with those he may purchase. Appendix VIII shows a system of producing plastic film or foil templates for use in standard print machines.

When plastic templates are used, the aisles, building features, and other frequently used items will be stripped in, using commercially available preprinted tapes or strips. These can be placed on the working sheet directly, as utility lines and similar fixed services. However, in the latter case, it is better to place lengthy utility and auxiliary lines on a separate clear overlay (or underlay) sheet to avoid confusion and difficulties in moving templates.

### Three-Dimensional Scale Models

Models in three dimensions are the clearest devices for layout planning. They afford the best possibilities in space visualization because they are more or less exact replicas of each piece of equipment, men, and material. They are obviously easy to "see," for they are self-explanatory—although they have limitations in exact identification. They allow very quick understanding of the planned layout, especially to non-engineering-trained people—often the very people who approve the layout plan, install it, and make it function as planned.

Models have these further benefits: They are helpful in explaining the layout to outside suppliers of equipment who are involved in the construction or installation. They are a genuine pitch kit, ideally made for getting attention and selling top management or operating people. As a device for instructing operating and support personnel in layout function, and for orienting all employees, models are ideal. Many people simply never can be made to understand 2-D plans. Besides, with models to move about, operating people really have a full opportunity to participate in making the layout. At the same time, they can un-sell themselves on quickly conceived or pet ideas which they had not had time to think through. Scale models frequently pay for themselves in these auxiliary areas alone, even when their contribution to layout planning itself may be in doubt.

As would be expected, the greatest value of models comes in reviewing, checking, selling, or explaining to others the layout. Therefore, the layout planner generally follows this sequence. He creates or captures layout ideas with rough sketches; he works out alternative layouts with templates; and when he has the layout arrangement he feels best, he prepares his model to review and check with others what he has planned. The model will be rearranged according to review suggestions that seem justified. He then revises his template layout and prepares to have it reproduced. From the print he prepares installation instructions; with the model he instructs installation and operating personnel. See Figure 12–7.

This is generally the way models are used. Planning the layout—as compared with review-checking, getting approval, or instructing others—is not often done with models. They are used to supplement templates and drawings (or sketches)—not to replace them. In fact, the layout engineer who uses models too early in his visualization process will find himself decidedly handicapped in practical maneuvering. Models suffer in comparison to drawings and templates, too, because of their cost, storage space, ease of breakage, the necessity of keeping the complete model horizontal and clean, and the fact that without special photographic equipment their arrangement cannot be quickly or clearly reproduced.

**Figure 12-7.** Use of drawings or two-dimensional template prints (on the wall) in conjunction with three-dimensional models in detail layout planning. The chief advantage of models is that they allow others to understand clearly what the layout engineer has planned. Here he is reviewing his layout with operating and staff personnel. This combination of drawing, or template print, and model generally affords the optimum in clearness of visualization and in ease of recording and communicating plans.

In general practice, models are a refinement justified in less than half the cases of layout planning. Economically, they are limited to high-investment installations and those involving problems of vertical complexity, new and unfamiliar products or processes, or especially high-cost floor space. However, on projects that involve complex vertical problems, such as chemical process plants or plants with diagonally curving overhead conveyors, it may even pay to do the layout in 3-D and then scale the model for dimensions and specifications. Particularly when engineering is done away from the actual location, it may be practical to deliver to the site a model, complete with piping and other details, rather than engineering drawings, and to scale the model for dimensions right on the job.

Still, partial use of key models is always recommended. As we plan in two dimensions, it is practical to keep a few models of workers, stock containers, or typical pieces of equipment handy for ready placing on the template layout or drawings. This keeps our dimension scale constantly before us. And when the layout involves new, different materials, products, and equipment, models help keep in mind the true dimensions involved.

### Working Practice

Industrial engineers recognize that the accessibility and convenience of a worker's tools are as important as his knowledge of how to use them. The same holds for the layout planner. As a practical matter, the organization of working materials is important. Therefore, we recommend:

1. A system of layout panels for detail layouts that integrates with the overall layout plan and the building or site arrangement (see Figure 12–8)
2. A systematic way of recording, obtaining, and filing templates and models
3. A method of storing layout panels for ready use
4. A system of identifying each panel
5. A written record of symbols or codes used on the layout
6. A specific way of including on each layout a north symbol, the scale being used, and identification of the plant or building
7. A way of identifying the project, date, and who planned each layout on each print
8. An established procedure for getting approvals of layout plans

Appendix IX describes a drawing numbering system, for a large company, into which layout plans may be incorporated.

### Conclusion

A summary guide explaining when to use drawings, templates, and models is given in Figure 12–9. We have not given any cost figures on comparative techniques of

**Figure 12-8.** A panel grid system for a company's total property. The site, with buildings, property lines, and main features shown, is divided into convenient blocks. The illustration shows how each block is then represented by a layout-planning panel, at an enlarged scale. Here, panel B-3 shows machine and storage rack details; the blocked out master on the wall shows where panel B-3 fits into the total property.

The size of panels is worthy of pre-planning thought. If too small (say less than 17 x 22), there is too much time and confusion in piecing together panels and prints made from them. If too long, there is a problem in storing; if too wide, there is inconvenience in leaning over to work on the layout and in lining it up to run through the print machine.

| GUIDE TO THE USE OF DRAWINGS, TEMPLATES, AND MODELS | |
|---|---|
| DRAWINGS | 1. When "paper and pencil" are at hand, and time does not warrant or allow getting templates or models.<br>2. When roughing out an arrangement idea prior to moving templates or models.<br>3. When you have no experience with templates or models and the project will not justify or allow time to gain it.<br>4. When you want a quick replica for someone while away from your layout planning office.<br>5. When you are refining your template or model layout and recording suggestions or capturing new ideas on a print of the alternative or proposed layout. |
| TEMPLATES | 1. When planning detail layouts for a large overall area or for repeated relayouts.<br>2. When template materials are readily available and materials or facilities for making drawings are not at hand.<br>3. When you have had prior experience with templates.<br>4. When project involves many alternatives--especially when preliminary to preparing 3-D model.<br>5. When equipment for print reproduction is readily available. |
| MODELS | 1. When the layout involves complex interferences and diagonally-oriented vertical equipment--that is, a 3-D problem.<br>2. When the layout involves new or changed processes, products, methods or procedures, or a completely different type of layout than previously familiar to the company.<br>3. When planning a layout involving substantial capital investment or a long-term (relatively permanent) layout where full detail planning makes good "insurance"--especially when the cost of floor space is high.<br>4. When layout planners and operating and support people have little ability or experience in layout planning so that review and check of layout by others is significant.<br>5. When the layout will involve many outsiders during the construction and/or installation planning, many problems in selling management or directors, and many new employees and supervisors to be trained. |

**Figure 12-9.** Choice of materials for detail layout planning.

## PROCEDURE FOR VISUALIZING DETAIL LAYOUTS

1. Prepare a Space Relationship Diagram for the detail area according to the S L P pattern.
2. Rough-out (on cross-section paper) logical arrangements of facilities, involving the various modifying considerations and their practical limitations.
3. Prepare templates and make a template layout for each worthwhile arrangement--or print each one, and re-arrange the next plan with the same materials.
4. Keep a 3-D model of a worker or typical key equipment handy, and have available models of new and different products or equipment. Place these on templates or move them around as appropriate so space dimensions are kept in mind.
5. When layout is ready for review-check and approval, reproduce it in 3-D for those areas that justify the extra time and expense.
6. Make review-check and approval from colored prints of the layout and/or the model; then bring template layout up to date prior to reproducing prints for installation.
7. Use the model thereafter for explaining, orienting, instructing, and promoting the layout; use the two-dimensional record for installation drawings and specifications of utilities, auxiliary services, mechanical or architectural details, and the like.

**Figure 12-10.** The SLP procedure for visualizing detail layout plans.

visualizing layout plans for these vary widely depending on how materials are purchased. We suggest each planner solicit current prices from those suppliers listed in Appendix VII which offer the items suitable to him.

By way of summary, Figure 12–10 sets down the SLP procedure for visualizing layout planning details.

# Part Four

# Introduction to Part Four

*Part One set forth the fundamentals of SLP—its framework of phases, its key input data, and its pattern of procedures.*

*Part Two exploded the pattern. We described in detail how to apply the procedure of each box, or step, for general overall layout planning and set forth a series of simple conventions, describing how to use them with various procedures in the pattern.*

*Part Three was devoted to describing detail layout planning. As Part Two covered Phase II of the framework, Part Three covered Phase III.*

*Part Four includes three chapters. These final chapters are devoted to Phase I and IV of the framework and to how layout projects should be managed.*

*Recall that Phase I, Location of the area to be laid out, comes in the four-phase framework, but that we postponed its discussion. We did so because Location does not always fall under the responsibility of the layout planner and because its SLP significance can better be understood after thorough study of the two strictly layout planning phases.*

*This postponement is not meant to belittle Phase I. Every layout project has its four phases, regardless of who is responsible for executing each one. Even a minor rearrangement of a small department involves four phases, though the entire project be but a small part of what would be considered detail layout planning were the whole company to be relocated. In minor relayouts, the planner must first know whether the area involved is to remain in its present location, and if not, where it is to be placed. This is Phase I, Location. Then the area is planned generally (II), in detail (III), and installed (IV). (Refer to* Figure 2–3 *for a pictorial recap of these four phases in action.)*

*Actually, the planner must look at the phases from the viewpoint of the area being laid out. Otherwise, he may not clearly see the four phases. From the standpoint of one building on a site, a departmental relayout occurs as Phase III—or III-A. But, if the department is relocated within the building, this location is Phase I for the department; yet it is a readjustment of the general overall layout (II) from the standpoint of the total building. Thus, the four phases involved in every layout project should be considered from the viewpoint of the area being laid out.*

# Chapter 13
# Location

In this chapter the techniques followed in Phase I, determining the Location of the area to be laid out, of the SLP framework are described. Yet this chapter will cover a broad range of interpretations, rather than devote itself to "Plant Location" as generally defined.

### Meaning of Location

The term *location* means the place something is. In the narrowest sense, detail layout planning involves the *location* of each piece of machinery and equipment. If relocating only one machine were the layout project, the phases would be:

I. Locate the area where the machine is to be laid out.
II. Place it generally for best flow and in relation to its supporting services.
III. Align and orient it with the specific features of equipment immediately adjacent to it.
IV. Plan and make the installation.

At the other extreme, *location* may involve an international problem of site selection. But the usual location problem is neither so simple nor so complex. It may be what is generally described as Plant Location. More often, it is finding where to locate an expanding department within a given building (or complex of buildings on a site), or finding a location in another building (or site) in the same community or near-by area.

As a result, location problems typically involve questions like these: Do we build a new building or stay where we are? Do we expand into the finished goods warehouse and rent new warehouse space or do we expand to the South and rearrange the whole plant? Do we consolidate the office area on the second floor or leave it divided? Questions like these, which involve determining the location of the area(s) to be laid out, should be answered before serious layout planning.

### Location Screening

One of the first things to do when starting to seek a new location, a different site, is to clearly assign responsibilities. In smaller companies this frequently involves a committee or team, though some member of top management remains as chairman. Even in large companies where a central staff of specialists may be available to provide technical analyses, a specific official, usually of the product or operating division directly involved, is in charge of investigations. Still, picking a new location for the company or for an expanded major branch is a problem involving practically all top-level management or their representatives. In large, multi-plant companies, this is true of the top management of the division involved.

Location projects involving a new site usually follow the same four-phase overlap-and-refine logic of most projects:

I. Set specifications, scope, restrictions.
II. Analyze and select general region or community.
III. Analyze and select specific site.
IV. Procure the site.

### Location Planning

In any case, each location problem boils down to essentially three features or elements; (1) what is wanted; (2) what is available; and (3) what is most suitable.

"What is wanted" is frequently termed *location requirements*, or specifications. These requirements typically include:

1. Area–size or amount of space required
2. Condition–nature and characteristics of the space
   a. shape, orientation
   b. topography and drainage
   c. subsoil and understructure
   d. prevailing wind
   e. site (or building) improvements or preparations
   f. relocation of existing utility lines or pipelines
   g. flood history
   h. site access ways
3. Relationship with sources and destinations
   a. raw materials
   b. suppliers
   c. customers or markets
   d. related transportation (rail, truck, water, air)
   e. related travel convenience
4. Contacts
   a. with personnel–supply, availability, type, labor history
   b. with utilities and auxiliaries–electric power, water, gas, coal, fuel oil, sewage and waste disposal
   c. with local services–local transportation, communications, banks, commercial and professional services, police, fire, trash
   d. with government–taxes, zoning, easements, restrictions, codes
5. Surroundings
   a. general–physical climate, state and community attitude, neighborhood, appearance
   b. housing, hospitals, health and welfare
   c. schools and education, recreation, cultural opportunities

   d. city planning, roads
   e. other business attitude
6. Investment
   a. land
   b. site improvements
   c. buildings, construction, rent
7. Profit Potential
   a. operating costs
   b. savings and return

With these requirements spelled out—usually in a longer, specific list—a search is begun for what is available.[1] As pointed out, this may involve regions first, communities next, and then specific sites. As we get closer to the actual site selection, the data sought becomes more specific. In any event, there is a matching or comparing of the requirements (what is wanted) with the areas available to decide what is most suitable.

Selection of the location is a matter of compromises and alternatives. This is logical: There is seldom a perfect location for a perfect layout available at a perfect price.

Preliminary research, to reduce the number of potential locations, is done with maps and economic data to eliminate areas which do not fall within desired population ranges or do not have adequate transportation routes.

Next, in finding available communities, we suggest turning to the many outside sources of information. State development commissions, chambers of commerce, public utilities, and railroads have a wealth of data they are only too happy to supply. Plant location consultants, organized industrial districts, and commercial realtors can help the planner learn about available sites and obtain specific information about their nature, characteristics, and suitability.

It is possible and frequently justified to make projected profit-and-loss statements for assumed plants at the locations under consideration. Or a comparison of the net delivered-to-average-customer cost per unit of product can be developed for each site and compared. On the other hand, intangible considerations generally prevail, so that a comparison of costs and intangibles is often more appropriate. The comparison takes the form of a cost analysis and evaluation of alternatives combined. Figure 13–1 illustrates one such study made in selecting a community.

### Site Selection

Within a given community, the selection of a site follows the same comparison of "wanted" with "available." But the factors are more specific and a more detailed analysis is made. For sites within the *same* community, many factors change their defi-

1. For a more detailed and comprehensive list of factors see: "Checklist of Site Selection Factors," *Site Selection Handbook,* Conway Research, Inc., Atlanta, Georgia.

| Site | Cost of Labor | Cost of Transptn. | Cost of Taxes | Cost of Fuel & Utilities |
|---|---|---|---|---|
| F | $325M | $109M | $21M | $ 8M |
| D | 417M | 100M | 37M | 10M |
| A | 442M | 98M | 43M | 13M |
| C | 634M | 90M | 62M | 17M |

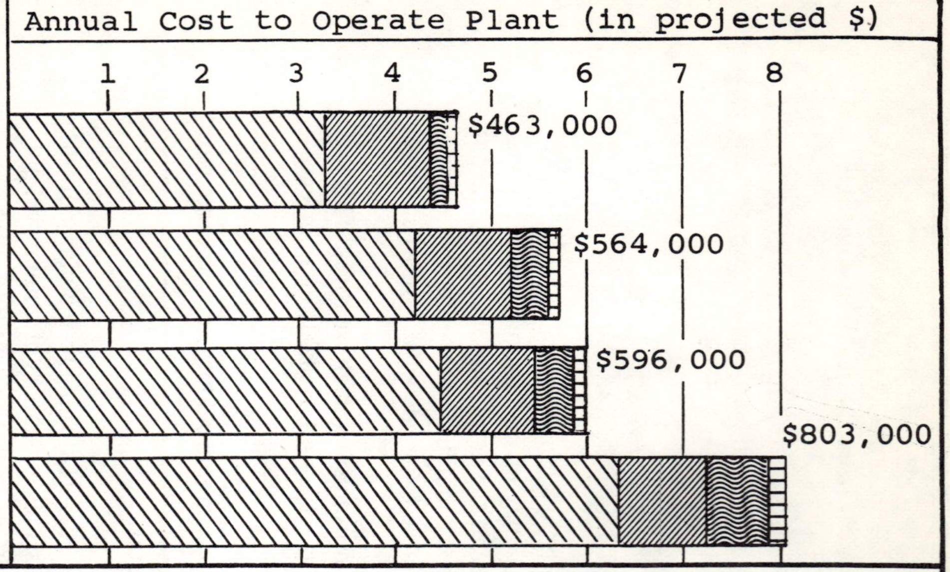


| Site | Proximity of Existing Plant | Proximity of Future Markets | Availability of Labor | Type of Labor | Union Activity | Community Attitude | Public Trans-portation Services | Reliability of Utilities | Recreation | Culture | Technical Schools | Hospitals | Hotels | Airport |
|---|---|---|---|---|---|---|---|---|---|---|---|---|---|---|
| F | E | E | I | E | E | A | E | O | E | I | E | I | U | I |
| D | A | I | O | O | U | E | E- | I | I | I | I | E | I | I |
| A | E | I- | E | I | I | E | I | I | I | E | I | A | O | O |
| C | E- | A | I | E | O | E | E | I- | U | I- | A | E | A | A |

nition—transportation may now become trucking, for example. And factors like community attitude and physical climate tend to disappear from the evaluation entirely.

Figure 13–2 shows the results of a specific study of transportation cost only as it affects site location. Similar investigations are made for all elements of expenditure. The cost comparison worksheets described in Chapter 10 can be applied here.

Likewise, the factor analysis method of evaluating alternatives can very well be employed to help choose a location, just as to help select a layout. Actually, the techniques of Chapter 10 apply here for evaluating locations almost equally well, the combination of cost comparison and cost justification with the factor analysis of intangibles generally being the most effective.

### Common Oversights

Perhaps the greatest single cause of trouble in location analysis is the failure to define sufficiently what is wanted. But there are many other oversights which otherwise astute managements commonly by-pass. A list of the most *common errors made in new-site location* is included here.[2] It can be used as a checklist to help avoid making similar mistakes.

1. Overlooked requirements and incomplete investigation
2. Too much attention to land costs
3. Underestimating the cost of moving
4. Hasty acceptance because of community inducements
5. Prejudices or preferences of key executives governing the selection, instead of impartially gathered facts
6. Resistance to change by key executives in moving from home ground to new location
7. Areas already or about to be overindustrialized
8. Judging the potential labor supply in terms of wage rate, while neglecting productivity, work standards, and labor history
9. Community's cultural and educational standards so low that key people accept employment elsewhere
10. Basing move on immediate or short-term benefits only; failure to look ahead adequately
11. Lack of proper coordination or control of the site-selection work, particularly getting all major groups within the company properly represented

2. This list of common errors is compiled in part from points suggested by A. Kingsley Ferguson, in *Pacific Factory*, April 1954, and by *Duns Review & Modern Industry*, April 1956 and March 1959.

**Figure 13-1.** Selecting a community for a new plant site. Costs have been determined for major items of expenditure. These are compared (numerically and visually) for each of four contending communities. Other less tangible factors are compared below, using the same vowel-letter rating system used throughout this book.

12. Failure or unwillingness to recognize changes or shifts in market areas, transportation methods, raw materials, and other major cost reduction elements

**Land-to-Building Ratio**

A typical mistake (one buried in the above listing) is selecting a site that is too small. This is true in times of both expanding economy and recession. Perhaps it is because managers who see the need for new facilities do not want to scare their direc-

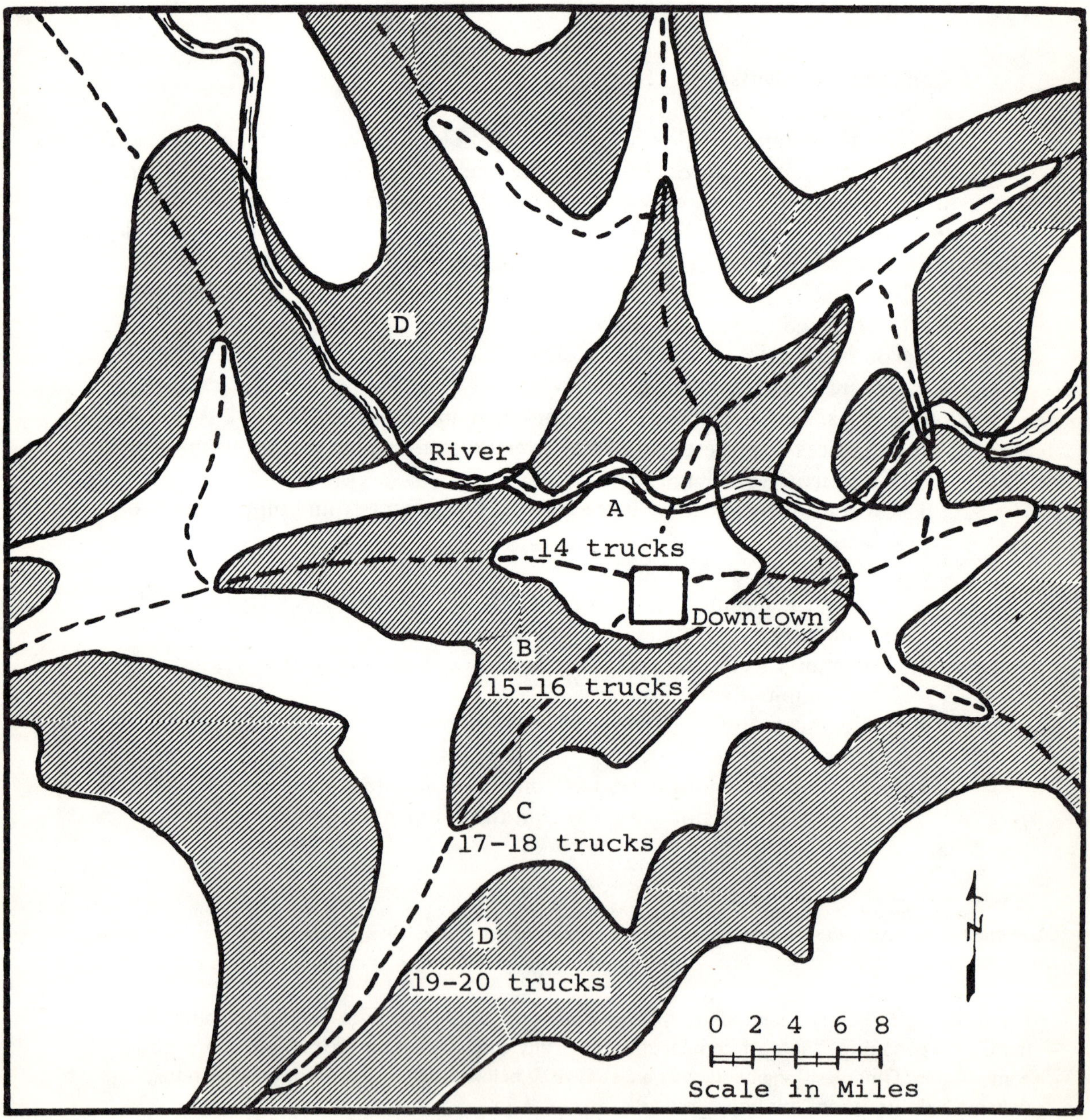

tors with too large a request. Perhaps it is because top managements as a whole are concerned with current problems and are not doing the kind of long-range planning they probably should. In any event, we do not want to "go in fresh" with too small a location.

If there is no possibility of expansion, either limit potential future products, processes, supporting services, and ability to meet demand, or move the plant again. If there is no extra building space, the planner cannot work out alternative layouts or hold occasionally needed expanded inventories. If there is no space for parking, the labor market is limited. Actually, in most situations the cost of land is relatively small compared to the total investment. Yet its unavailability, once a company is committed to a site, hinders the operation from then on and eventually causes premature abandonment of the company's investment.

Actually, much of the extra land "required" is protection—insurance against future increased capacity requirements and also against neighboring property owners who buy and build around the site. What was once prairie or back woods can become a residential community in a surprisingly short time, and when this happens the plant that originally started the area's development often finds itself an obnoxious or at least out-voted member of the new community.

**Figure 13-2.** Location and trucking costs. Before building a new warehouse a distribution company made this study of how the number of over-the-road trucks depended on the site location. The figure shows that a location in zone A nearest the downtown area would require 14 trucks. On the other hand, any site in zone D would require 20 trucks, a difference of six trucks or approximately $80,000 per year. Notice how the zone lines tend to follow the major highways, which are shown by dotted lines. By comparing the difference in trucking cost and the cost of land in the different districts, it is possible to select the district(s) which give(s) the optimum cost of land and trucking.

The total trucking time per day was derived as follows: The warehouse receives truckloads from a number of freight terminals and delivers to different shopping centers located throughout the area. The trucking time per day to each of the freight terminals or shopping centers is:

$$T = N\left(L + \frac{2D}{V}\right)$$

T = Trucking time per day (hours)
N = Number of trips per day
L = Loading and unloading time per trip (hours)
D = Distance between warehouse and freight terminal or shopping center (miles)
V = Average travel speed (miles per hour) on this route

The total trucking time per day, which is used for deciding the number of trucks needed, is the sum of the total time from the warehouse to and from each terminal and each shopping center. This is simplified, for in most cases the additional conditions have to be included.

(a) Delivery trucks with less than full load for one destination will visit more than one shopping center per trip.
(b) Trucks may first deliver to a shopping center (s) and then pick up a load at the freight terminal on the return journey to the warehouse.

Moreover, probably it is desirable to have access to the site from two or three, maybe all four, sides. Probably it is desirable to be able to drive all around the plant within its property boundaries. Plenty of room to bring in future rail sidings, to spot waiting vehicles, or even to place direct air-receipts or air-delivery facilities is doubtless wanted. As mechanization increases, a relatively increasing amount of supporting-service facilities probably will be demanded. Water tower, cooling pond, filter beds, and similar facilities for reclaim, waste treatment, and central distribution or collection not only require special or very different equipment from that in the operating departments, but are notoriously overlooked by management and layout planners alike.

As a result, the stated policy of many larger companies has come to read something like this: "New sites for plant facilities should be at least 100 acres or have a ratio of usable open land to under-roof, built-upon land of ten to one."

The land-to-building ratio is a guide to relating a total site to anticipated or existing buildings. Appendix X gives a table of needed land areas based on various ratios for a given building size. "Building" is ground area under structure or roof, not the area of usable floor space. "Land" is the open ground area, not the total site. The ratio wanted by any company at a particular time varies significantly. And from industry to industry, there are many extremes. Therefore, this ratio is a guide only—but an exceedingly helpful one for comparing or planning the structural density of various sites.

The problem of too little space is not confined to new sites; we discussed it earlier when balancing area requirements versus area available. (See Chapter 7.) The same type of space balancing is involved in finding where to put the layout to be planned.

It must be remembered that planning is even less specific in the Location phase than in Phase II. Frequently, what happens is that the planner gets a "feel" of the amount of space needed and then gets agreement from the important people involved. The planner can calculate himself to death trying to get some figures to stand behind, when actually there may not be an effective way to calculate them with any degree of accuracy. As a result, one is likely, in Phase I, to use the ratio-trend-and-projection or the converting method of space determination.

#### Overlapping Phases

We pointed out that Phase I, Location, of the SLP framework of phases should overlap Phase II. Before choosing a specific location, therefore, the planner must know what kind of a general overall layout can be fitted into or onto the location. This is particularly true when there is limited flexibility in the location, such as in existing-building locations.

This means that the planner does in fact have to know what can be done layout-wise with a location if it is selected. For example, in Figure 13–3 any of the existing buildings would satisfy the space requirements, but the rail siding affects the general

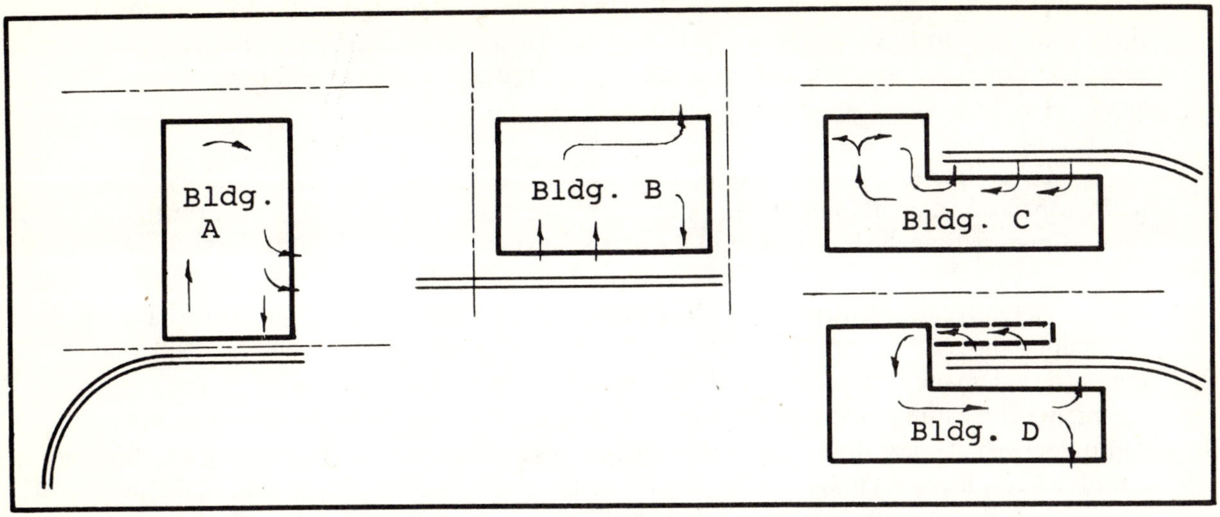


**Figure 13-3.** Alternative locations and the effect of rail siding access on each insofar as what can be done with the general overall layout of each location. Buildings A and B give a pretty good layout. Building C/D doesn't lend itself as well to the operation and therefore should probably be passed by as a prospective location, assuming other considerations are equal.

overall layout possible within each building. This means making a rough overall layout of what could be done layout-wise for the three alternative locations as a part of deciding on the best location.

This raises the question of whether to work from the outside in or the inside out. Some industrial engineers believe in starting at each workplace. Improve the methods there; expand to an improved department layout; and build this up to an overall layout which can be used as a basis for plant-location or building-selection studies. They quote the principles "plan the layout around the machinery and equipment" and "plan the building around the layout" as their support.

To be sure, a good program of methods improvement may be needed. It might well be a lot more productive of reduced costs than a relayout of the area. In fact, in a company where no attention has been given to methods work recently, it is not uncommon to save both the cost of relayout and the need for expansion by a good program of improving operating methods. Certainly, at the time a new layout is made, the planner should not miss this golden opportunity to sweep away old, bad practices and get rid of loose, poor methods.

On the other hand, if the planner waits for a full methods study program, and then a detailed layout, and then an overall plan, before moving ahead with location, in all likelihood the project will be set back too far to help the immediate need. As a matter of fact, he may never get to the bigger layout project if he waits for the perfect methods.

If, however, by "methods" is meant the selection of the processes used to produce the products and the selection of the type of operating machinery and equipment, these must be known before proceeding with layout planning. Recalling the key input data, (R) routing or process, and its sequence and equipment must be known before an effective flow of materials can be planned.

Or, if "methods" are taken as work simplification analysis, then certainly eliminating operations, combining them, or changing their sequence will influence routing, flow-of-materials analysis, and, therefore, the layout.

Actually, the framework of phases does follow the two principles quoted a few paragraphs above. It says further, "Start with the whole, and work to the details." And the pattern demands, "Plan the ideal and from it the practical." Start with rough measurements and estimates and refine these as the ground becomes more certain and the end objectives plainer. Rough tools and techniques are used for early planning and fine ones for later, more accurate details. Decisions are made in each phase, each phase overlapping into the next phase as warranted, but full details are postponed until the basic direction is reasonably defined.

As Figure 1-6 illustrated, the planner deals with more intangibles and less specific details and techniques in Phase I than elsewhere in layout planning work. For example, space requirements used to solve a location problem are estimated and seldom as detailed as figures calculated according to number of machines times space per machine (as in Chapter 6).

On the other hand, where ample space is available so the plan can be fitted and adjusted within it in many different ways, the planner can move ahead with overall planning—even get into detail planning—before the location is actually settled.

Occasionally, it is true, full details are preplanned. For example, where typical buildings or office layouts are more or less standarized for use by branch shops or offices, several optional layouts can be prepared to meet the conditions of each likely location. See Figure 13-4. This is a practical way of reducing the cost of planning. But even here, it is necessary after the location is determined, to select the layout that best fits overall and then adjust the details to meet local conditions.

### Locations on Existing Site

The problem of locating the area to be laid out more often than not falls within the confines of an existing site.

When relocating a department or small area, take the same approach as with plant location: Establish what is wanted (requirements); determine what is available; analyze to decide what is most suitable.

As with larger location problems, set up a list of *requirements* or specifications. Figure 13-5 shows such a list, using the same classifications of factors listed earlier in this chapter. The importance of each factor in the selection of the desired location is

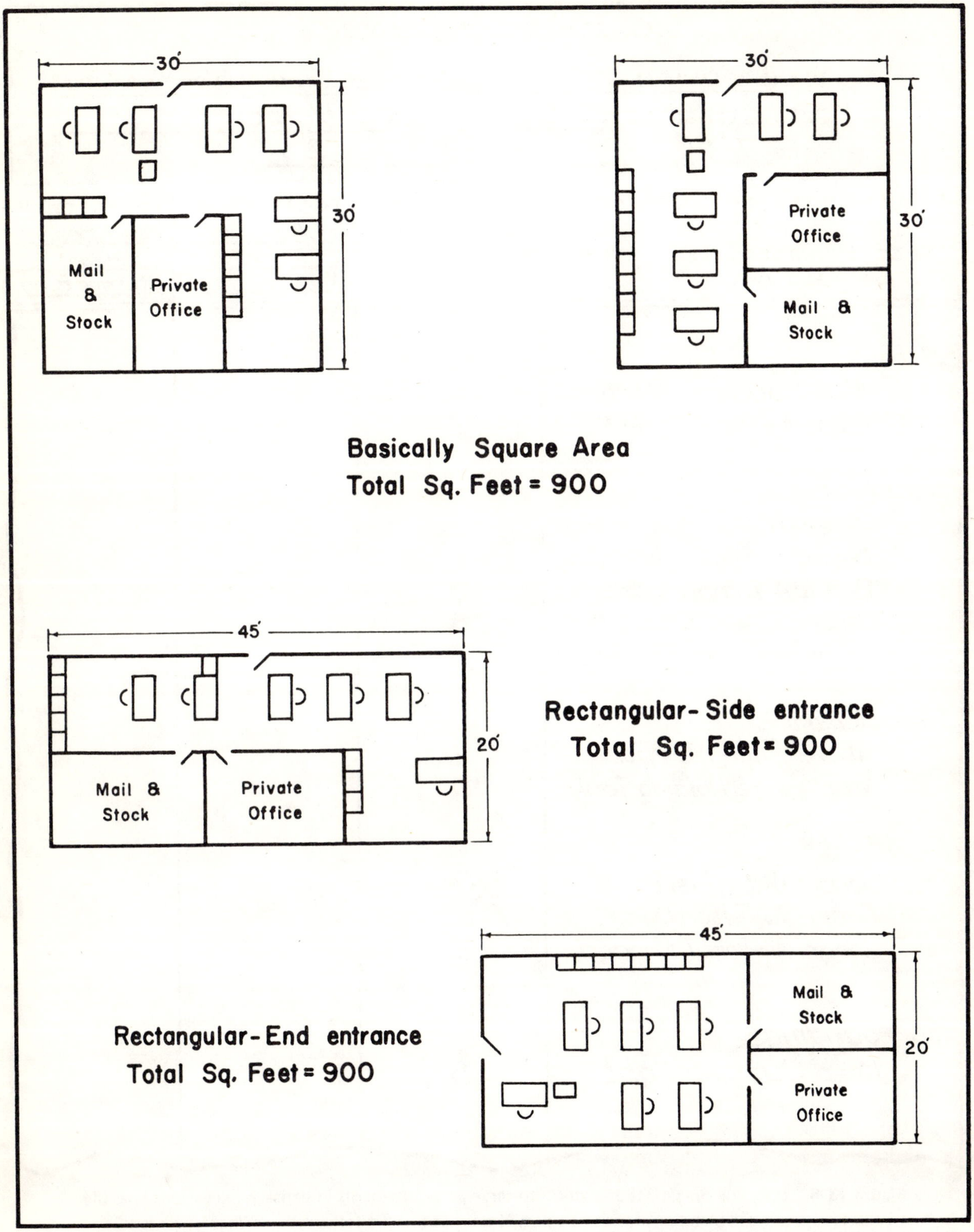


**Figure 13-4.** Typical preplanned standardized layouts. These are developed for branch sales offices. Layouts like this are planned long before the location is selected. Then quick, brief adjustments only become necessary when the specific office area is being considered—or after it is selected—for lease.

LOCATION SPECIFICATIONS

Plant *Zurich A* Project *Expand Blending Area*
Established by *R.L. Egli* With *J. Habich*
Date *7/6* Sheet *1* of *2*

| Requirements of Location (What is wanted from area to be laid out.) | Importance of Requirement to the Project | Location A *2-4/A-C On 2nd Floor, New Wing* | Location B *Shed, Beside Lubricant Storage Bldg.* |
|---|---|---|---|
| *Size* | | | |
| *4000 Square Meters* | 9 | | |
| *Expansion Possibility To 7500* | 7 | | |
| *Condition* | | | |
| *Second (Or 3rd) Floor* | 3 | | |
| *Basically Rectangular (Can Be 1 To 4 Proportions)* | 3 | | |
| *Concrete Floor 800 kg/M² Min.* | 8 | | |
| *4.8 Meters Clear Overhead* | 5 | | |
| *Relationship* | | | |
| *Above Filling Dept.* | 2 | | |
| *Near Aging & Holding Tanks* | 10 | | |
| *Contacts* | | | |
| *Normal Water, Electrical* | 10 | | |
| *Drains To Filter Beds* | 8 | | |
| *Special Air Ventil. Needed* | 5 | | |
| *Surroundings* | | | |
| *Keep Out Sunlight* | 8 | | |

**Figure 13-5.** Location Specifications set up to serve as requirements in considering various possible areas where expanded blending area can be put. Note that this list follows the "What is wanted" section of the text: items are filled in; their importance to the project is weighed; and each prospective location available can then be rated against this list of specifications to select the most suitable. This is similar to the factor analysis method of evaluating alternative layouts.

called out in the column next to the factor description. Available locations are then considered and rated in adjacent columns. The form explained in Chapter 10 for evaluating alternatives by factor analysis could be used here equally well.

Determining what is *available* (or can be made available) within the existing site does not involve the extensive search necessary for locating a new site. But essentially the same sequence is followed:

1. Finding possible locations that reasonably meet the requirements
2. Matching the requirements against each possible location
3. Determining the restrictions or limitations and screening out all but the logical alternative locations

Point 3 is important. Every location has some restrictions or limitations on its use or availability. These must be established before a likely looking location can be considered truly "available."

We have already touched upon the methods of determining the most suitable location:

1. Advantages versus disadvantages, for rough or screening decisions
2. Factor analysis, for intangibles or when lacking adequate financial figures
3. Cost comparison (and justification) for final decision or when reasonably accurate and complete cost figures are on hand

Frequently, the problem on an existing site is one of juggling or "playing checkers" with areas. Thus, relocating one department or area can start a chain reaction of rearrangements with attendent costly disruptions. And, the more divided the space, such as with multi-story buildings or buildings with fixed or load-bearing walls, the problem of finding suitable fits of areas is even greater. This kind of shuffling of space is illustrated in Figure 13–6. The figure also illustrates the importance of logically planning future space needs to avoid frequent and serious relocation problems.

We again point out that what is a "location" problem for a departmental area is a "rearranging of the general overall layout" problem for a building of several departments. That is, what is a Phase I problem for one area to be laid out, is a Phase II rearrangement for the overall plant or building. Similarily, placing an individual desk is a location problem as to the desk alone, but it is part of a Phase III rearrangement of the office in question.

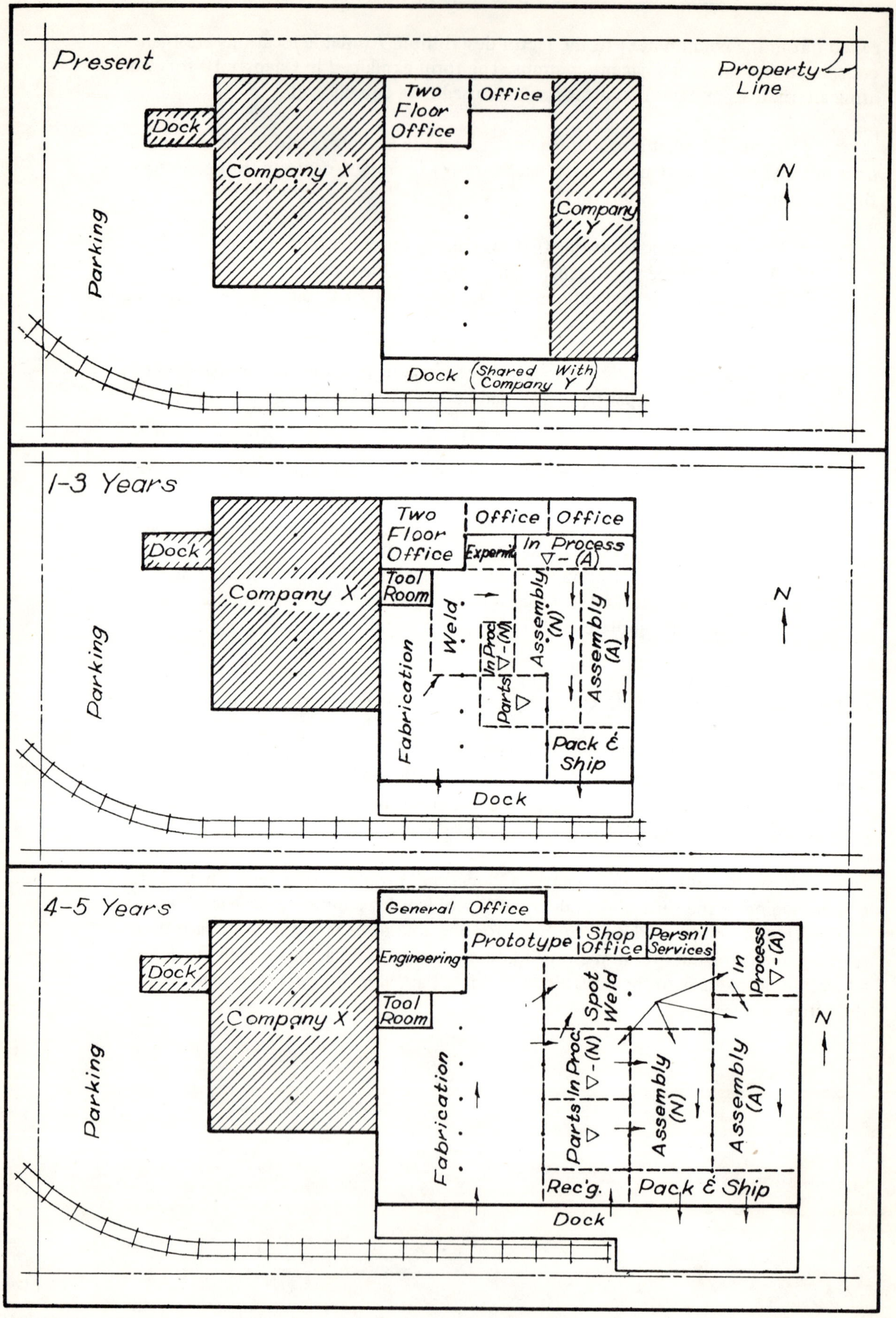

Present
Property Line
Dock
Company X
Two Floor Office
Office
Company Y
N
Parking
Dock (Shared With Company Y)
1-3 Years
Dock
Company X
Two Floor Office
Office
Office
Experim't
In Process ▽-(A)
Tool Room
Weld
In Proc ▽-(N)
Assembly (N)
Assembly (A)
Parts ▽
Fabrication
Pack & Ship
Dock
N
Parking
4-5 Years
General Office
Engineering
Prototype
Shop Office
Persn'l Services
In Process ▽-(A)
Dock
Company X
Tool Room
Spot Weld
In Proc ▽-(N)
Parts ▽
Assembly (N)
Assembly (A)
Fabrication
Rec'g.
Pack & Ship
Dock
N
Parking

**Figure 13-6.** Shows the problem of adjusting locations as a company looks ahead and plans future growth on its present site.

The top layout shows the present building features and property lines. Three companies are located in the same building. The company under consideration is located in the center with Company X and Company Y located respectively in the West and East portions of the building.

The other layouts show what can be done with the layout and flow pattern for one to three and four to five years in the future. Space occupied by Company Y is absorbed in the projected one-to-three-year expansion. Enlarging the building is planned for the four-to-five-year projected expansion. Contemplated expansion (not shown) for 10 years in the future, incorporates the area now used by Company X.

Planning future area locations provides for logical property development at a minimum of expense, and often prevents a company from being "boxed in". Equally significant is the question of what can be done to make effective layout plans as new locations (areas to be laid out) become available or as management plans its negotiations to make added areas available.

# Chapter 14
# Installation

Installation of the layout is Phase IV of the framework of phases. Phase IV is like Phase I in that the SLP pattern of procedures does not apply to it. Also, it, too, may not fall under the responsibility of the layout planner. Ordinarily, it is chiefly the work of the plant engineering or maintenance group.

The fact that machinery and equipment is to be moved offers a real opportunity to make changes and improvements. This is the time to take such steps as the following:

1. Repair, rebuild, or repaint equipment.
2. Add new fittings, attachments, feeds, deliveries.
3. Initiate new working methods, procedures, controls.
4. Abandon bad operating practices of all kinds.
5. Convert to new materials or new product specifications.
6. Realign the manpower assignments, balancing, skills, and time standards.
7. Incorporate better safety practices.
8. Repair the floors, walls, ceilings, and do other major maintenance work.
9. Realign supervisory responsibilities.

### Installation Make-Ready

The physical act of picking up, moving, and placing the machinery and equipment in its new location is only part of Installation. *Make-ready* and *follow-up* are also important. The preparation for the physical move is the really big job. Included in the make-ready may be securing final approval and appropriation of money for the expenditure (with the necessary installation costs as back-up information); preparing installation plans and drawings; arranging for changes in the building or the auxiliary service lines and utilities; scheduling the move; and notifying everyone involved.

Similarly, the *follow-up* steps, after the move, are important and time-consuming. They involve hooking up the equipment; try-out; release to the operating people; and cleaning up the areas—both new and old.

A necessary part of the request for financial appropriation is backing it up with cost information—a breakdown of estimated costs for the installation of the new layout. Not that costs are overlooked in Phases II and III—no, cost information is necessary there, too. In fact, the planner may obtain the installation funds when getting Phase III approval . . . or the funds for footings and structure at Phase II approval, if the project is rushed. However, if new construction, major renovation, or costly new equipment is involved in the project, a more definite, more accurate cost breakdown can be made after the detail plans and installation drawings are prepared.

As an aid to management in deciding whether to proceed with the layout, the planner may update and formalize for presentation earlier cost analyses. He itemizes the savings that are expected as the result of the new layout, as well as the capital investment and expenses of the installation. The analysis should be factual and on a measurable basis so that after the installation results and estimates can be compared. Still, in many cases it is necessary to list the intangible results expected which are not included in the actual expenditure or savings figures.

Figure 14–1 shows a method of compiling estimated installation costs. This sheet—or these figures, along with their back-up information—together with investment sheets for any capital equipment involved in the new layout, is the basis for the appropriation or allocation of funds.

Because the events prior to making the actual moves are so significant, an Installation Coordination Worksheet has been developed. Figure 14–2 shows this. The form is designed to serve as a check list of items likely to be overlooked, as a guide in installation planning, and as a status check. Establishing what is to be done, when it is needed, and who is to do it, achieves the biggest part of coordinating the installation. Therefore, make definite assignments for the completion of each step, calling out when it should be done and who is responsible for the action. This means names and dates on each line of the form. Then make periodic status checks to be sure the new installation goes in on time and with a minimum of disturbance.

### Installation Drawings

The ability to pass instructions along to the various persons concerned with the installation is important. It can best be accomplished with installation drawings and written instructions.

The development of the detail layout is explained in Chapters 11 and 12. The easiest way to make these layout plans available for installation is by some reproduction process. Regardless of the reproduction method used, every "drawing" should show north direction (N), the scale being used, identification of the plant or building, and the sheet or panel. The project identification date and who prepared the drawing can be added to each print, rather than to the original drawing. These are minimum requirements. The drawing numbering system shown in Appendix IX may be helpful here. Moreover, symbols for utilities and other layout features should be coded and used on the installation prints. See Appendix XI.

**Figure 14–1.** Installation Cost Summary for a small departmental rearrangement of a toy factory. The form is used to estimate installation costs. It is subdivided into specific steps by nature of work to be done. This example shows the recording of an outside bid (line 2)—used when it is not practical or economical to do the job with company people. On the other hand, the outside bid can be compared with the company estimate (line 3). The outside bid for painting was not acceptable and therefore is circled to so indicate. Using the outside bid for building repairs and the company estimate for painting, total cost of the installation work becomes the sum of the total cost column.

## INSTALLATION COST SUMMARY

Description: Relayout Stuffed-Animal Department

Plant: Toddle Toys, Inc. Project: 76

Estimated by: H.D.Howard Extension by: M.C.

Date: 3/12 Sheet 1 of 1

| NATURE OF WORK | Estimated Mat'l. | Estimated Hours | Rate/ Hour | Labor Cost | Total Cost | Outside Bid |
|---|---|---|---|---|---|---|
| 1. Clear and prepare new area - - - including marking aisles, columns, locations. | $100. | 25 | $3.80 | $ 95. | $195. | |
| 2. Building repairs, alterations, or construction | -- | -- | -- | -- | 650. | $650. |
| 3. Paint new areas, before and after move. | 80. | 20 | 4.10 | 82. | 162. | (256.) |
| 4. Clean up and repair machines and equipment now in storage. | -- | -- | -- | -- | -- | |
| 5. | | | | | | |
| 6. Disconnect utilities - - - electric, water, air, gas, etc. | -- | 5 | 4.40 | 22. | 22. | |
| 7. Disconnect auxiliaries - - - vents, drains, ducts, conveyors, other handling equipment, etc. | -- | 3 | 4.30 | 12.90 | 13. | |
| 8. Move out operating machines and equipment. | -- | 20 | 4.60 | 92. | 92. | |
| 9. Move out all service and miscellaneous equipment. | -- | 5 | 4.60 | 23. | 23. | |
| 10. Move out materials, work-in-process, stores, tools, supplies. | -- | 3 | 3.60 | 10.80 | 11. | |
| 11. Remove operating machines and equipment not to be relocated. | 20. | 5 | 4.60 | 23. | 43. | |
| 12. Remove service and miscellaneous equipment not to be relocated. | -- | -- | -- | -- | -- | |
| 13. Prepare non-movers for storage, sale, or other disposition. | -- | -- | -- | -- | -- | |
| 14. Fill-in pits, close holes, and clean up old area. | -- | -- | -- | -- | -- | |
| 15. | | | | | | |
| 16. Install pits, foundations, openings, special enclosures. | -- | -- | -- | -- | -- | |
| 17. Install conveyors, hoists, other handling equipment, racks, shelving, storage equipment. | 200. | 25 | 4.60 | 115. | 315. | |
| 18. Install electric power equipment, leads, outlets, lighting fixtures, etc. | -- | 5 | 4.60 | 23. | 23. | |
| 19. Install utility lines and outlets - - - water, air, gas - - drains and sewers. | -- | 5 | 4.60 | 23. | 23. | |
| 20. Install heating, ventilating, air conditioning, ducts, fans, filters, dust collectors, etc. | -- | 3 | 4.60 | 13.80 | 14. | |
| 21. Move in operating machines - - - spot, level, lag, mount. | -- | 20 | 4.60 | 92. | 92. | |
| 22. Move in all service and miscellaneous equipment (not already installed). | -- | 8 | 4.60 | 36.80 | 37. | |
| 23. Move in all materials, work-in-process, stores, tools, supplies. | -- | 5 | 3.60 | 18. | 18. | |
| 24. Hook-up or connect, straighten-up, try out prior to operating. | -- | 4 | 4.60 | 18.40 | 18. | |
| 25. | | | | | | |
| Total | $400. | -- | -- | $701. | $1751. | $650. |

NOTES:

RICHARD MUTHER & ASSOCIATES – 302

## INSTALLATION COORDINATION WORKSHEET

Plant: Sturdy, Ltd. — Project: A 965-3
By: T. Lynch — With: G.C.
Date: originated 8/2 — Sheet 1 of 1

| Phase | Step | WHAT | WHEN | WHO | STATUS 10/2 | STATUS 10/16 | | | |
|---|---|---|---|---|---|---|---|---|---|
| MAKE READY | PLAN | 1. Start planning the installation | 8/20 | J.Heller | Complete | – | | | |
| | | 2. Establish sequence of moves | 8/20 | J.Heller | Complete | – | | | |
| | | 3. Inventory everything to move | 8/20 | H.Dunlap | Complete | – | | | |
| | | 4. Get disposition of non-movers | 8/26 | V.Daniel | | Complete | | | |
| | | 5. Schedule moves in detail | 9/10 | J.Heller | Complete | – | | | |
| | | 6. Assign move numbers; check vs. inventory & equipment (tag) number | 9/20 | J.Heller | Complete | – | | | |
| | PROVIDE | 1. Decide who will make moves | 10/5 | C.Jacks | Complete | – | | | |
| | | 2. Secure bids as necessary | 8/20 | A.Printz | Complete | – | | | |
| | | 3. Determine and reserve moving equipment required | 9/10 | A.Printz | This Week | Complete | | | |
| | | 4. Set up communications for both ends of move | 11/2 | J.Heller | | | | | |
| | | 5. Appoint key men for each area | 10/10 | D.Bell | | Complete | | | |
| | | 6. Get work order (s) for move (s) | 10/10 | C.Jacks | | By 10/20 | | | |
| | PREPARE | 1. Prepare new locations - - physical area, conditions, auxiliaries | 10/20 | C.Jacks | | | | | |
| | | 2. Broadcast plans | 10/10 | Z.White | | Briefing 10/26 | | | |
| | | 3. Brief personnel specifically involved | 10/10 | Z.White | | ditto | | | |
| | | 4. Mark everything to move: identification, move no., destination | 10/15 | J.Heller | | Resched. to 11/3 | | | |
| | | 5. Disconnect or ready equipment | 10/27 | C.Jacks | | | | | |
| | | 6. Check out equipment release to movers | 10/27 | J.Heller | | | | | |
| DO | MOVE | 1. Move equipment intact to reduce reassembly time | 11/10 | C.Jacks | | | | | |
| | | 2. Move close to spot to reduce line-up and hook-up time | 11/10 | C.Jacks | | | | | |
| | | 3. Post move performance as accomplished | 11/10 | C.Jacks | | | | | |
| | | 4. Keep moving crew informed, coordinated | 11/10 | C.Jacks | | | | | |
| | | 5. Be on hand - - layout interpretation | 11/10 | J.Heller | | | | | |
| | | 6. Be on hand - - auxiliaries interpretation | 11/10 | J.Heller | | | | | |
| PUT AWAY | HOOK-UP | 1. Spot equipment; check location | 11/10 | J.Heller | | | | | |
| | | 2. Temporary hook-ups where needed | 11/10 | J.Heller | | | | | |
| | | 3. Check and release for permanent connections | 11/10 | C.Jacks | | | | | |
| | | 4. Inspect the installation & release for tryout | 11/12 | C.Jacks | | | | | |
| | | 5. Maintenance tryout | 11/13 | C.Jacks | | | | | |
| | | 6. Release to operating group; secure acceptance | 11/14 | J.Heller | | | | | |
| | CLEAN-UP | 1. Survey-inspect old and new areas | 11/15 | C.Jacks | | | | | |
| | | 2. Schedule & assign clean-up - - old and new areas | 11/18 | C.Jacks | | | | | |
| | | 3. Verify layout as installed | 11/18 | J.Heller | | | | | |
| | | 4. Verify auxiliary service as installed | 11/18 | J.Heller | | | | | |
| | | 5. Verify or adjust layout & service-specification records | 11/20 | J.Heller | | | | | |
| | | 6. Recap installation costs and performance | 11/25 | J.Heller | | | | | |

RICHARD MUTHER & ASSOCIATES - 310

Several copies of the layout plans are necessary. If the plan is made on opaque back-up (or undersheet) material, some form of photography is required, or the layout is redrawn on tracing paper. In the latter case, as with all forms of flexible, clear back-up (or undersheet) material, the layout plan can then be run through a standard print machine and a print prepared by light passing through the sheet. This is one of the biggest advantages of clear, transparent plastic layout planning sheets.

Actually, when many prints are needed it is better to use a transparent intermediate or brown-line sepia print. From this transparency, prints can be made on regular blueprint paper. This intermediate step has several advantages: It saves the original layout sheet from further exposure and possible damage. More important, it allows marking on the transparency instructions to the various trades involved without cluttering the drawing with too much information. For example, one transparency can be marked with electrical data, another with mechanical requirements, another with piping and drains, another for conveyor or fixed handling equipment, still another for building modifications. The transparency(ies) is run through the print machine with the written information thereon. As many copies as wanted, or sets of the various copies, can be reproduced in this way, clearly and without tying up the original planning sheet or panel. And prints can be sent out to contractors for bidding or used locally—everyone working to the very same layout plan, which forms the basis of the original transparency.

Another method of instructing the various parties is to use clear, plastic overlays. Each overlay can be made for a particular function or trade such as electrical, mechanical, and the like. The overlays are run through the print machine in conjunction with the layout tracing or plastic back-up sheet, thus enabling each group to see what it is supposed to do with respect to the entire project.

The planner can get full "mileage" out of his 3-D model by using it to instruct, train, and generally orient operating people and installation crews. This is one of the big advantages of 3-D. Use it to aid installation, both by briefing people from it directly and by photographing the layout model at an angle and releasing photoprints to the installation and operations people.

### Work Orders and Instructions

In addition to the prints, written explanations are usually required. This means that the prints and instructions must be prepared and then covered by a work order. Each plant has its own design of form and procedure for preparing and issuing work orders. However, Figure 14–3 shows one form of move notice that can be used effectively within any work order system.

**Figure 14–2.** Installation Coordination Worksheet is used to assign responsibilities and check status of installation planning. This sheet shows the timing established for each step in coordinating a move. Along with the due times go assignments to specific individuals for the completion of each step. A status report is issued periodically (information on the right side of the form) to prevent detail "fall downs" from affecting the completion date scheduled for the entire project.

## MACHINE & EQUIPMENT MOVE NOTICE

Plant *East* Project *B-659* Notice No. *633*

Project Description *Relocate & Relayout portion of Machine Shop* Work Order No. *60-375*

Prepared by *V. Reynolds*

Moves to be completed by *11/1* Approved by *A. J. R.*

Date issued *10/5*

| Description Asset or Tag No. eqpt. to be moved | | Bldg. | Floor | Dept. | Col. | Explanations Check "OK by" and Date |
|---|---|---|---|---|---|---|
| 1. *Brake Press #1346* | F | | *2* | *44* | *D5* | |
| | T | | *1* | *43* | *E8* | |
| 2. *Punch Press 87 Ton #1113* | F | | *2* | *44* | *D5* | |
| | T | | *1* | *43* | *E8* | |
| 3. *Lathe #1262* | F | | *1* | *Receiv.* | *A2* | |
| | T | | *1* | *43* | *E8* | |
| 4. *Lathe #1237* | F | | *2* | *52* | *L4* | |
| | T | | *1* | *43* | *D8* | |
| 5. *Lathe #1238* | F | | *2* | *52* | *L4* | |
| | T | | *1* | *43* | *D8* | |
| 6. *Drill #1706* | F | *Dead* | *eqpt.* | *storage* | | |
| | T | | *1* | *43* | *D8* | |
| | | *Tool Room to Recondition* | | | | |
| 7. | F | | | | | |
| | T | | | | | |
| 8. | F | | | | | |
| | T | | | | | |
| 9. | F | | | | | |
| | T | | | | | |
| 10. | F | | | | | |
| | T | | | | | |

Instructions: *Punch Press #1113. Disconnect electrically after 4PM 11/1. Necessary wiring at new location to be done without interruption of service to production line. Installation of this machine to be complete by 7AM 11/3.*

*Lathe #1262 (new Gisholt)-Move in place and hook up soon as possible. Lathes #1237 and 1238 disconnect on 11/8 and be in operation by 11/10.*

*Brake Press #1346-Disconnect after 4PM 11/6 and in operation by 11/18.*

Drawing/Plan No. *M-2563* Photo/Sketch No. ______

Distribution:
- White--Movers
- Blue---Maintenance
- Pink---Plant Layout
- Yellow-File

Return white copy to *A. James* when installation is complete.

Where many copies of installation instructions are needed, it is practical to reproduce the instruction sheet right with the layout prints. The Installation Instruction Summary, Figure 14–4, is an example. After the instructions are entered on the form, it is attached to, and along side of, the layout plan or sepia transparency and the two run through the print machine together.

On the Installation Instruction Summary, the item numbers can be keyed to similar numbers marked on the layout plan or the sepia transparency. This makes it easier to find the piece of equipment under consideration.

One of the simplest aids to clear understanding of the installation work to be done is to post in the new area a print of the installation work instructions and colored identification of various items. This should be attached to something firm, mounted up off the floor or on the maintenance supervisor's desk. Significant items should be hand colored for easy visualization; this is simple to do and very often extremely helpful.

### Locating Equipment

There are basically two ways to aid installers in lining up machines and equipment. One is to print the layout planning grid sheet or to superimpose a grid on the layout drawing. This allows anyone to count the number of grid lines and convert it to distance. The other way is to mark the dimensions on the drawing or sepia transparency before run-off of the prints. Either method allows the machine to be located in the proper position.

Providing the installation people with the actual dimensions has proven to be more accurate, though it is more time-consuming for layout planners. Figure 14–5 shows various methods of lining up equipment. Obviously, it would be difficult for the installation crew to place such machines accurately if dimensions were not given. Moreover, when black-line grids are used, the grid lines show through the clear templates, causing confusion and making interpretation difficult. Remember that the total system of templates, undersheet, overlay, reproduction method, and installation instructions must be considered when setting up materials to visualize detail layout plans.

### Who Should Make the Move

The layout planner or analyst may plan and coordinate the installation, although more often than not someone else is responsible for planning and coordination of the installation phase. The physical installation usually falls to the maintenance group. For large or complex jobs, outside movers and installation contractors are fre-

**Figure 14-3.** Move Order used in conjunction with a work order system. Each item of machinery or equipment is identified on one line; from and to locations are placed opposite; and instructions entered below. Space at right is for follow-through notes such as installation delays, movers' explanations, and completion sign-off.

# INSTALLATION INSTRUCTIONS SUMMARY

Description *Relayout of Kitchen*

Plant *Nortot* Project *343*
By *O.M. Cramer* With *D.F. Addis*
Date *11/10* Sheet *2* of *2*

OTHER WORK—(Code who and describe): C-Carpenters, M-Methods, S-Safety Eng'r., F-Fire Marshall, P-Plant Engrg., *-Out. Contrtr.

| ITEM No. | DESCRIPTION / MACH./EQPT. No. | MOVE From / To | UTILITIES: Air | Water | Gas | Drain | Vent | Steam | MILLWRIGHTS—MECHANICAL | ELECTRICAL: Volts | Phase | KVA | Amps. | H.P. | ELECTRICIANS | OTHER WORK: WHO | WHAT |
|---|---|---|---|---|---|---|---|---|---|---|---|---|---|---|---|---|---|
| 1 | Sink D-1 | -- / -- | – | ✓ | – | ✓ | – | – | Reorient | | | | | | | | |
| 2 | Food Prep. Table D-2 | A-5 / B-10 | – | – | – | – | – | – | Relocate | 110 | | | 15 | | Install | C | Revise table per drawing T6538 |
| 3 | Steamer (stag) D-3 | A-5 / B-10 | – | ✓ | ✓ | ✓ | ✓ | ✓ | Relocate | 110 | | | 15 | | Hook up | | |
| 4 | Proof Box D-4 | A-7 / B-10 | – | – | – | – | – | – | Relocate after repairs | 110 | | | 15 | | Install | | |
| 5 | Bake Oven D-5 | A-7 / B-10 | – | – | ✓ | – | ✓ | – | Relocate | | | | | | | F | Check for fire regulations |
| 6 | Portable Mixer D-6 | A-7 / B-10 | – | – | – | – | – | – | Relocate | 110 | | | 15 | | Install | | |
| 7 | Range F-7 | C-5 / B-12 | – | ✓ | ✓ | – | ✓ | – | Relocate | | | | | | | F | Check for fire regulations |
| 8 | Range F-8 | C-5 / B-12 | – | ✓ | ✓ | – | ✓ | – | Relocate | | | | | | | F | Check for fire regulations |
| 9 | Portable Slicer F-10 | C-5 / B-12 | – | – | – | – | – | – | Relocate | 110 | | | 15 | | Install | S | Certify present guards |
| 10 | Exhaust Hood | -- / -- | | | | | ✓ | | Rearrange and tie-in with 3,5,7,8 | | | | | | | | |
| 11 | Partition | A-5 / 76 | | | | | | | Remove | | | | | | | C | Disassemble and salvage |

To accompany Drwg. Nos. *M-29548*

Work Order *60-936*

quently called in. On the other hand, for simple relayouts, operating personnel themselves often make the change-over.

Whenever the installation work is divided among two or more groups, the work each group will do should be clearly defined. In addition to its own assignments, each group should have information on what the other group is supposed to do. In this way, one group cannot "assume" that another group will cover a certain portion of the job because it does not happen to be listed on the instructions of the first group.

Actually, most firms make their own layout installations for the following reasons:

1. The cost is likely to be less when the company has its own installation crew. This is especially true if operating people are idle anyway and can be used effectively, although their effectiveness is often very limited.
2. The maintenance men become familiar with the installation when they put it in and, therefore, find it easier to maintain and repair.
3. There is less need for elaborate paper work on contracts, prints, specifications, installations drawings, and the like—that is, less staff work.
4. Where speed and time are important, as in the case of hurried change-overs, it is frequently not practical to wait for an outside contractor.
5. By having the company's own men do the installation work, the presence of maintenance men is consequently ensured in case of emergency.

There are, however, many advantages in employing outside contractors to make the installation.

1. Contractors are often more skilled and familiar with layout installation work and techniques; they have the proper equipment available and can do a safe, efficient job.
2. Frequently a company does not have a qualified or adequate maintenance group to handle the rather infrequent task of layout installation. For new layouts, there may be no company employees available at all.
3. A company's own construction and maintenance crews will have many other details to attend to during a relayout, and, therefore, may not have time to handle the installation.
4. Labor union agreements may restrict the use of company personnel.
5. The installation can be a long drawn-out sequence, or a continuous process plant may have to be kept operating, in which cases, outside crews are practical.

**Figure 14-4.** The Installation Instructions Summary is a condensed way of issuing instructions. By having it on tracing paper, it can be fastened to an installation drawing or sepia transparency of the layout, and both run-off together. The item number is keyed to the layout plan for easy reference. Description refers to the equipment being moved. A check or an X in the columns under Utilities indicates services that must be provided or removed. The Electrical column calls out the electrical requirements. The columns labeled Millwrights-Mechanical and Electricians will hold a brief description of the work each trade is to perform. The remainder of the form allows room for instructions or special assignments for other people involved with the installation.

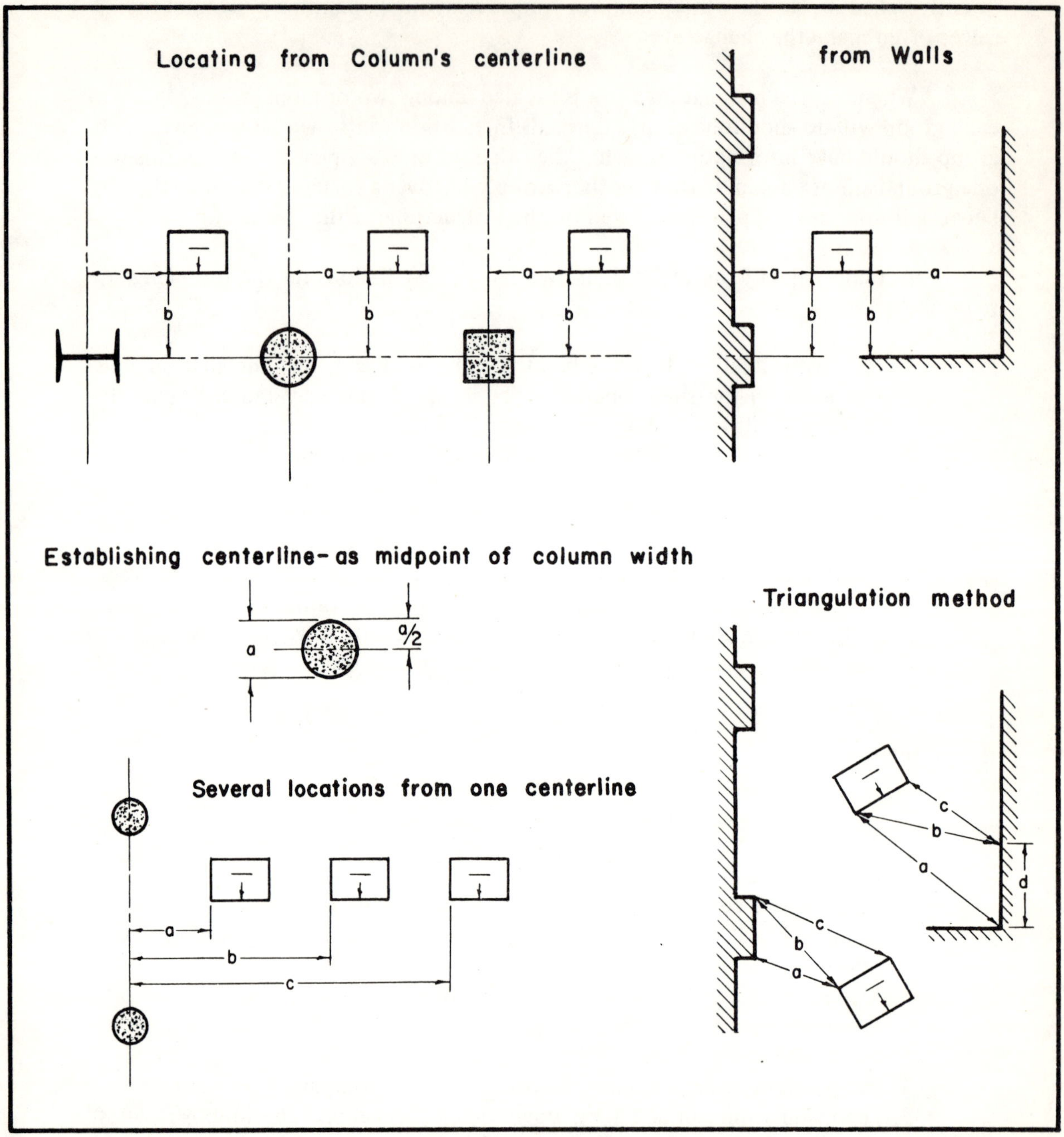


**Figure 14-5.** Methods of indicating machinery and equipment placement. The centerline method is most commonly used in locating equipment. It may be necessary to dimension to both sides of the machine (or machine template) if the reference point is not clearly defined.

The use of triangulation requires three dimensions to be specified for proper location of equipment.

Machines can also be located from aisle markers. The method is the same as if the measurements were given with respect to a wall, with the outside edge of the aisle marker customarily used as the measurement line.

Where outside contractors are employed, it is often practical and usually advantageous to have one or more company men work closely with them. Again, be sure to spell out for both groups which details of the installation each will do.

### When to Move

The time chosen to install the layout may be most important. The planner should schedule the move during periods of low production or changes in product, process, or equipment. Scheduling moves in this manner helps avoid interruptions in shipping schedules, losses in production time, and disrupting effects on employees.

It is usually impossible to find a time which will completely satisfy all parties. An attempt can be made to maintain production schedules during the move. Or, the move can be made on weekends, holidays, vacations, or during seasonal lulls when production is normally down. It is often advisable to suspend operations and make the move all at one time, rather than tangle with everyone during the move.

When the move time is critical, the installation phase can dictate the schedule for the entire planning project, earlier work being scheduled against the "on stream" or "in operation" dates. Regardless of the planning schedule(s), the installation work should have a specific schedule—finely broken down and closely controlled—especially when closely synchronized and complex construction, utilities, and equipment are involved. Critical path or PERT network analysis and scheduling are most meaningful under such conditions.

Once the schedule is set, it is often better to go ahead than to hold up if everything is not quite ready. That is, make the installation before being ready with every layout detail. If the installation were postponed until everything was completely ready, the move might never be made. And if we tolerate delays, we may find in the future our installation planners never will meet schedules.

When the move takes place, the layout planner must be on the floor, or at least readily available, for consultation with the movers. No matter how well the layout planning has been done, there are layout adjustments at installation time. Therefore, the layout man must be available to answer questions, interpret plans, inspect for completeness, and secure as early a resumption of production as possible.

Regarding changes, the policy should be firmly established that the installation people can not deviate from the prints and work orders. The layout man or installation coordinator can make changes on the prints or work orders and must sign and date such changes before the installation group can honor them.

### Sequence of Moves

If a small move is being made, move things closest to the door first. This clears space. And if the equipment can be moved in one load, the first item in the vehicle is the last item out. But it is probably much more important that the new areas are set up and completed in orderly sequence—so let this objective control the sequence of moves.

Before the move, pillars or columns should be clearly identified, preferably by painted letters and numbers on each. This allows ready orientation for equipment spotting. Also, paint, tape, or otherwise mark aisles before starting the move. This prevents the movers from setting equipment in the aisles and clogging the flow of other equipment. Besides, most of the equipment will be located with respect to the columns and aisles.

Whenever floor space is clear, the location of each piece of equipment in the new layout should be marked on the floor. The equipment can be moved as close as practical to the point where it is to be used. Then it can be lined up with a minimum of further moving.

Making a relayout is often not unlike a game of checkers. One move may call for a move into a spot presently occupied by some other piece of equipment. Hence the sequence of moves must be planned, perhaps even with extra moves merely to clear space.

### Condition Employees for Change

Prior to the move, operating and supporting-service supervisors should be briefed. Their understanding of the installation plans can materially aid a smooth move. By having prints, photos, or the 3-D model available to them, supervisors can explain to their employees what is going to happen and where each worker will be located in the new set-up. Actually, elaborate training of operating people may pay off handsomely, depending on the nature of the project and the relative "newness" of product, process, and personnel.

Keeping employees "in the know" improves morale, especially during the time when there is a certain amount of disruption. And, of course, don't overlook personnel staff; give out new locker numbers, explain the new parking or time-clock procedure which has developed as a result of the new layout. New practices of this type, when improperly introduced to employees and supervisors, can give a black eye to the entire layout project—with the layout planner frequently catching most of the blame.

### Release to Production

Machinery and equipment should be checked over and tried out before being released by maintenance to the operating departments. Figure 14-6 shows a form help-

**Figure 14-6.** The Installation Record serves as a progress sheet of each piece of equipment's installation. By checking-off and/or writing-in as necessary, the sheet gives work assignments and completion due dates at left, installation completion sign-offs at right, and acceptance of the installation below. The sheet will be kept with the machine until its installation is completed and accepted. Any reason for delays in installation or hook-up can be entered in the comments section or as a footnote. Return of the form to the issuing office indicates completion of this machine's installation.

## INSTALLATION RECORD

Plant St. Louis Project 60-36
Bldg. B Area 1st Floor Dept. L
Nearest Column Location E 5
Sheet issued by* J. Rothman
Date issued 12/7 Date returned

Mach. or Eqpt. No. 5
Mach./Eqpt. Type 16"Wire Drawing Mac.
Mfgr. Federal Wire Rope Corp.
Size/Model
Signif. Identfn. 7B--7H

**MECHANICAL, MAINTENANCE, MILLWRIGHTS**

| Item | | Sign | Date |
|---|---|---|---|
| Level | Yes | A.Butts | 12/15 |
| Lag | | | |
| Mount | | | |
| | | | |

Complete (sign) A. Butts Date 12/15
Comments Everything OK

**ELECTRICIANS**

| Item | | Sign | Date |
|---|---|---|---|
| Power | 440 3 phase | L.Carter | 12/17 |
| Control | X | " | " |
| Lights | X | " | " |
| Connect spooler | | " | 12/18 |

Complete (sign) L. Carter Date 12/18
Comments OK

**PIPING & SHEET METAL**

| Item | | Sign | Date |
|---|---|---|---|
| Water | X | F.Royer | 12/17 |
| Steam | | | |
| Drains | | | |
| Comprsd. Air | X at front | F.Royer | 12/17 |
| Gas | | | |
| Coolant, Lube | | | |
| Hoods, Ducts, Fans | | | |
| | | | |

Complete (sign) F. Royer Date 12/17
Comments Will need new valve installed when available.

**PAINT and CLEAN-UP**

| Item | | Sign | Date |
|---|---|---|---|
| Paint | | | |
| Clean-Up | X | G.Art | 12/18 |
| | | | |

Complete (sign) G. Art Date 12/18
Comments OK.

**SAFETY**

| Item | | Sign | Date |
|---|---|---|---|
| Guards | X | J.Wolf | 12/18 |
| Fire Regs. | | | |
| Fumes, Acids | | | |
| | | | |

Complete (sign) J. Wolf Date 12/18
Comments Old style guards—replace with new style soon as available.

**OTHER INSTALLATION WORK SIGN OFF**

| | Sign | Date |
|---|---|---|
| | | |
| | | |
| | | |
| | | |

Complete (sign) Date
Comments

**FINAL ACCEPTANCE** (by installation coordinator or operating supervision)

Name H. Sharp Title Layout Engineer Sign H. Sharp Date 12/19
Comment Working OK
Name J. O'Conner Title Dept. Foreman Sign J O'Conner Date 12/19
Comment O.K

(After final acceptance, return this sheet to person originally issuing it. See above.)

NOTES, Special Instructions, Reasons for Delay

RICHARD MUTHER & ASSOCIATES - 320

*Return this sheet to person originally issuing it, after final acceptance.

ful in this *installation check-out*. It is used to certify that the piece of equipment is installed and functioning properly. This form has a place for each craft to check off the work it has completed. It includes a place for the safety engineer to sign, indicating that the machine conforms to the plant's safety regulations.

### Follow-Up

For the sake of future layout projects and adequate space control the planner should follow up the layout to see that it is correct and functioning as planned. Check with the people using the new layout to determine what gripes may have developed. Immediate attention to these details will aid in making the new layout successful and at the same time develop the reputation of the layout planning group.

If the layout engineer finds the layout is not functioning as planned, he must analyze the problems. The operating people may have misunderstood what was intended. The layout engineer may have planned something impractical. Or the service people for some reason can not or will not follow the procedures or use the equipment as planned. This is a serious reflection on the layout planner, and he should correct the situation quickly. If he is wrong he should OK the change. After all, some oversights are to be expected; it is too costly to plan every detail perfectly. If he is right and cannot induce the operating people to work the layout as planned, he may have to involve "higher ups" in the organization.

When the layout project is wrapped up, it is wise to make a post-installation audit of the savings and costs that were estimated. This increases the accuracy of installation estimating and gives a sounder basis for future operating-cost estimates. Also, where relayouts are frequent, the layout plans should be updated so an exact record of the current layout is always available for future use. Some plants find it practical to keep two sets of layout panels, one for planning and one for permanent current record.

## Chapter 15

# Managing Layout Projects

In this final chapter, layout projects are considered from the standpoint of the person responsible for supervising plant layout activities and from the viewpoint of the individual charged with the responsibility for a specific project. The reader is addressed here accordingly.

### Management Interest

The degree of management attention to facilities layout varies with the relative amount of the capital investment involved, with the extent of owner or stockholder interest, and with the degree of change in products, processes, and sales volume. Top management is generally interested in Phases I and II of layout planning but delegates to others execution and approval of Phases III and IV. However, as explained in the Introduction to Part Four, the rearrangement of a small departmental area has a Phase I and a Phase II, but they may be of no interest at all to top management. Moreover, what is a major Phase II problem for a branch plant, may have little or no significance from the viewpoint of headquarters top management.

Therefore, as shown in Figure 15–1, the degree of management attention is *relatively* greatest in the early phases and less important in the later phases of layout planning. On the other hand, the total manhours increase with each phase. There are more people involved and more time is spent in the later phases. But, since the work is generally at lower levels of the total employee group, management delegates the planning and control of these manhours to others.

In the complete schedule of any project, the total manhours figure increases while the total elapsed-time figure generally remains about the same for each phase. This means that more important deliberations by fewer, high-level management personnel take place in the early stages of projects while the reverse is true later in the project.

### Organization

The organization for layout planning depends on a host of factors. In industrial plants, the layout function falls under Industrial Engineering, Plant Engineering, or their equivalents in 60% of the cases.[1]

In highly engineered process plants, men with high technical experience meld

1. See "Who Lays Out the Plant," a survey conducted by the Technical Services Division of A.M.H.S., and reported in *Modern Materials Handling,* April 1964.

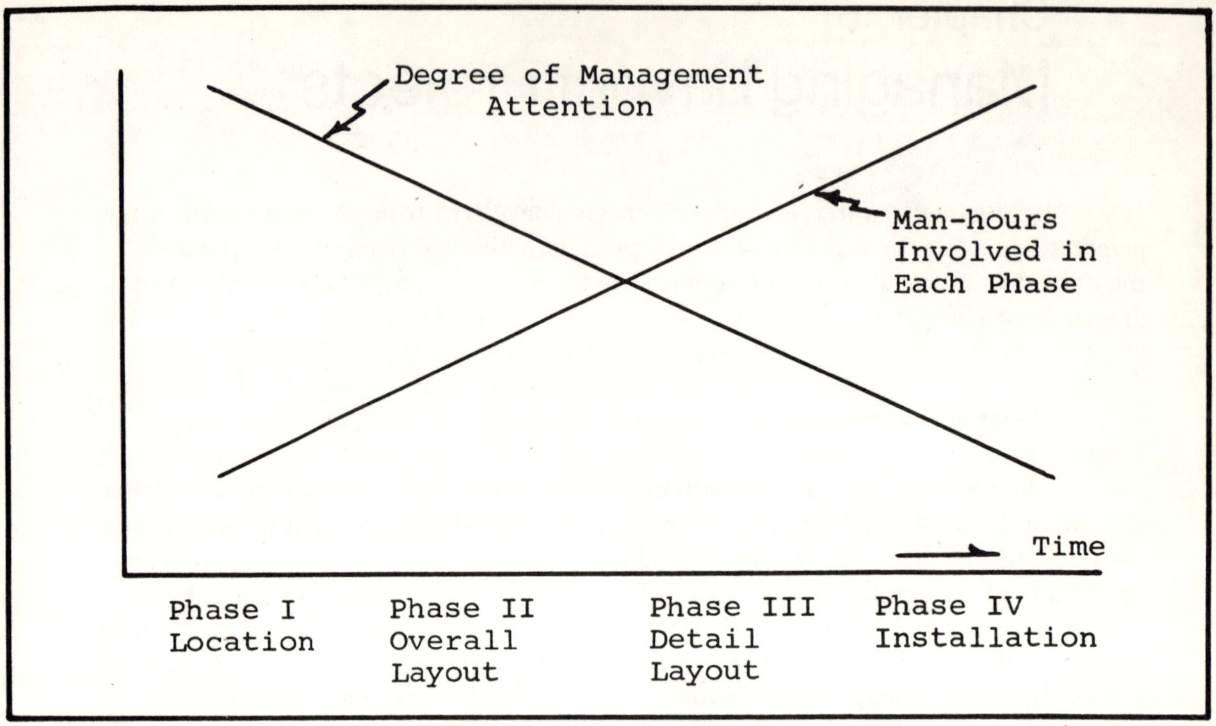


**Figure 15-1.** Management interest and total man-hours vary with layout-planning phases. Although calendar time tends to remain relatively the same in each phase, it is because management deliberates more on the more-important early phases, and delegates and spreads the work of detail planning and installation.

the layout planning right into the process planning. For office areas, on the other hand, where fixed investment is not high and rearrangements are easy to make, skills in procedures analysis, as well as an understanding of layout planning, may be of value. Where there is no formal layout planning function, the planning falls to the manager of each operating area, be he office manager, superintendent, service manager, assistant to one of these, or some other person designated for the job.

The main point is to have somebody responsible for planning the layout and to equip this somebody with the know-how necessary to do it. His position in the organization structure is more academic perhaps than it is significant. Teams, committees, and other investigating groups or outside consultants can definitely participate, and with good overall benefits; but one person, or group with a designated person in charge, should be responsible for getting the layout planning executed.

### Get Ready for Planning

Getting set up properly for layout planning is important. Too often, the person responsible for the project is handicapped by lack of working space or suitable materials, by ignorance of the total company program, or by lack of recognition or authority. The planner should get set up with suitable and sufficient materials. Don't be afraid to waste a few extra prints or grid sheets. Such items are meant to be used—and they are relatively inexpensive in comparsion with the time of the layout planner. Get an adequate place to work, with table-top area or wall space to spread out plans.

Get an overall layout of the site or building plans which identify the buildings and floors and the assignment of current departmental areas. Such plans need not be greatly detailed, but they should show major features of the site or buildings. An example of such an orientation plan is shown in Figure 15–2. In detail relayouts, detailed plans of existing structures and utilities are used—and, in the average company, they are notoriously inaccurate if they are available at all. Make it easy for the layout planner to get at the information he will need: projected input data—product designs (P), anticipated sales forecasts (Q), the process routing (R), supporting services (S), and timing considerations (T). He needs to know what changes are expected in these during the life of his layout. And, he should know about any long-range master plans for the site, buildings, or space allocated, so that his layout is integrated with the larger space-planning program for the company. If these are not available see that other division officers or department managers involved are alerted to the fact that they will be expected to supply such information to the layout planner. Failure to plan or to communicate the basic criteria or specifications to which the layout must be designed is one of the most common oversights by top management in initiating layout planning projects. Knowing this, the planner should not be hesitant to develop his own input data and submit it to others for approval.

### Facilities Planning

The term *facilities planning* is sometimes used almost interchangeably with plant layout. Some companies use facilities planning to describe Phases I and II of major projects, and plant layout to mean Phase III only. Regardless of individual company terms, facilities planning is a broader term and encompasses the total company program of real estate, buildings, process machinery, and utility equipment.

In reality, a layout is only part of a total industrial or commercial facility. The total facility can be compared to a living organism. The different parts of the anatomy are different components of the facility. Figure 15–3 draws this comparison.

Following the concept of planning in phases, the layout planner integrates each of these components at each phase—overlapping his planning into the following phase as necessary. But both the planner and the approver of his plans should recognize that all five components need to be considered as an integrated whole when each phase of planning is submitted for approval.

For overall planning of these activities, a long-range capital investment budget is recommended. This not only forces management to plan ahead—which any reader of this book will subscribe to as being important—but also sets up an organized routine for screening and appraising various projects that must be justified in the competition for available investment money. For effective total facilities planning, this capital invest-

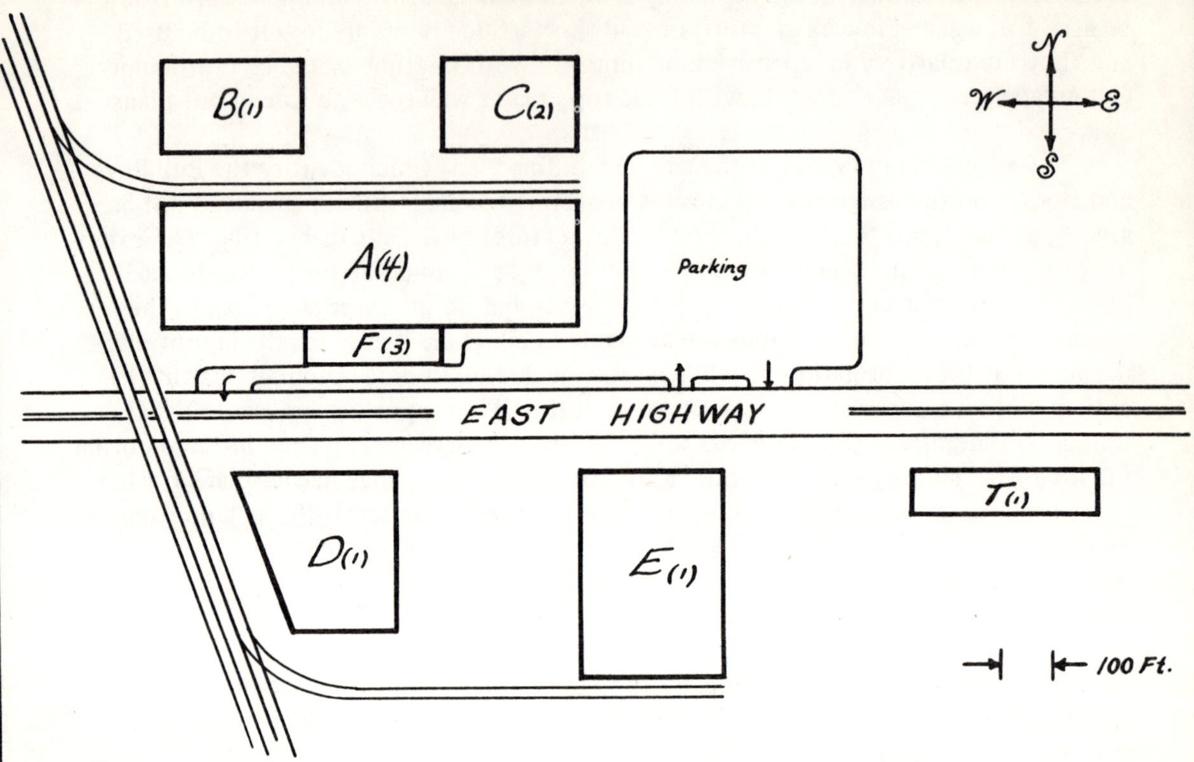


FACTORY

| Department | Bldg. | Floor |
|---|---|---|
| Large Press | A | 1 |
| Small Press | A | 1 |
| Screw Machines | B | 1 |
| Gear Shop | B | 1 |
| Assembly Dept. | A | 2 & 3 |
| Tool Room | C | 1 & 2 |
| Heat Treating | D | 1 |
| Experimental | C | 3 & 4 |
| Forge Shop | E | 1 |
| Test Range | T | 1 |

OFFICE

| Department | Bldg. | Floor |
|---|---|---|
| Accounting | F | 1 |
| Billing | F | 1 |
| Engineering | F | 3 & 4 |
| Payroll | F | 1 |
| Cafeteria | A | Bsmt. |
| Plant Engineering | F | Bsmt. |
| Purchasing | F | 1 |
| Industrial Engr. | F | 2 |
| Tool Design | C | 2 |

**Figure 15-2.** Overall orientation plan to aid layout planning. This diagram of the buildings involved and list of allocated floorspace is valuable to the layout planner as he works to adjust and rearrange space. He should always have at hand—or at least obtain early in his project—this record of existing departmental space and where it is located. Here the buildings are indicated by letter and the number of floor levels (including basement) by number in parenthesis.

ment budget should be based on a five-year or ten-year plan of overall company action. The budget itself essentially reduces to dollars against a time schedule of anticipated plans for facilities alteration or expansion.

While all companies may not go so far as to follow the procedures set forth in

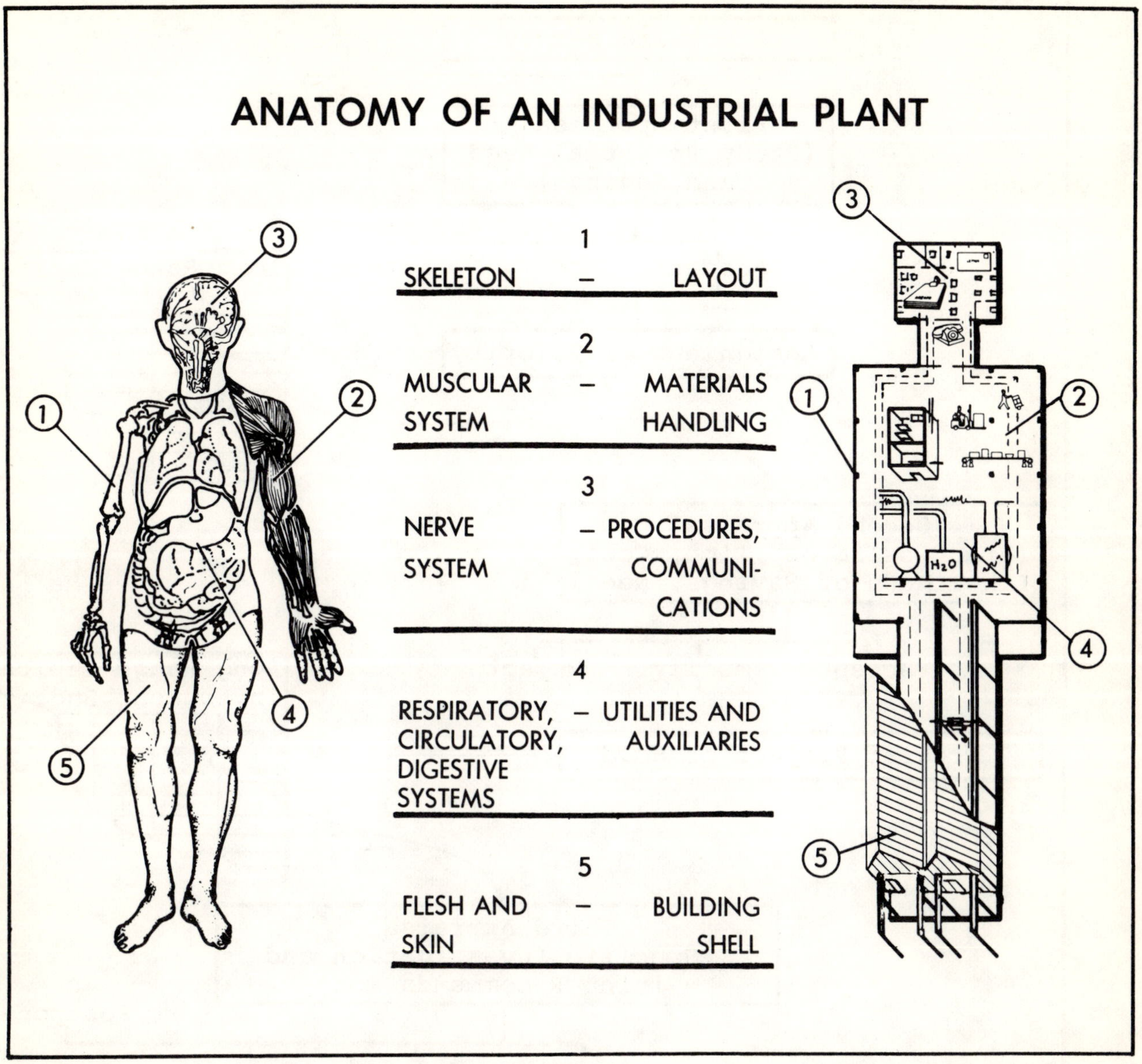


**Figure 15-3.** An industrial plant is a dynamic combination of physical systems. It may be likened to a living animal wherein each component system has its counterpart as a component of the industrial plant. The layout is only one of five components making the combination that constitutes an industrial plant.

Figure 15-4, some form of it is helpful—if not invaluable—to every organization expecting to grow or expand.

### Doing the Planning

Once a project is activated, make sure it is defined and identified. Every project should have a descriptive name that is generally understood. Each project should prob-

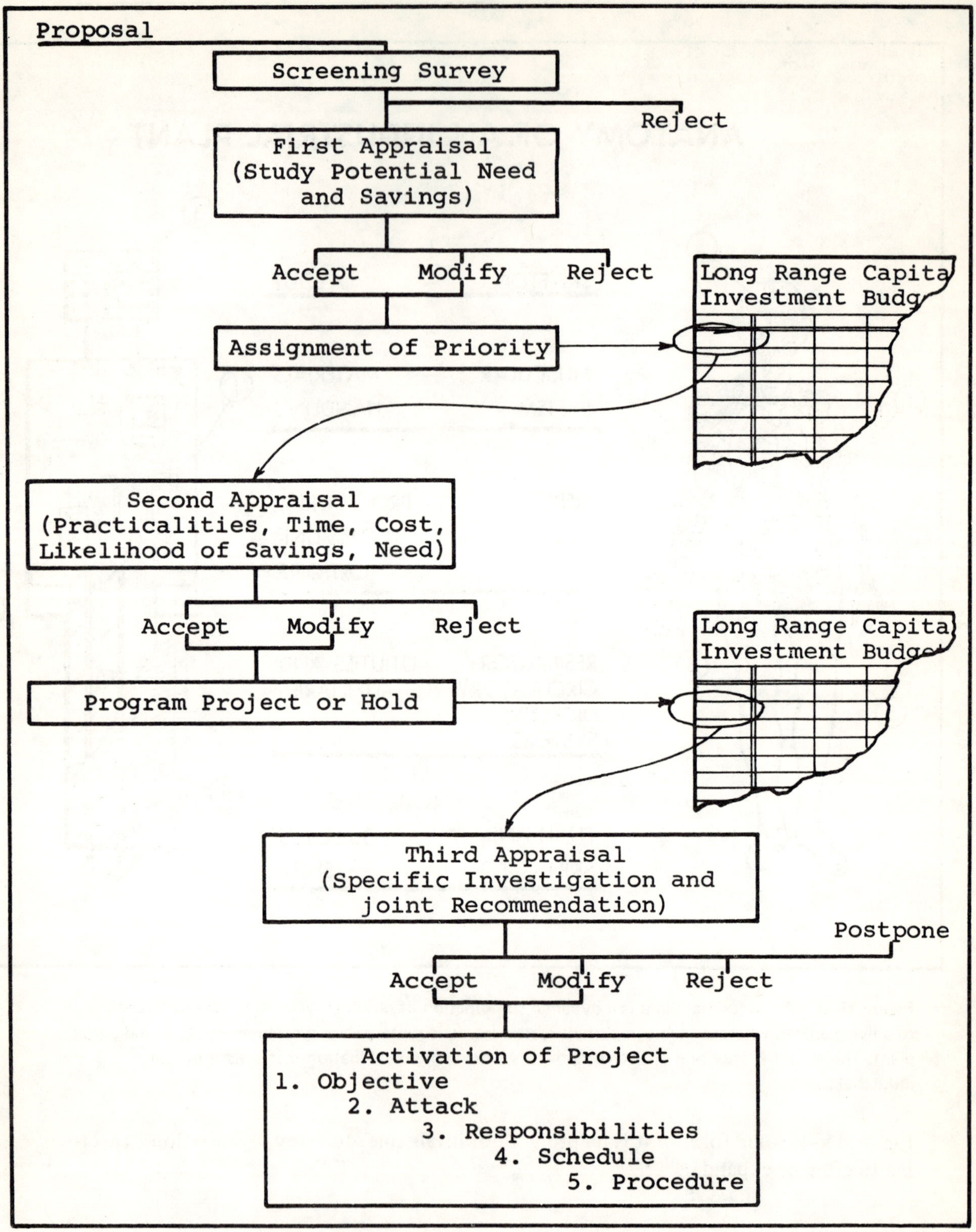


**Figure 15-4.** Screening and programing capital expenditures. Major layout projects—especially those involving new processes or new buildings—fall into the area of long range capital expenditure planning. They must first be worthwhile, then be compared with other requests for funds, and wait their turn for logical activation.

ably also have some identifying project number, where a series of projects are involved. Moreover, a clear statement of the objective of the project—*what* is to be done and *why*—should be set down in writing and agreed upon.

Next, make clear *who* is to handle the project. This means a clear assignment of responsibility for the project. And, it should involve an understanding or indication of what other groups or individuals the layout planner must coordinate with; what the limits of his project are; and where he overlaps with others who must supply him with information or to whom he will feed his results.

Every layout project should be scheduled. *When* the project should be done may be as significant as what is to be done and who is to do it. Scheduling alone can improve the performance of the layout planner, for most humans strive to meet the challenge of deadlines, especially ones they set or agree they can meet. Before proceeding with the project scheduling study, the reader should pause to examine Figure 15–5. It sets forth a procedure for starting any layout.

Probably the best way to schedule a project is to break it down into the four phases of layout planning; then subdivide:

Phase I into P, Q, R, S, T, and activities analysis
Phase II into steps (or Boxes) of the pattern
Phase III into the number of areas to be laid out in detail.

In this way, each project can be scheduled and its files kept according to this systematic breakdown.

There are no generally agreed upon time standards for layout planning. Under certain circumstances, an entire area can be laid out by one man in one day following the SLP methodology. On the other hand, under other conditions, a complete plant may involve several planners' working more than a year just to execute one phase. Still, by reasoning out how much data is available, the phases involved, the steps of the pattern, the detailed areas involved, and the complexity of the installation, a project can be planned and scheduled according to a logical and systematic sequence of events.

Moreover, scheduling this way keeps the project within the SLP framework and pattern. If the planner understands them and can make them work to his advantage, he can work faster and more soundly. If he has to stop and think about which to follow and what materials to use, then time is lost. In the final analysis, if we save time in planning, we can afford or justify having a better man do the planning.

A project scheduling sheet with an example broken down by SLP phases is shown in Figure 15–6.

Recognize, however, that sometimes it is expedient not to follow the framework too strictly. For example, in crash programs it may be advantageous to move ahead with detailed layout plans before the general overall layout or even the location is agreed to. This usually results in much redoing of the detailed planning and is, there-

PROCEDURE FOR STARTING ANY LAYOUT PROJECT WITH S L P

1. Establish in writing:
   a. The identification and description of the project.
   b. Its objective(s) and reasons why needed.
   c. Its breakdown in phases and appropriate steps.
   d. The name of the person responsible for the project.
   e. The time(s) when planning should be done and/or installation should be complete.

2. Determine where the project lies:
   a. In what phase?
   b. Are previous phases complete--fluid, fixed, approved?
   c. If not, how far along is each step in each phase?

3. Set a schedule for the project.
   a. The date for completion of each phase.
   b. The date for completing the appropriate steps.

4. Get, or verify, projected input data on P, Q, R, S, T.
   a. What is known?
   b. Is data approved by those in authority?
   c. If not, get basic input data.

5. Move into the appropriate step or box of S L P and follow forward from there.

**Figure 15-5.** Getting started on layout projects using SLP.

fore, more costly. Still, if getting the total project accomplished on time is the most important consideration for the company the extra cost is justified. By having more time to negotiate the site procurement or the lease arrangements, moreover, top management may have saved far more than the extra layout planning manhours cost.

Layout planners must realize that top management's common criticism of them is that the planning takes too long. Maybe management contributes to this. But planners too often drag out projects, meticulously seeking perfection. Most managers realize that engineers tend toward too much detail. Management may, therefore, do well to push projects along toward completion before all details are fully resolved.

On the other hand, a layout is not a good layout so far as the operating man who uses it is concerned, unless he believes it is good. Therefore, it is well to be sure that reasonable time is allowed to let operating people sell themselves or become convinced of the benefits of the new layout. Then, they will support it as a "good layout" for the company.

In this connection, a word should be added about the use of computers in layout planning. For processing data on P and Q, for extending flow-of-materials intensities, for inventory of machinery and equipment, for calculating machinery requirements, and the like, the computer can be of great value to the planner. Unfortunately, the computer does not qualify as a thinking device, and, as a result, computer programs for making layouts thus far have been largely an aid, rather than a reliable planning tool.[2] The use of a computer in parallel with mental/manual SLP is described in Appendix XIV.

### Recording Data

For the layout planner himself, certain practical routines can avoid confusion and save time. As a matter of habit, he should identify every data or analysis sheet with which he works. This identification should include:

1. Project name and/or identifying number
2. Name of person(s) preparing the data sheet
3. Date the sheet is begun
4. Sheet number, in sequence for each step of the analysis or study (is helpful but not mandatory)

It is quite common in layout planning to have one project set aside in favor of another of higher priority. By the time the planner returns to the first original, he may not be able to identify his own prior data. Or, if the planner is transferred or promoted whoever inherits the project may be at a complete loss to identify what has been done, or whether or not the sheets even apply to the particular project.

A project can be set back two or three years and the original planning completely wasted if it is not properly identified. One layout project which went as far as the 3-D review-check was completely abandoned when the layout planner was killed by an automobile. More than two and a half years of layout planning and seven thousand dollars worth of models were wasted. The job had to be completely started from scratch when it was reactivated. Trying to untangle and interpret such records and analyses as had been kept was not worth the time involved, for no identification or filing of the work sheets had been made.

### Balance of Elements

One important guide to bear in mind in working with the great variety of features and people likely to be involved in any layout is a proper *balance* of all elements.

---

2. See "Four Approaches to Computerized Layout Planning," by Richard Muther and Kenneth McPherson, in *Industrial Engineering,* February 1970.

**PROJECT SCHEDULE SHEET**

Covering (Plant) Louisville

As of ~~8 Dec.~~ 22 Dec.

Distribution: Ind'l Engng (3), Facilities, Ind'l Rel., Prodtn., R.T.Walker

Originating Dept. Industrial Engineering

Prepared by E.F.Coulter

Date 20 Nov. Sheet 1 of 3

Project No. 25731 Wax Paper Relayout

Week of: Nov. 20, 27; Dec. 4, 11, 18, 25; Jan. 8, 15, 22, 29; Feb. 5, 12, 19, 26; March 5, 12, 19, 26; April 2, 9; Further Schedule

| Proj. No. and/or Description | Work to do; Action to take | Total time |
|---|---|---|
| 1. | Establish location | 3 |
| 2. | P-Q Analysis | 2 |
| 3. | Flow of Material & Activity Relationship | 5 |
| 4. | Space Allocated | 5 |
| 5. | Alternative layouts | 10 |
| 6. | Evaluation and Selection | 5 |
| 7. | Approval of General Overall Layout | 3 |
| 8. | Detail Layout: Wax Machines | 8 |
| 9. | Wrapping Machines | 3 |
| 10. | Boxing Machines | 3 |
| 11. | Raw Paper Storage | 3 |
| 12. | Finished Product Stor. | 3 |
| 13. | Service Areas | 10 |
| 14. | Installation Cost Estimates | 5 |
| 15. | Approval of Detail Plans | 3 |
| 16. | Installation Drawings and Instructions | 15 |
| 17. | Move and Cleanup | 5 |
| 18. | | |

Gantt Chart Code: ┌ Date work scheduled to start ┐ Date work scheduled to finish ┌─┐ Total time scheduled for work ▬ Amount of work done

(Each vertical period represents one unit of time. Use Gantt Chart Code or enter numbers and/or other indicators.)

It is easy to be led astray by statements based on preconceived thinking or personal preference of individuals likely to be affected by the layout. Their quick opinions based on too sudden or too early solicitation of their thinking may turn out to become stubborn "facts," merely because they were expressed or stated earlier.

Moreover, a common mistake is to overcorrect short-comings or limitations in an existing layout, when planning a new one. For example, perhaps the present layout has no adequate provision for maintenance personnel. This has become a major bone of contention for an outspoken maintenance supervisor. So in the new layout the planner goes overboard to provide access for maintenance personnel—at the same time perhaps wasting floor space and adding to housekeeping problems. The latter problems are taken for granted and not emphasized in the new layout planning because they were satisfactory in the original layout. Therefore, perhaps one of the most difficult mental adjustments for the layout planner is to maintain a realistic balance of all elements and factors affecting the layout plans.

Seeing the layout project in its entirety, which can be done with the SLP framework of phases and pattern of procedures, helps the planner give balanced attention to various elements. It helps him get the total picture—and get it with a depth of perspective—for he can readily see just how any project fits into the larger framework and into the pattern of attack.

Similarly, middle and top managers should be cautioned about blaming "poor layout" for failure to produce competitively. Changes in layout can make dramatic savings. But frequently, the basic cause of high costs lies hidden in a poor production control system or poor services. Moreover, poor layout can be an overworked excuse by some weak managers. There is no sense in making a new layout, or relayout, or being concerned with how best to make it, if it is not necessary. We should look for other ways to "skin the cat" besides simply "planning a new layout."

**Figure 15-6.** The Project Schedule Sheet is especially effective for scheduling various steps of any layout project or a number of projects. The Gantt chart brackets show the span of calendar time alloted for each step. The number above the bracket indicates the number of planning man-days estimated to accomplish that step. Note that line 1 represents Phase I; lines 2 through 7 cover Phase II . . .

Periodically, as work is completed on each step scheduled, a heavier accomplishment line is filled in between the appropriate brackets. This is drawn from the left-hand bracket towards the right, thus showing the amount of work completed and still to be done.

As of 22 December, this project is running ahead of schedule except on lines 9, 10 and 11. A record can also be kept on the chart of the actual number of planner days spent. Thus, one sheet is used for both project scheduling and performance reporting.

The form can also be used to indicate plan-and-performance-against-plan for several projects; another way, to indicate the assignment of engineers—or the loading of engineering staff—by entering in each vertical column the number of days (or hours) allocated each week or two-week period.

**COORDINATION AND PROGRESS SUMMARY**

Covering (Plant) South #4 As of 6/1

Distribution: Groner, Angelo, Hanson, Tucker(2), File

Originating Dept. Methods & Standards

Prepared by P.R.Slocum PRS

Date 5/1 Sheet 1 of 1

| | Project No. and/or Description — Work to be done; action to be taken | Dept. Acting | Code | Status Remarks |
|---|---|---|---|---|
| 1. | M-121 Bracket Line Rebalancing | Methods | 1 | Delay due to flooding of area. |
| 2. | M-126 Hermetic Winding & Assembly | Plant Enginrg. | 4 | Maintenance still waiting on shaping press. Need drinking fountain. |
| 3. | M-128 Consolidate Core Iron Shearing | Plant Enginrg. | 3 | Bridge crane electrification complete. On schedule. |
| 4. | M-129 Relayout Large Motor Department | Methods | 1 | |
| 5. | M-133 Relayout Auto.Screw Mach.Dept. | Methods | 0 | Scheduled to begin planning August 1. |
| 6. | M-134 Air Brake Assembly and Storage | Plant Enginrg. | 3 | Need new test benches on valve line to replace obsolete temporary benches. |
| 7. | M-137 W-1996 Washer from D-46 to D-40 | Plant Enginrg. | 3 | Equipment installed. Adjustment still required. |
| 8. | M-138 Move W-6922 Press | Person'l | 4 | Installation complete--waiting for approval from safety engineer. |
| 9. | | | | |
| 10. | | | | |
| 11. | | | | |
| 12. | | | | |
| 13. | | | | |
| 14. | | | | |
| 15. | | | | |
| 16. | | | | |
| 17. | | | | |
| 18. | | | | |

Status Code; 0, not yet begun; 1, Preliminaries underway; 2, Prelim. complete & work in process; 3, well along; 4, almost done; Compl.,complete.

RICHARD MUTHER & ASSOCIATES – 710

### Coordination of Projects

Every layout project involves coordination with other departments and people. In a small, one-project situation, it may be possible to keep these relationships in mind. But, as a general rule, it is best to get these relationships down on a coordinating sheet of some kind. This not only gives formal recognition to allied responsibilities, but also allows an organized method of reporting status. A self-policing type of form like that shown in Figure 15-7 can be of significant help in coordinating layout planning and installation when several groups are involved—as they practically always are.

### Conclusion

In speaking of performance and accomplishment, Frederick W. Taylor, the recognized early leader of scientific management, has said "The greatest production results when each person is given a definite task to be done in a definite time and in a definite manner." Modern management theory would alter the statement to read ". . . is given and accepts a definite . . . " or "is committed to . . . "

If his statement is broken down and applied to managing layout projects, a relationship similar to that in Figure 15-8 is derived. By clearly specifying what is to

| "Greatest production results when.... | Key Word | Application to Layout Planning Project |
|---|---|---|
| "each PERSON is given a | Who | Layout Planner |
| definite TASK to be done in a | What | Specific Project at Hand |
| definite TIME and in a | When | Project Schedule |
| definite MANNER." * | How | Systematic Layout Planning |
| *Frederick W. Taylor | | |

**Figure 15-8.** Effective performance in executing of layout projects reduced to fundamentals of production output.

**Figure 15-7.** Keeping abreast of facilities installation. A periodic progress record informs the various interested parties of the progress being made on each layout installation project. Each project is listed as an aid in keeping the layout planners and the layout installers (in this case Methods & Standards and Plant Engineering) moving along together. Projects falling behind can be highlighted; the cause of delay and the party responsible for corrective action can be identified and reported.

be done, who is to do it, when it is to be done, by getting acceptance of the program for accomplishment, and by following an organized, systematic method of planning layouts, productivity of planning must improve.

I encourage you to organize your activities and conduct your projects with a clear-cut *what, who,* and *when*. I trust that Systematic Layout Planning can give you the *how*.

# Appendixes

## Appendix I—Description and Tables for Mag Count

Various units are used in analyzing the intensity, or capacity, of materials movements. In a blast furnace operation, tons are commonly used to measure the amount of materials involved; in an oil refinery, barrels or gallons suffice; in a flour mill or grain-handling operation, tons again is adequate. Even in a warehouse of mixed items, truckloads or palletloads may be adequate. The ton-mile, kilogram-meter, and the palletload-foot are then used to express the total work done in a given period of time—that is, the intensity of movement × the distances moved. And, of course, the actual time per trip or per distance traveled is highly important in the end result.

However, in diversified manufacture where a considerable variety of materials is involved, neither weight nor volume nor piece count alone truly measures the "transportability" of the material on a comparative basis. And, for overall layout planning—before any layout has been established and before the handling methods have been determined—there is no common container or unit load that can be used as a common denominator. In these cases, the so-called "Mag" is a unit of real assistance.

"Mag" is an abbreviation for magnitude. It is a unit created to measure the transportability of materials. It measures the material, product, or material load for its difficulty-of-transport characteristics, largely independent of how it is to be moved or transported. When the mag count of any item is multiplied by the quantity of pieces moved per unit time, the intensity of materials movement, or intensity of flow, is established.

Obviously the mag concept has limitations, for the transportability of the piece or item itself may depend in part on the method of handling used, both in container and in handling equipment. Still, before these are known—as is true in practically all cases of theoretical analysis of flow—the mag is perhaps the best way to measure and compare the relative difficulty of the materials handling problem in a series of transportations involving highly diversified items.

### Factors Affecting Transportability

There are many factors that can affect the ease or difficulty of transporting or handling materials. Basically, they can all be reduced to the following:

A. *Size* of the item
B. *Density* or bulkiness of the item*
C. *Shape* of the item
D. *Risk* of damage to the material, facilities, and employees
E. *Condition* of the item
F. *Value* or cost of the item (included only in certain cases, and usually included in factor E)

*The weight of the item is not included because, for any given material, weight is related to size. Density or bulkiness is therefore substituted for weight.

Because of these many factors, it is difficult to measure and compare the transport or handling effort for different materials. However, by classifying and rating these factors in a measurement system, the relative ease or difficulty of material transport can be quantified.

### Definition of Mag

The Mag Count *establishes a base value for the size of an item and then reduces or increases this base value by means of modifying values for the other influencing factors.* Thus, the mag count of the item is derived. The magnitude of this end value is a measure of the transportability of the item in question.

By definition, one mag equals a piece of material that

a. can be held conveniently in one hand;
b. is reasonably solid;
c. is compact in shape and has some stacking qualities;
d. is slightly susceptible to damage; and
e. is reasonably clean, firm, and stable.

A typical example of one mag is a block of dry wood measured and roughly cut to 10 cubic inches in size—or a shade larger than two inches by two inches by two inches. Thus, if ten pieces of an item could be held conveniently in one hand, the item would have a value of approximately one tenth mag. One pack of twenty cigarettes is one half mag; a piece requiring two hands cupped together to hold it conveniently has a value of approximately two mags. Starting with this crude, simple scale, a range of base values for typical benchmark items can be set up.

Note that the definition of base value in terms of what can be held in one hand does not mean that the transport will be made by hand. The handful measure is merely convenient and has nothing to do with the materials-handling method or equipment. Also, it does not refer to the number that can be picked up by one hand, but to the number that can be held in the hand on a non-prearranged or nonstacked basis.

The mag unit is used instead of actual cubic inches or cubic feet for several reasons.

1. There is a difference in units between the English and metric systems.
2. The "volume" is difficult to measure for very flat sheets or very long items.
3. There is not a linear relation between transportability—or handle-ability—and volume, so that a conversion or other calibration would be involved.
4. The use of actual cubic measurement would imply more accuracy in numbers than the precision to which items could be practically measured.

### The Base Factor-Size

Figure 1 gives the benchmark values for the size of an item. Notice that the base value is not directly proportional to the volume or cubic size of the item.

Figure 2 shows typical items and their respective base values. In practice, any company can list its own typical and readily recognized items.

### The Modifying Factors

Figure 3 gives the recommended modifying values. These factors are expressed in degrees, one degree equal to 25 per cent increase or reduction in the base value.

In terms of a formula, the mag count of any item equals:

$$A + [\tfrac{1}{4} \times A(B + C + D + E + F)]$$

or

$$A\,[1 + \frac{(B + C + D + E + F)]}{4}$$

where A is in mags and B, C, D, E, and F are number of degrees positive or negative

Example:

A four-drawer file cabinet, formed and welded, but unpainted and without drawers has a mag count as follows:

| | | | |
|---|---|---|---|
| A | 30 base value | D | 0 |
| B | –2 degrees | E | 0 |
| C | +1 degree | F | 0 |

Total: $30 + [\tfrac{1}{4} \times 30(-1)] = 30 + [\tfrac{1}{4}(-30)] = 30 - 7\tfrac{1}{2} = 22\tfrac{1}{2}$, or

$30\,[1 + (-\tfrac{1}{4})] = 30[\tfrac{3}{4}] = 22\tfrac{1}{2}$

This would probably be rounded off to 23 in practice.

### Basis and Accuracy

Mag Count is a simple procedure. With a little practice, one can make the calculations in his head rather quickly. Some modifying factors do not apply at all for many items.

First-time users of this measure of transportability must note that the mag

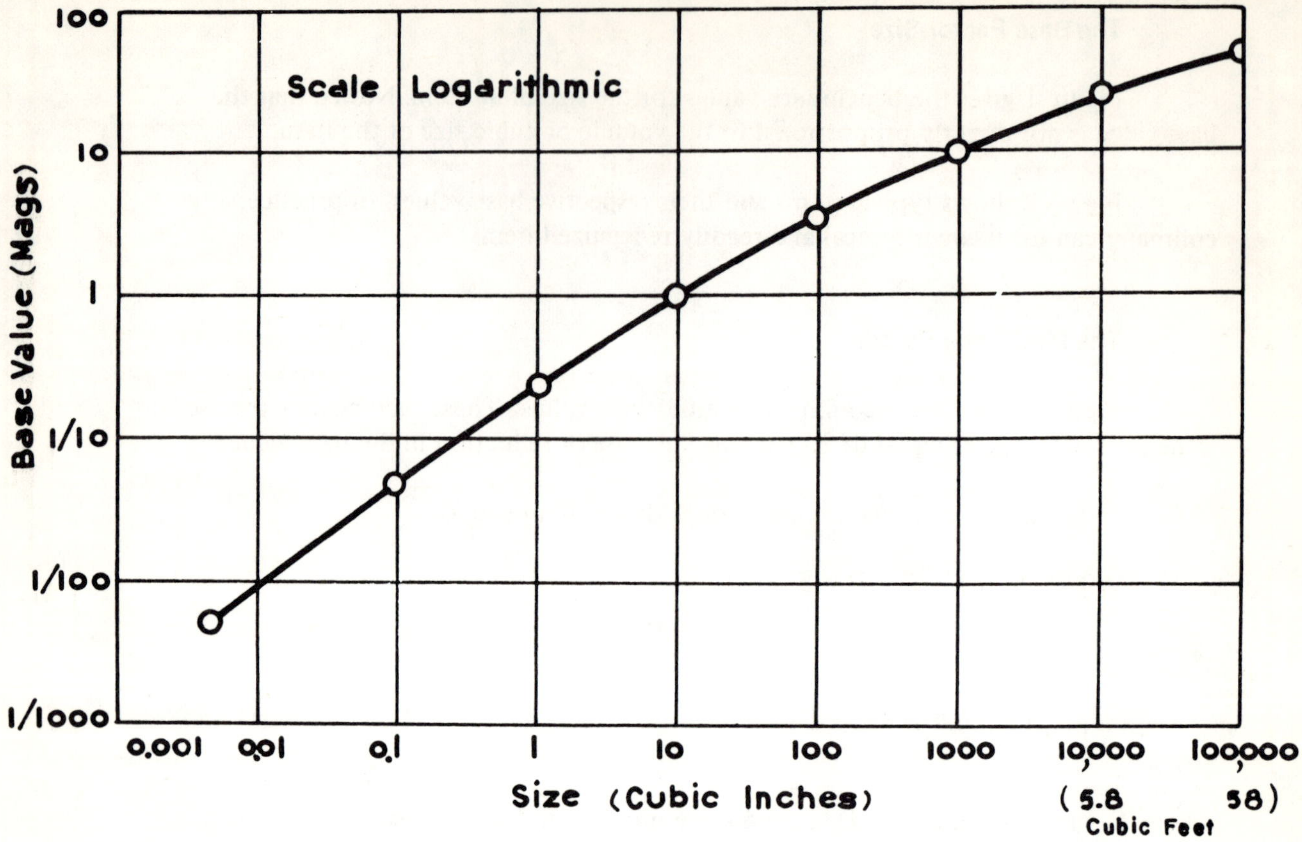


Curve showing relationship between size (volume) and base value. The base value is not directly proportional to the size; rather, it becomes relatively easier to transport an object per unit of its volume as it becomes larger. When measuring the volume, use outside dimensions and do not subtract for inside cavities or irregular contours.

| Benchmark | Volume | Mag Count |
|---|---|---|
| | 0.005 cubic inch | 1/200 |
| | 0.1 cubic inch | 1/20 |
| | 1.0 cubic inch | 1/4 |
| | 10 cubic inches (approximately 150 cubic centimeters, metric) | 1 Mag |
| | 100 cubic inches | 3½ |
| | 1000 cubic inches | 10 |
| | 10,000 cubic inches | 25 |
| | 100,000 cubic inches | 50 |

**Figure 1.** Base Value for Size-Factor A

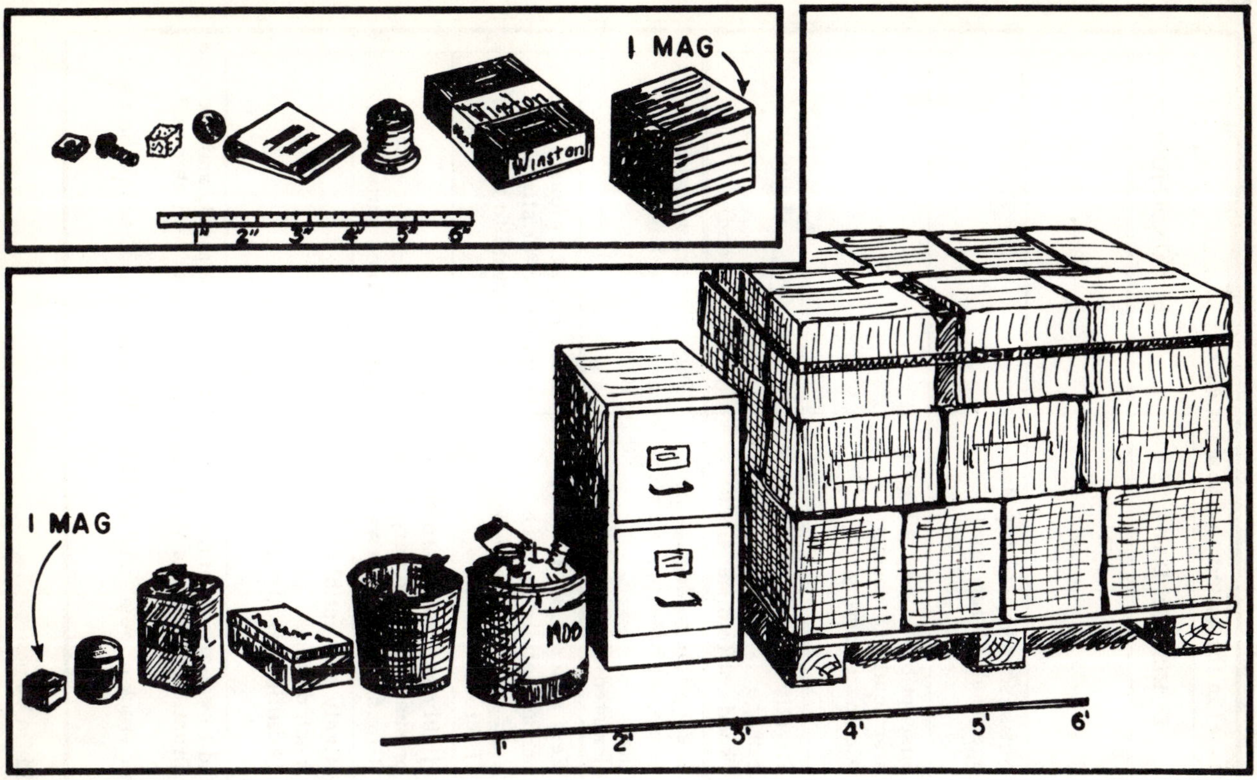


| Examples (*Items shown ) | Base Value |
|---|---|
| Cotter pins 1/4" long | 1/200 |
| Washer 1/4" diam. hole | 1/150 |
| *Nut 7/16" x 3/16" | 1/50 |
| *Stove bolt 1/2" x 1/4" | 1/50 |
| *Sugar cube, small | 1/16 |
| *Glass marble | 1/16 |
| *Penny book of matches | 1/6 |
| *Household fuse | 1/4 |
| *Pack of cigarettes | 1/2 |
| *Wood block approx. 2" x 2" x 2" | 1 |
| *Quart paint can | 3 |
| *Gallon container (incl. handle) | 5 |
| *Retail shoe box | 6 |
| *Household pail | 9 |
| Desk-type letter size file | 11 |
| *Five-gallon gasoline can | 12 |
| File drawer, filled | 18 |
| *Two-drawer file cabinet | 24 |
| 4' x 4' standard wood pallet | 28 |
| Four-drawer file cabinet | 30 |
| Household bath tub | 36 |
| Pallet load (4' x 4'—low stack) | 38 |
| *Pallet load (4' x 4'—medium stack) | 44 |
| Pallet load (4' x 4'—high stack) | 48 |
| Telephone booth | 50 |

**Figure 2.** Examples of Size Values—Factor A

Each "degree" is equivalent to 25% increase (or decrease) in the base value (Factor A)

| DEGREE | B. BULKINESS OR DENSITY | C. SHAPE* | D. RISK OF DAMAGE | E. CONDITION |
|---|---|---|---|---|
| -3 | --- | Very flat and stackable or fully nestable* (Flat sheet of paper or metal) | --- | --- |
| -2 | Very light and empty (Bulky sheet metal) | Readily stackable or nestable* (Pad of paper, soup bowl) | Not susceptible to any damage at all (Scrap iron) | --- |
| -1 | Light and bulky (Knocked-down corrugated carton) | Fairly stackable* or slightly nestable (Book, tea cup) | Susceptible to negligible or practically no damage (Compact casting) | --- |
| 0 | Reasonably solid (Block of dry wood) | Basically square with some stacking qualities (Block of wood) | Slightly susceptible to some damage (Wood cut to dimensions) | Clean, firm, and stable (Block of wood) |
| +1 | Fairly heavy and dense (Cored casting) | Long, rounded, or somewhat irregular (Sack of grain, short bar) | Susceptible to damage by crushing, breaking, scratching (Painted items) | Oily, flimsy, unstable or awkward to handle (Oily chips) |
| +2 | Heavy and dense (Solid casting, forging) | Very long, spherical, or irregular (Desk-model telephone) | Very susceptible to some or susceptible to much damage (TV tube) | Covered with grease, hot, very flimsy or slippery, very awkward to handle |
| +3 | Very heavy and dense (Die block, solid lead) | Extra long, curved, or highly irregular (Length of steel beam) | Highly susceptible to some or susceptible to very much (Glass crystal stemware) | (Tacky glued surface) |
| +4 | --- | Extra long and curved, or especially irregular (Formed tubing, wooden armchair) | Highly susceptible to very much damage (Acid in glass, explosives) | (Molten steel) |

F. VALUE (or cost) of each item is not rated here because frequently it does not lead to variations in transportability within a plant, and because care in handling is already incorporated in the "risk of damage" factor. But if your situation requires a Value modifier, set your own zero and scale.

* When FLAT or NESTABLE items are commonly handled in a stack or nest, this unit stack or nest, and not the individual piece, should be used when measuring all six factors.

**Figure 3.** Modifying Factors and Add-On Values

count for each item may vary from operation to operation—as when a piece of sheet metal is formed, a product is painted, or a package is filled. Note also that when one item is assembled to another, its mag count ends and is included thereafter in the mag count of the assembled article.

Actually, the accuracy of this system is no closer than a range of 25 percent, so there is little reason not to round off the values. A half degree can be used for any of the modifying values if wanted, but this implies a greater precision in the modifying values than perhaps the base size value of the item itself involved.

Moreover, it is possible (though it seldom occurs) to calculate a negative mag count, which is realistically impossible. In this case, the lower limit is arbitrarily set as 25% of the A value.

In this same connection, it should be recognized that the mag count system is itself not based on scientific research or mathematical analysis. Rather, it was developed by practical materials handling and plant layout engineers, representing three different countries, to meet a specific need on a particular project. Later, it was further reasoned through and refined.

One application of mag count is explained in Figure 4. It is typical of cases where there are diversified products or materials which have to be moved—especially when significant changes in the product and, therefore, in its transportability, take place during the process. This is where mag count has its greatest use: in diversified fabrication and assembly, where no universal handling equipment and containers are employed. Mag Count fills the need for a common unit of measuring the transport characteristics of materials and of quantifying them in a realistic way.

### Adaptation and Extension of Mag Count Concept

The Mag Count is based on a unit of such size that it can be conveniently held in one hand. There are many industrial situations, however, in which the physical sizes of the objects being moved are appreciably larger. Palletized materials, large tote boxes, containers, and crated goods are typical. In projects where these relatively larger loads predominate, it is often practical and convenient, when making an analysis of materials flow, to select, for Factor A (size), a larger "base unit," one closer in size to that of the objects being handled. The analyst then has a more meaningful yardstick for measuring the comparative size of his materials. Moreover, he can explain his findings to other people associated with the project in terms they can more readily understand.

### Macromag

The Mag Count concept lends itself readily to such adaptations. Indeed it may be less important to have a definite base value than to grasp the very real significance of (a) the shape of the curve of Factor A (size) and (b) the impact or influence of the other factors (density or bulkiness, shape, risk of damage, and condition—the latter

One application involved a firm producing kitchen, restaurant, and hospital utensils. The fabricating shops had to be relocated. In planning the new layout, engineers found that most of the diverse products followed different routings or paths of flow. Therefore, process groupings of equipment were in order and the layout was planned on a basis of the intensity of flow between each pair of process departments.

A From–To Chart (Cross Chart) was made and postings for each transportation were entered thereon. Here is how the engineers determined the intensity of movements between the various departments.

Item: Nine-inch diameter metal mixing bowl.

Routing:
- Shear
- Press
- Trim

Plan–for Quantity:
10,000/year

Mag Count for move "From Press to Trim"
- A = 4 Mags
- B = –2 degrees
- C = –1 degree
- D = 0
- E = +1 degree
- F = –––

Net Modifier = –2 degrees

Mag Count = 4 + [¼ x 4 (–2)]
= 4 – 2 = 2 mags/piece

Intensity of movement
= 2 x 10,000 pcs./year
= 20,000 mags/year

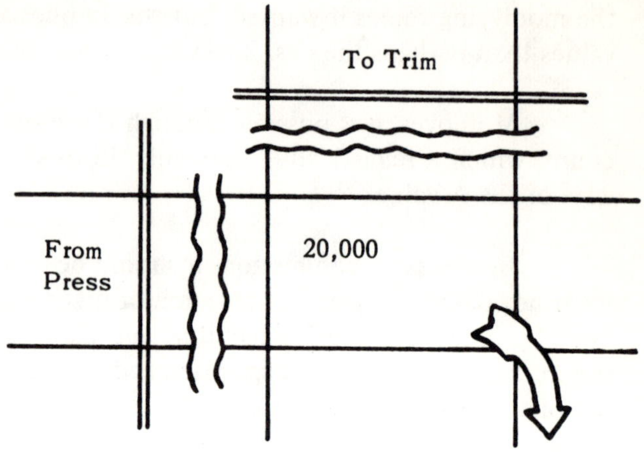


Mag Count times plan-for quantity was entered on From-To Chart in its respective "box" as a measure of the movement between the two departments, so far as that item was involved.

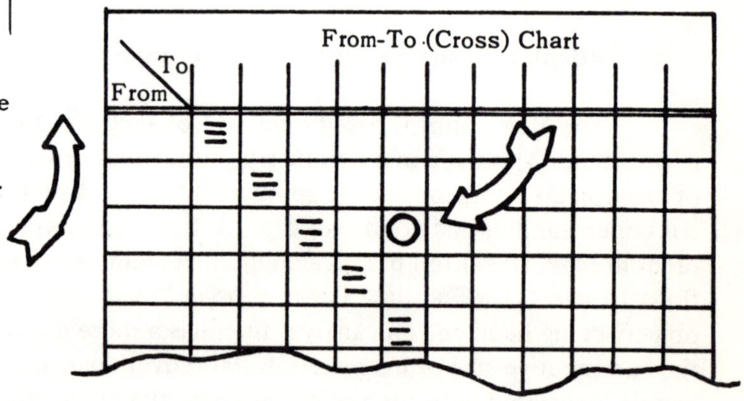


The total entered amounts were then posted to a final From-To Chart, representing the total intensity of flow between all departments for all items.

**Figure 4.** Application of Mag Count

including the value factor). It is logical, therefore, for planners and analysts to use their standard shop boxes or industrywide containers as base values, and then proceed to apply the other Mag Count factors. This leads to intensity-of-flow (magnitude per period of time) measurements in terms of "equivalent shop boxes per day," "equivalent palletloads per month," and the like.

A typical adaptation is to convert to a cubic meter as the base size unit. Experienced planners have used the term *Macromag* when measuring transportabilities based on this size unit.

A *Macromag* is thus equivalent to a cube of wood one meter on each side. In English units, this approximates a block 40 x 40 x 40 inches in size. Such a block approximates 44 *Mags*. If the Macromag is then called "1" on the size scale, the other factors can be applied to it just as they would when the Mag is used as the base value.

### Application of Macromag to an Actual Industrial Project

The planning of a new factory layout for a manufacturer of steel office furniture called for an analysis of the proposed materials flow in the new plant. The relatively large size of many of the parts to be manufactured led to the selection of the Macromag (one cubic meter) as the base value for unit size. Size measurements of materials and containers used throughout the proposed plant were expressed in the metric units of cubic decimeters (1 cubic meter = 1000 cubic decimeters) and converted via equation to Macromags. The equation relating base value of transportability in Macromags and size in cubic decimeters has been determined empirically to be:

$$\log_{10}(\text{Macromags}) = -0.03553\,(\log_{10}\text{dm}^3)^2 + 0.51104\,(\log_{10}\text{dm}^3) - 1.21336$$

With this equation, a table of secondary benchmarks was established. These were plotted in the form of a curve, as shown in Figure 5.

In this project, as with most similar applications, the physical properties of the Macromag unit other than its size (i.e. density, shape, risk of damage, condition and/or value) were taken to be basically the same as the original wooden Mag block. This had numerous advantages. One cubic meter of very heavy wood with a specific gravity of 1 weighs 1 metric ton (1000 kilograms), which is the order of magnitude of many loads commonly encountered in materials handling. Thus the Macromag has both size and weight characteristics which are quite typical for industry.

In addition, use of a base unit whose other physical characteristics were the same as those of the Mag block, permitted the Table of Modifying Factors to be used unchanged.

These facts made it possible to draw up a direct-reading table (Figure 6) to help the analyst quantify each of his typical existing transport units (containers).

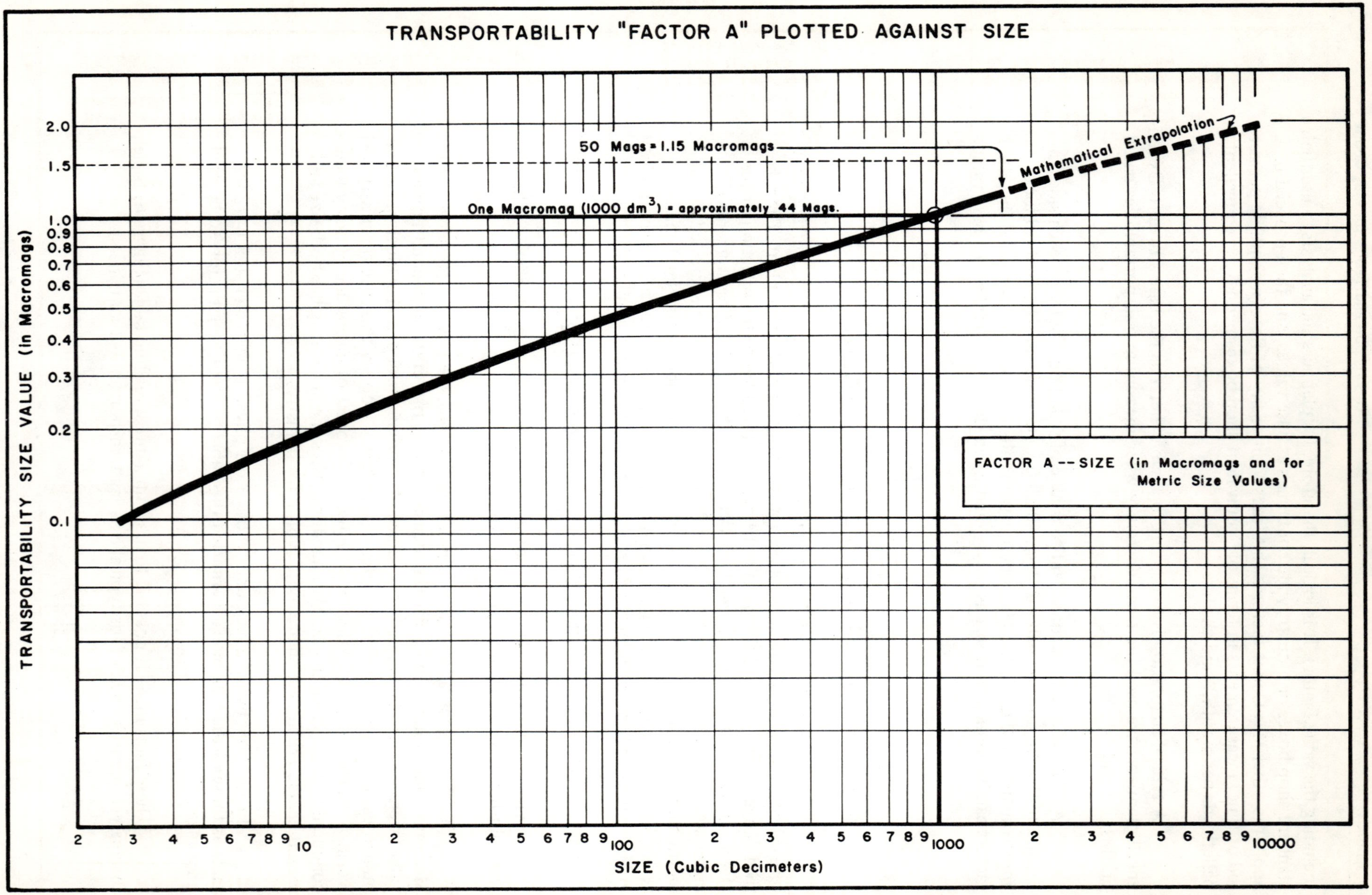


**Figure 5.** Transportability "Factor A" Plotted Against Size

# DIRECT - READING CHART FOR SIZE FACTOR IN MACROMAG UNITS

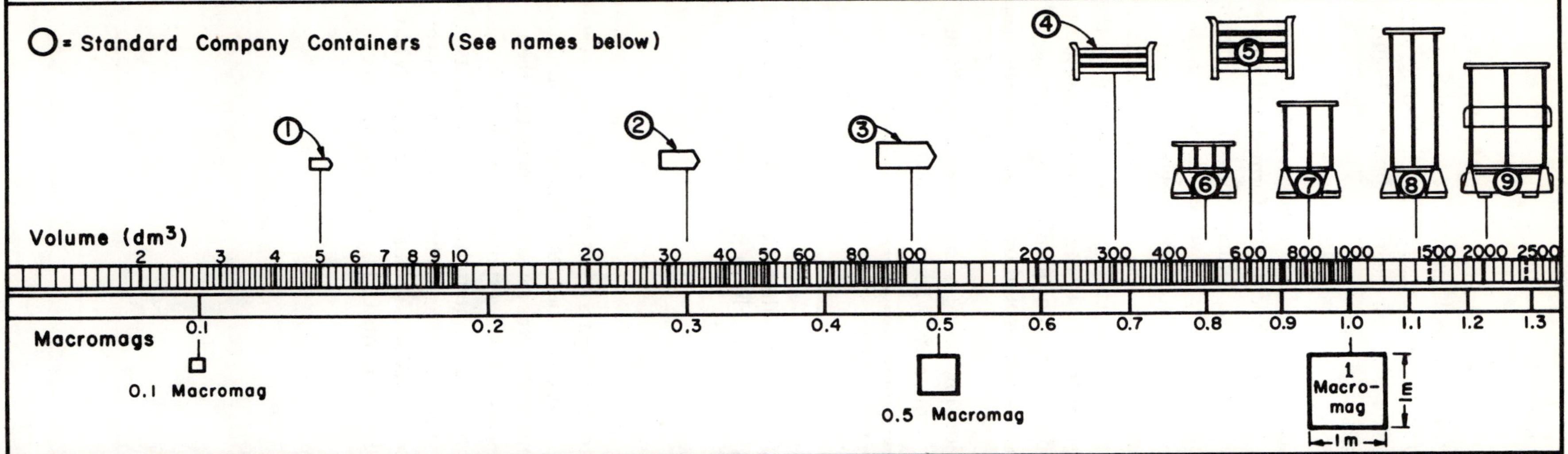


## Modifying Factors

B. Bulkiness or Density

C. Shape

D. Risk of Damage

E. Condition (including value)

The Macromag values for the Base Size Factor A, derived from the above chart, are modified to incorporate the influence of Factors B through E at left. Use the same modifiers and add-on values tabulated in Figure 3 of Appendix I.

## Calculation Example

Coil of sheet steel

O.D. = 1.00m = 10.0 dm (radius = 5.0 dm)

Width = 1.25m = 12.5 dm

Volume = $\pi \times (5)^2 \times 12.5 = 982\ dm^3$

| | |
|---|---|
| Size | Factor A = approximately 1.0 Macromag |
| Density | Factor B = +2 degrees |
| Shape | Factor C = +1 degree |
| Risk | Factor D = ±0 degree |
| Condition | Factor E = ±0 degree |
| | Total = +3 degrees = +75 percent |

Transportability Value = 1.0 x 175% = 1.75 Macromags

## Further Typical Examples of Modifying Factors

B. Bulkiness or Density
- -2 degrees - Empty standard desk body
- -1 " - Knocked-down cartons
- 0 " - Wood spacers
- +1 " - Pallet of Part No. 72691
- +2 " - Coil of steel
- +3 " - Stack of flat steel sheets

C. Shape
- -3 degrees - Desk covers (stacked)
- -2 " - Waste baskets (unpacked)
- -1 " - Drawer front panel
- 0 " - Two-drawer file cabinet
- +1 " - Coil of steel
- +2 " - Bracing rods
- +3 " - Lengths of angles & shapes
- +4 " - Executive chair (unpacked)

D. Risk of Damage
- -2 degrees - Scrap from presses
- -1 " - Heavy steel plate
- 0 " - Unpainted drawer panel
- +1 " - Painted drawer panel
- +2 " - Plate glass inserts

E. Condition (Ignore Value)
- 0 - Normal finished part
- +1 - Oily steel sheet
- +2 - Freshly pickled sheet
- +3 - Freshly glue-coated desk top

(Standard Company Containers (as sketched above)

1. Small Rackable pan
2. Assembly bench stack pan
3. Large tote pan
4. Low skid box
5. High skid box
6. Low carrier frame
7. Medium carrier frame
8. Extra high carrier frame
9. Large carrier frame

Figure 6.

### Visualizing Flow in Mag Count Measurement

Besides helping to construct a realistic layout, the concept of Mag Count can be used in checking or explaining the flow plan. For example, when the layout planning for the office-furniture factory was completed, the materials flow in Macromags was superimposed on the proposed layout plan. The resulting diagram, shown in Figure 7, exhibits the following interesting characteristics:

1. As the sheet steel passes through the plant, its form changes progressively from a simple compact steel coil, to relatively complex formed and painted furniture components. During this process, the tonnage moved from department to department does not change markedly. However, the size, shape, risk of damage, and condition are altered as each item progresses through the production process. The progressive changes in the corresponding materials flow intensities, expressed in Macromags and shown on the diagram by the width of the flow line, reflect the real changes in the amount of transport effort required to handle the product. If tons or numbers of pieces were the basis for measuring the flow, these changes would be revealed only in a limited way, if at all.
2. Note the large increase in the major flow lines (as the parts emerge from painting) and the reduction in flow intensity when the drawers are assembled into the main desk body and the combined assembly is packed (thereby reducing the "risk of damage" factor of its transportability).

The above application of the Macromag and its measurement of "transportability," demonstrates how this unusual yardstick can provide a more realistic picture of the transport effort than calculations based on size, weight or piece count alone.

### Modification for Weight-Based Items

Often a company has good data as to the weights of its materials, but because of the various sizes, shapes, and conditions of its end products, it can well afford to use mag count measurements. A rubber company that mixes its raw material but then molds or extrudes endless varieties of shapes and sizes is a typical example. It keeps its mixing records by weight. Often it is easier to weigh an item than to establish its cubic size. In such cases, the analyst should start with weight as his base value and convert via density estimates (representing an add-on or deduct for bulkiness or density) to reach a modified Mag (or Macromag)—to which Factors C, D, and E can then be applied.

MATERIAL FLOW INTENSITY (IN MACROMAGS)
SUPERIMPOSED ON PROPOSED LAYOUT PLAN

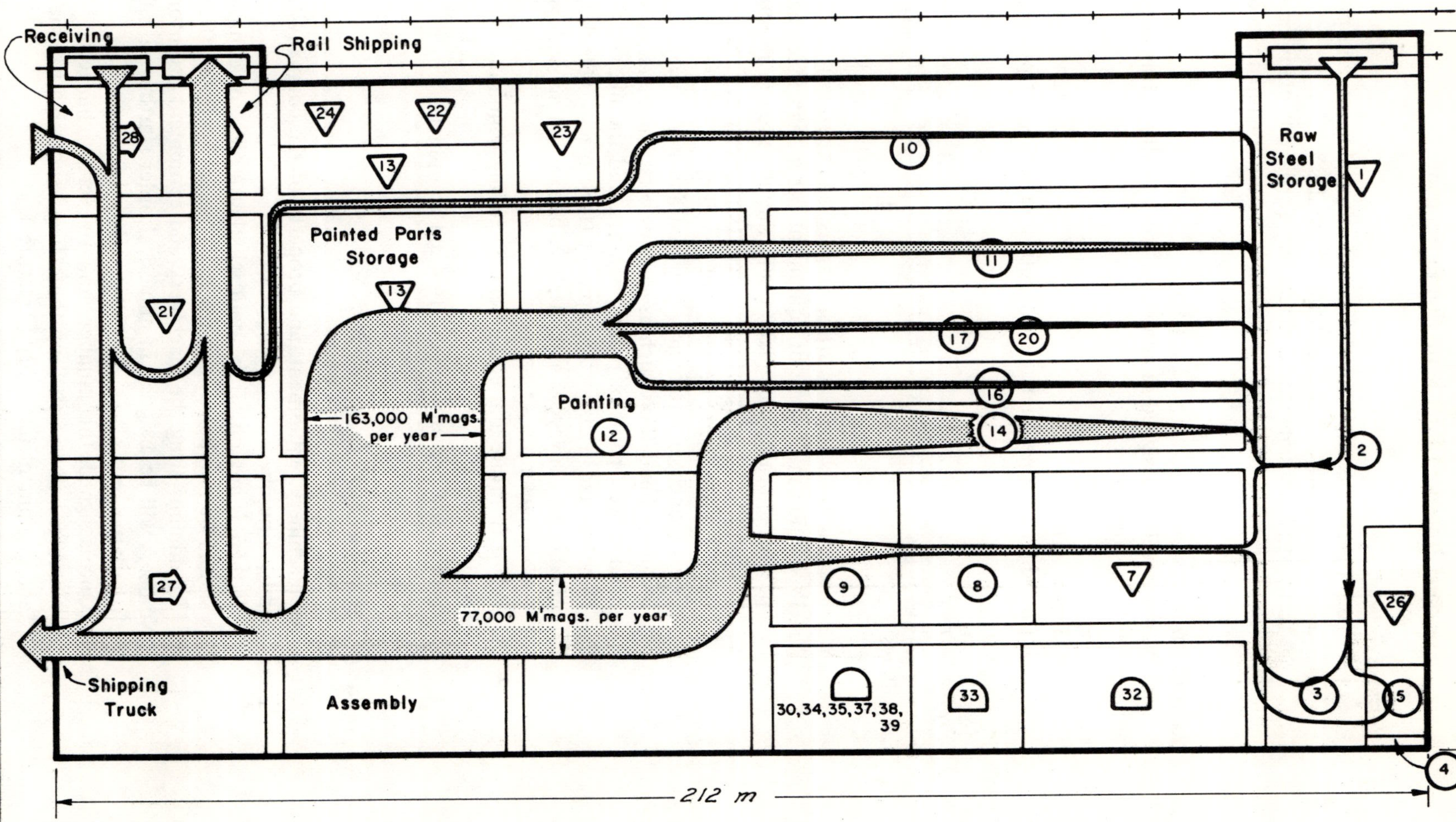


Figure 7.

## Appendix II—Identifying Machinery, Equipment, and Property

### Standard Procedure for Identifying Equipment

*Purpose*

1. To properly identify capital equipment, machinery, and property for asset accounting and depreciation schedules.
2. To identify equipment and machinery for layout planning and installation (or rearrangement), and for maintenance and/or repair records.

*Identification System*

Numbers will be assigned in numerical order to all existing and new machinery and equipment as follows:

1. Prefix letter
2. Four-digit number (starting with 0001 regardless of prefix letter)

*Prefix Letters*

Selected prefix letters indicate the general type of machinery and equipment as follows:

M – Machine tools (lathes, drills, shapers, and the like)
P – Production machinery or equipment other than machine tools (welders, presses, anneal ovens, platers, and the like used in production)
E – Equipment for supporting services (compressors, fans, lockers, tool cabinets, transformers, heaters, and the like)
T – Tools (bench, maintenance, powered and non-powered types, but not including hand tools such as screwdrivers, hammers, or wrenches, nor hand gauges, paint or plating hangers, and the like)
H – Handling and transport equipment (fork trucks, vehicles, major containers, hoists, conveyors and the like, but excluding skids, pallets, and general containers)
O – Office furniture and equipment, including communications and filing equipment
S – Safety, fire prevention, and first aid equipment

*Number (Tag or Machinery Inventory Number)*

1. All machinery, equipment, and property of the company must be identified by a number.
   a. If there is any question about whether a piece of equipment or a tool should be numbered, consult the Treasurer, whose office is responsible for administering the machinery and equipment inventory.
2. Numbers will be assigned in numerical order starting with 0001, regardless of the letter prefix.
   a. Once assigned, this number should never be changed.
   b. Numbers will be issued by the office responsible for the machinery and equipment inventory (Treasurer).

*Tags*

1. Brass tags with the Company Name, Letter Prefix, and the Numerical Inventory Number should be attached to each piece of machinery and equipment that is company owned.
2. The Maintenance Department should affix this tag—in a conspicuous place for all shop machinery and equipment; not too prominent a place for office equipment.
   a. Example:

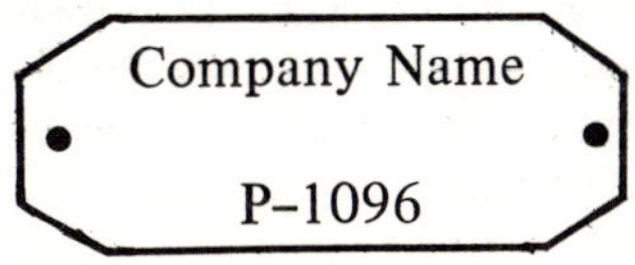


P = Production Equipment (other than machine tools)
1096 = Chronological company number

*Other Property*

1. Portable jigs, fixtures, and dies are not included in this inventory. They will carry the stamped or painted part or assembly number(s) on which these tools are used.
2. Buildings or installed building features like plumbing fixtures, partitions, and ventilating ducts are also excluded.

## Appendix III—Classification of Material Handling Equipment and Containers[1]

This breakdown is carried to only two digits; the full classification further extends each two-digit subdivision to several three-digit types of equipment.

*1. Conveyors*

(All fixed and portable equipment for conveying material between two fixed points with continuous or intermittent forward movement—continuous drive.)

11. Belt Conveyors
12. Elevating Conveyors (Capable of Vertical Lift)
13. Carrier Chain and Cable Conveyors (Load Carried on Moving Conveyor)
14. Haulage Conveyors (Weight of Pushed or Pulled Load Carried by Stationary Troughs, Surface, or Rails)
15. Roller Conveyors
16. Screw Conveyors
17. Pipeline Conveyors
18. Vibrating and Reciprocating Conveyors
19. Special Conveyors and Auxiliary Equipment

*2. Cranes, Elevators, and Hoists*

(All equipment for moving material having a reversing vertical or lateral movement—drum windup and payoff or reciprocating plunger drive.)

21. Fixed Cranes and Derricks (Stationary Mount)
22. Traveling Cranes (On Rail Runways)
23. Portable Cranes
24. Elevators
25. Cableways, Drag Scrapers, and Cableshifters
26. Hoists
27. Winches
28. Crane, Elevator, and Hoist Auxiliary Equipment
29. Special Cranes, Elevators, and Hoists not otherwise Classified

*3. Positioning, Weighing, and Control Equipment*

(Equipment used for local positioning and transferring, for weighing, and for controlling material movement.)

31. Manipulators and Positioners
32. Up-enders and Dumpers
33. Positioning Tables and Platforms
34. Positioning Transfers
35. Fixed Positioning Bridges and Ramps
36. Portable Positioning Equipment
37. Weighing Equipment
38. Handling Control Devices

*4. Industrial Vehicles*

(All types of industrial trucks, rail cars, trailers, tractors, excavating and grading equipment, off-highway carriers and agricultural vehicles used in materials handling.)

41. Powered Industrial Trucks
42. Hand Trucks
43. Industrial Trailers
44. Industrial Cars
45. Industrial Tractors and Locomotives
46. Bulk Materials, Handling Vehicles and Portable Equipment
47. Agricultural Vehicles (Not Including Motor Vehicles)
48. Special Military Vehicles (Not Including Motor Vehicles)
49. Special Industrial Vehicles and Components

---

1. From *Materials Handling Handbook*, Edited by Harold A. Bolz (New York: The Ronald Press Co., 1958, Pages 1.5 to 1.16).

*5. Motor Vehicles*

(All types of highway passenger and cargo vehicles including trucks, tractors, and trailers whose dimensions and capacities permit use on public highways.)

51. Passenger Automobiles
52. Motor Busses (Capacity for Carrying 10 or more passengers)
53. Trucks and Truck Tractors
54. Truck Trailers
55. Passenger-Car Trailers
56. Motorcycles

*6. Railroad Cars*

(All standard and narrow-gauge public carrier railroad rolling stock used on the country's railroad system.)

61. Passenger Cars
62. Baggage Cars
63. House Cars
64. Flat Cars
65. Special Railroad Cars
66. Hopper Cars
67. Tank Cars
68. Locomotives

*7. Marine Carriers*

(All water-borne vessels, including ocean, lake, river, and canal boats, small craft, and other miscellaneous marine equipment used in the handling of materials.)

71. Ocean Vessels
72. Lake Vessels
73. River and Canal Boats (Self-Propelled)
74. Barges and Lighters
75. Towboats
76. Small Craft
77. Mobile Floating Equipment
78. Suction Dredges
79. Special Marine Carriers not otherwise Classified

*8. Aircraft*

(All types of aircraft, including passenger and cargo airplanes, rotary-wing aircraft, gliders, and lighter-than-air airships.)

81. Passenger Airplanes (Fixed-wing Airplanes- Combination Cargo)
82. Cargo Air Transports (Fixed-wing Airplanes)
83. Military Combat Airplanes
84. Rotary-Wing Aircraft
85. Gliders
86. Balloons and Airships (Lighter-than-air Aerostats)

*9. Containers and Supports*

(All types of pressure, tight, loose, and open-top containers; also platform and coil supports, and all types of securements such as strapping, cinches, bulkheads, dunnage, etc.)

91. Pressure Vessels
92. Tight Containers (Hermetically Sealed or with Breather)
93. Entirely Enclosed Containers (Loose)
94. Open-Top Containers
95. Platform Supports
96. Coil Supports
97. Securements
98. Auxiliary Packaging Equipment
99. Containers and Supports Not Otherwise Classified

## Appendix IV—Calculating Equipment Requirements

Establishing the number of machines required must include several considerations other than the operating capacity of the machine itself. Such things as working hours available to operate, set-up time and frequency of set-ups, number of machines operated by one person (or number of persons per machine, in some cases), lost time for various reasons, yield or quality loss, peak requirements, and inventory-balancing policies all affect the number of machines required.

To help better explain these calculations let us follow an example.

$$\text{Number of machines required} = \frac{\text{pieces per hour to meet production requirements}}{\text{pieces per hour per machine}} = \frac{\text{time per piece per machine}}{\text{time per piece to meet production requirements}}$$

Suppose that a drug company's requirements call for making five varieties of tablets each week on a Model EG3 tabletting machine. The plant operates one shift of 40 hours. The quantity of granulated powder for each tablet variety and the machine capacity for each variety is as follows:

| Tablet Variety | Lbs. per Week Required | Machine Capacity Lbs. per Hour |
|---|---|---|
| A | 4,560 | 30 |
| B | 3,040 | 40 |
| C | 760 | 50 |
| D | 570 | 37½ |
| E | 190 | 50 |

Previous studies show that on an average the Model EG3 tabletting machines actually produce at 80 percent of capacity (idle-time interferences, set-up allowance, and downtime losses) and that acceptable tablets are obtained 95 percent of the time (yield or quality loss). How many Model EG3 tabletting machines are required to produce the five tablet varieties?

*Tablet A*: 4,560 lbs. per week ÷ 0.95 yield = 4800 lbs. per week requiring tabletting. 30 lbs. per hour machine capacity x 0.80 machine efficiency = 24 lbs. per hour average production.

*Tablet B*: 3,040 lbs. per week ÷ 0.95 yield = 3,200 lbs. per week requiring tabletting. 40 lbs. per hour machine capacity x 0.80 machine efficiency = 32 lbs. per hour average production.

*Tablet C*: 760 lbs. per week ÷ 0.95 yield = 800 lbs. per week requiring tabletting. 50 lbs. per hour machine capacity x 0.80 machine efficiency = 40 lbs. per hour average production.

*Tablet D*: 570 lbs. per week ÷ 0.95 yield = 600 lbs. per week requiring tabletting. 37½ lbs. per hour machine capacity x 0.80 machine efficiency = 30 lbs. per hour average production.

*Tablet E*: 190 lbs. per week ÷ 0.95 yield = 200 lbs. per week requiring tabletting. 50 lbs. per hour machine capacity x 0.80 machine efficiency = 40 lbs. per hour average production.

*Tablet A*: 4800 lbs. per week ÷ 40 hours per week available = 120 lbs. per hr. required.
1 ÷ 120 lbs. per hour = 0.00833 hrs. per lb. required.
1 ÷ 24 lbs. per hour = 0.04166 hrs. per machine per pound produced.

$$\text{Number of machines required} = \frac{120 \text{ lbs. per hr. required}}{24 \text{ lbs. per hr. per machine}} = 5 \text{ machines}$$

$$\text{or} = \frac{0.04166 \text{ hrs. per lb. per machine}}{0.00833 \text{ hrs. per lb. required}} = 5 \text{ machines}$$

*Tablet B*:

$$\text{Number of machines required} = \frac{3{,}200 \div 40}{32} = 2.5 \text{ machines}$$

*Tablet C*:

$$\text{Number of machines required} = \frac{800 \div 40}{40} = 0.5 \text{ machine}$$

*Tablet D*:

$$\text{Number of machines required} = \frac{600 \div 40}{30} = 0.5 \text{ machine}$$

*Tablet E*:

$$\text{Number of machines required} = \frac{200 \div 40}{40} = 0.125 \text{ machine}$$

Total number of machines required to produce week's requirements = 5 + 2.5 + 0.5 + 0.5 + 0.125 = 8.625 or 8 5/8 machines.

Obviously, we cannot use less than a whole machine. Therefore, nine Model EG3 machines will be required. However, the possibility of obtaining the production of five-eighths of a machine by working overtime or additional shifts should be considered. Note that it is easier to make eight machines do the work of 8-5/8 machines (by overtime, extra shifts, or reduced idleness) than to make one machine do the work of 1-1/4 machines. In the first case the additional work is only just over seven per cent for each machine, whereas in the second case the overload is 25 per cent.

When we established 80 per cent of capacity as our anticipated production, this might have been based on one man operating two machines. If we decide that one man can actually operate four machines, we may save labor cost, but it likely there will be more downtime or perhaps lower quality yield. In that case, we would have to recalculate the number of machines using say 70 per cent of capacity and/or 93 per cent acceptable tablets.

In this example we have assumed average requirements. If our peak-period requirements were 25 per cent greater for each tablet, we would have to add additional machines. We could add enough to take care of peak conditions directly, or we could build an inventory of finished tablets during the slow period and use up the inventory faster than we produced during the peak period. Theoretically, on this latter basis we could balance our machines by adjusting inventories, thus requiring no more equipment than for level-production, average conditions. This does not always work out in practice, however. But the chief point is that our policy on inventory-building will make a significant difference on machinery investment and therefore on space.

Still one other consideration is how much insurance against failures to meet customer requirements can be tolerated. Extra or stand-by machinery may be added over and above that required for the normal expected losses in order to protect the company and its customers' needs. The amount of such protection will vary depending on the importance of this consideration, the availability of outside equipment that can be used or leased for short periods, the critical nature of the machine, the likelihood of prolonged downtime, and whether or not the work can be done on other types of equipment in the company, even if less productively.

## Appendix V—Use of Color in Layout Planning

One of the most helpful conventions used in layout planning is color coding. Colors can identify various departments or activities or different types of area or space. Color can also be used to indicate degree of closeness desired or magnitude of flow intensities.

Essentially, color coding is used to aid interpretation and understanding—that is, as an aid to clear visualization. Color adds a decorative and artistic quality, too, and can therefore add appeal to, and aid more ready acceptance of, one's layout plans. It can be used to appeal to the "reader's" heart as well as his eye and his mind.

For layout planning, using color reaps the greatest benefits when each color has a significant meaning. Map-makers long ago discovered that water-blue, grassland or forest-green, desert-tan, and iceland-white are readily recognized. In 1969, the International Materials Management Society approved a set of Standard Color Codes for Use in Layout Planning and Materials Handling Analysis. The SLP color conventions are almost identical with the IMMS Standard.

The SLP color conventions are basically *two:*

1. Colors for the types of space (activities, areas, or equipment)[1]

| | |
|---|---|
| Red | – Assembly, including Subassembly and Disassembly Areas. |
| Orange/yellow | – Transport-related Areas; Storage, Set-down, or Hold Areas |
| Green | – Forming or Treating Areas |
| Blue | – Inspect, Test, or Check Areas; Service or Support Areas |
| Brown or Gray[2] | – Office-Type Areas |

2. Colors for rating the relative closeness between activities follow the order of colors of the spectrum:

| | | |
|---|---|---|
| Red | (A) | Absolutely necessary |
| Orange/yellow | (E) | Especially important |
| Green | (I) | Important |
| Blue | (O) | Ordinary closeness |
| Uncolored | (U) | Unimportant |
| Brown or Gray[2] | (X) | Not desirable |

Note that only five colors have been included in the conventions. This limits the number of colored pencils (or slots in a multi-color mechanical pencil) to a practical number. Degrees of measure in between these color ratings can be indicated by dots or cross-hatching to represent a half degree.

1. Black-and-white shadings may be substituted for color when rapid, inexpensive reproduction is desired. The old heraldic tincture code for black-and-white shading has been adapted by both the IMMS Standard and SLP as a conventional interchange with type-of-space colors. This black-and-white shading code is indicated on the SLP Capsule Summary (back cover).

2. Gray, while not an IMMS Standard, is sometimes used in place of brown when conventional lead-pencil gray is more readily available.

Other comments on the use of color in layout planning include the following:

1. Modifications of the standard type-of-space color convention can be made—typically for the operating areas. For example:
   - red for hot operations and green for cold work
   - red for primary and green for secondary operations
   - green for forming and red for finishing.
2. When coloring a large area, the entire area may be covered, or a band of approximately a quarter or a half inch outlining the area along and just inside its boundary line can be used.
3. On site surveys, major contour lines can each be colored differently to indicate the topography of the site.
4. Different colors can be used to indicate degrees of intensity of materials flow.
5. Different colors can be used in diagraming to identify various paths of flow for different products, or product groups, or materials.
6. Different colored lines on light-sensitive clear plastic sheets make it possible to visualize a complex of several products or several routes through several layers of transparent sheets.
7. Color can be used to indicate the degree of attention that should be given to planning the layout of each area, or to indicate areas to be moved, to stay, to be renovated.
8. In detail layout planning, color can identify individual templates according to type of function the equipment performs; whether the equipment is new or old; or whether it is to be moved or left behind.
9. Colored lines or colored tapes for various utilities or service lines aid clear planning.

Certain problems in using color should be recognized.

1. Hand coloring individual drawings or prints is time-consuming work and always opens the possibility of error. Still, it may be well worthwhile on key installation drawings.
2. Color does not print easily or reproduce readily. Therefore, templates of individual machines should be outlined in black, red, or orange, which are all opaque colors. Inasmuch as other colors fade in varying degrees, a template which prints the equipment outline and equipment information data in color (other than black, red, orange or dark brown) is generally not satisfactory.
3. Light tints on individual templates are ideal. These disappear when the layout plan is printed or photographed, yet aid visualization and do not obliterate the machine or equipment outline and other information.

## Appendix VI—Factors or Considerations in Selecting the Layout

When evaluating alternative layouts there are many factors which can affect the selection of the most suitable plan. The factors most frequently identified are listed below. A definition of each is also given. Then key points to consider in making an evaluation are set forth. These factors, definitions, and key points—in a consolidated form or together with others that may be added as appropriate for a particular company or project—can be extremely helpful in keeping a clear understanding and meaning for each factor during the process of evaluation.

1. EASE OF FUTURE EXPANSION (*The simplicity of increasing the space employed.*)

a. Tie-in with long-range potential use of the space,
   with the future plans for building or property development,
   with the basic overall allocation of space, and
   with the overall flow pattern(s).

b. Ability to spread out to adjacent areas-beside, above, below,
   to encroach on readily moved storage or service areas, or
   to add vertical storage equipment, balconies, mezzanines.

c. Freedom from fixed or permanent building features,
   from divided or honeycombed areas, and
   from space blocked-in by physically long equipment, property lines, natural obstructions or limitations, and the like.

d. Regularity of allocated space amounts in terms of
   readily exchangeable amounts and types of areas,
   modular units of layout space,
   multiple unit areas.

e. The amount of disruption or rearrangement of areas other than the one(s) specifically being expanded.

f. Shrinkability- ease of contracting the layout economically, to cut down the size if necessary.

2. ADAPTABILITY AND VERSATILITY (*The ease of accommodating, in the layouts as planned (without rearrangement), changes (normal or emergency) in, and variety (or number) of, items like the following.*)

a. Product, materials, or items
b. Quantity or volume
c. Frequency of delivery
d. Process equipment
e. Operation sequence
f. Working methods and operating time
g. Handling or storing methods
h. Utilities or auxiliaries
i. Other services
j. Type or classification of employees
k. Time-keeping or count system
l. Hours of work
m. Material dispatching procedure
n. Inspection controls
o. Rework procedures
p. Standby equipment
q. Additional space for stock
r. Alternate routes
s. Test runs, pilot lots, experimental engineering.

3. FLEXIBILITY OF LAYOUT (*The ease of physically rearranging the layout to accommodate changes.*)

a. Mobility of machinery and equipment
b. Relative size and fixity of equipment
c. Standardization of equipment, containers, work places
d. Freedom from fixed building features or walls, unmatching floor levels, other barriers
e. Overly dense saturation of space
f. Independence or self-sufficiency of facilities (not dependent on central coordination or centralized service tie-in)
g. Ready accessibility of service lines, piping, power distribution, heating and ventilating, service holes, etc.
h. Access to the area laid out at more than one point or side

4. FLOW OR MOVEMENT EFFECTIVENESS (*The effectiveness of sequenced working operations or steps—without unnecessary back-tracking, cross flow, transfers, long hauls—of materials, paper work, or people.*)

a. Greatest flow intensities with minimum distances
b. Basic regularity or consistency of flow pattern(s)
c. Proximity of related areas to each other where movement of material, people, or major paper work is involved, or where frequent, urgent or significant personal contact takes place
d. Access to, away from, and between major areas (like receiving, shipping, key operating areas)
e. Flow of auxiliary or service materials: supplies, tools, scrap or waste, and other service materials
f. Accessibility for delivery and pick-up, visitors, or employed non-company service personnel

5. MATERIALS HANDLING EFFECTIVENESS (*The ease or simplicity of the handling system, equipment, and containers to move materials into, through, and out of the areas laid out.*)

a. Ease of tie-in with external handling methods and equipment: rail line, docks, highway, and other accessways
b. Necessity for re-handling, extra handling, delays, awkward positioning, undue physical effort, undue dependence on frequency or urgency of moves, undue amount of jury-rig or non-integrated equipment
c. Traffic congestion and interferences other than due to flow pattern
d. Balanced variety of handling systems, equipment and containers
e. High utilization of handling equipment and containers
f. Simplicity of handling devices
g. Equipment integrated for multiple use
h. Dependence on M. H. equipment on maintenance, repair, replacement parts
i. Avoidance of synchronizing two or more people at same time or place
j. Ability to move completely around buildings on company property
k. Take advantage of gravity
l. Combined purposes of handling equipment for storing, pacing, sequencing, inspecting, work-holding, weighing and the like, as well as moving

6. STORAGE EFFECTIVENESS (*The effectiveness of holding required stocks of materials, parts, products, service items.*)

a. Inclusion of all storage—raw, in-process, finished goods, supplies, tools, scrap or waste, trash and equipment or materials not in current use
b. Accessibility of items stored
c. Ease of locating or identifying items stored
d. Ease of stock and inventory control
e. Ability to make stored items available according to urgency of demand
f. Protection of material (fire, moisture, dust, dirt, heat, cold, pilferage, deterioration, spoilage.)
g. Adequacy of storage space(s)
h. Suitably close to points of delivery and use

7. SPACE UTILIZATION (*The degree to which floor area and cubic space is put to use.*)

a. Conservation of floor space, property, or land—or most desirable portions thereof
b. Utilization of overhead space in terms of cubic density
c. Ability to share or exchange space among similar activities, and balancing of areas with seasonally complimentary space requirements
d. Effectiveness of aisle space:
   to serve areas adjacent to them,
   to lead to areas needing access,
   to handle traffic without wasting space or without excessive aisleways (too few, too many, too wide, too narrow, too cornered or crooked, too angular.)
e. Waste or idle space, caused by split, divided, cornered, scattered or otherwise honeycombed structures, too-close columns, too-frequent partitions or walls.
f. Less desirable or out-of-way space utilized for slow, dead areas; convenient space for fast, active areas

8. EFFECTIVENESS OF SUPPORTING SERVICE INTEGRATION (*The way supporting areas are arranged so as to serve the operating areas.*)

a. Ability of existing (or planned) systems, procedures, and controls to work effectively with the layout, including: production planning, scheduling and control, timekeeping, material or stock issuing, work count, tool control, personnel records, receiving and shipping system
b. Ability of the layout to integrate with desired or effective pay plans, performance measure, cost reports, lot size, order quantities
c. Physical closeness of service areas according to each area's need for the service (actual versus desired relationships).
d. Ability of the utilities, auxiliary service lines, and central distribution or collection systems to serve the layout. (Compressors, steam generators, transformers, chargers, and the like, and their accompanying pipes, ducts, wiring, etc.)
e. Service convenience of baler, salvage equipment, reclaim, incinerators, filter beds, scrap collection, and similar waste control areas or equipment
f. Ability of engineering groups, and technical advisors to support the layout effectively.

9. SAFETY AND HOUSEKEEPING (*The effect of the layout and its features on accidents or damage to employees and facilities, and on the general cleanliness of the areas involved.*)

a. Basic regularity of the aisles and work areas, and degree of freedom from equipment protruding into aisles or work areas, congestion, blind corners
b. Degree to which all safety codes and regulations are satisfied
c. Risk of danger to people or equipment
d. Availability of adequate exits and clear escapeways
e. First-aid facilities and fire extinguishers nearby
f. Floors free of obstructions, spillage, and mess, and not overly congested
g. Adequate protection or segregation for dangerous or unsightly operations
h. Workers not located under or above unprotected hazards;
   workers not located too near moving parts, unguarded equipment, and other hazards
i. Workers able to get benefit from special safety devices or guards
j. Effectiveness of ways to clean or clear area of waste, offal, trimmings, trash
k. Ease of keeping areas clean, sanitary, snow-white, under controlled conditions

10. WORKING CONDITIONS AND EMPLOYEE SATISFACTION (*The extent to which the layout contributes to making the area(s) a pleasant place to work and free from inconveniences, awkwardness, or disruptions for employees.*)

a. Effect of layout on attitude, performance, or general morale of employees
b. Working conditions suitable to the type of operation
c. Suitability of the layout's arrangement and allocated space to the personnel
d. Convenience for employees—access, distances, interruptions, delays, and adequacy and convenience of parking, lockers, rest rooms, food facilities, etc.
e. Freedom from features causing workers to feel afraid, hemmed-in, embarrassed, discouraged, discriminated against
f. Noise, distractions, or undue heat, cold, drafts, dirt, glare, or vibrations
g. Utilization of employee know-how and skills
h. Balanced man-power allocations

11. EASE OF SUPERVISION AND CONTROL (*The ease or difficulty for supervisors and managers to direct and control the operations for which they are responsible.*)

a. Ability to see the area fully and easily
b. Ability to get around the area conveniently
c. Ease of controlling quality, quantity counts, schedules, inventories in process
d. Ease of controlling waste time, lost materials, or supplies
e. Ease of moving or reassigning personnel to other work

12. APPEARANCE, PROMOTIONAL VALUE, PUBLIC OR COMMUNITY RELATIONS (*The ability of the layout to afford engaging or attractive facilities, having value in promoting the company name or reputation in the community and territories served by the company.*)

a. Attractiveness of external or viewable features, yards, main structure, out buildings
b. Ability to serve as show-place or reflect reliability, progressiveness or other company qualities
c. Regularity, symmetry, clean-lines, and organized appearance
d. Fit with community appearance, tradition, character
e. Effects on neighbors (benefits and irritants)

13. QUALITY OF PRODUCT OR MATERIAL (*The extent to which the layout affects quality of the product, material, or their workmanship.*)

a. Damage or risk to materials caused by nature of the layout or its transport facilities
b. Contamination, corrosion, spoilage, or other detrimants to the product's nature or condition as caused by the layout
c. Convenience and inter-relationship of quality control activities: inspection areas, Q. C. office, test facilities, control laboratories, engineering office, sample room, gauge crib, and the like

14. MAINTENANCE PROBLEMS (*The extent to which the layout will benefit or hinder maintenance work, including building and machine repair as well as day-to-day service.*)

a. Adequacy of facilities for maintenance and repair work
b. Sufficiency of space for access to machinery and equipment to be lubricated, checked, cleaned, adjusted, on-spot repaired, or otherwise maintained
c. Appropriate janitor and cleaner facilities

15. FIT WITH COMPANY ORGANIZATION STRUCTURE (*The degree to which the layout matches or disrupts the planned or desired organization structure.*)

a. Eliminate, combine, or streamline supervision, or effectiveness with which the layout helps otherwise improve the alignment of managerial personnel
b. Areas having the same supervisory responsibility are adjacent or convenient to each other
c. Staffing or manning of layout fits with job classifications and salary schedules

16. EQUIPMENT UTILIZATION (*The extent to which machinery and equipment, both operating and service, is used.*)

a. Degree of utilizing all equipment: operating, utility and auxiliary handling, storing, servicing or otherwise supporting
b. Necessity for duplicating equipment caused by layout versus use of common equipment and services
c. Over-capacity equipment necessitated by the layout
d. Man-machine efficiency planned into the layout

17. PLANT SECURITY AND THEFT (*The ease or difficulty of safeguarding company proprietary or security-classified information, and of controlling theft or pilferage.*)

a. Ease of controlling and/or monitoring access to the plant:
   during working hours
   during off-shift hours
b. Ability to provide guard-controlled traffic access for all pedestrian and vehicular
c. Effective control of visitor or casual access to secure areas or information within the plant
d. Provision of adequate vaults for safe, secure storage of valuable or confidential records and documents
e. Ease of patrolling building(s) and/or grounds
f. Ease of controlling access to and dispensing of: drug supplies, tools, and expendables or easily pilfered items

18. UTILIZATION OF NATURAL CONDITIONS, BUILDING OR SURROUNDINGS (*The extent to which the layout takes advantage of or capitalizes on the natural conditions of the site, physical surroundings, building structure, or neighboring areas, and the suitability of the layout to these features.*)

a. Slope, topography, foundation, drainage
b. Direction of sun, prevailing wind
c. Rail line, highway, waterway, bridges, accessways, crossings
d. Building features, structure, shape, height, construction, docks, door locations, elevator(s), windows, walls, columns
e. Zoning of site and restrictions of community or neighborhood
f. Fit of the area(s) laid out onto the natural site or into the existing building or area allocated

19. ABILITY TO MEET CAPACITY OR REQUIREMENTS (*How well the layout actually meets the planned needs or output wanted from the installation.*)

a. The right products or materials, properly meeting specifications
b. The right quantities of each variety or item in the operating time planned, without overtime or premium pay
c. The right yield in terms of projected quantities and qualities of product

20. COMPATIBILITY WITH LONG-RANGE COMPANY PLAN (*The ability of the planned layout to fit with long-range growth projections and with long-range master site plan or total facilities development plan(s).*)

a. Degree of tie-in with long-range projections of products and/or materials, sales or operating quantities, process sequence and equipment, services, working hours and operating times
b. The ease of complete renovation, rehabilitation, modernization, or change in function
c. Ease of integration with other buildings, plants, or sites of the organization
d. Effect of the layout on the re-sale value of the property

## Appendix VII—Suppliers of Layout Visualization Materials

| Name and Address of Company | Items Supplied | | | |
|---|---|---|---|---|
| | Tape | Grid Sheet | Template | Model |
| ACS TAPES, INC.<br>480 Neponset St.<br>Canton, Massachusetts 02021 | x | x | Std.<br>Plas.,Adh. | |
| APPLIED GRAPHICS CORP.<br>58 Shore Road<br>Glenwood Landing, L.I., N.Y.11547 | x | x | Std.,Ctm.<br>Plas.,Adh. | |
| CHART-PAK (1)<br>One River Road<br>Leeds, Massachusetts 01053 | x | x | Std.,Ctm.<br>Plas.,Adh. | |
| F. WARD HARMON, ASSOC. (1)<br>P. O. Box 206, Halesite<br>L.I., New York 11743 | x | x | Std.,Ctm.<br>Plas.,Adh. | Std.<br>& Ctm. |
| LABELON CORPORATION, INC.<br>10 Chapin Street<br>Canandaigua, N.Y. 14424 | x | x | Std.<br>Plas.,Adh. | |
| PLANPRINT COMPANY (1) (2)<br>Chalfont, Pa. 18914 | x | x | Std.,Ctm.<br>Plas.,Adh. | |
| REPRO-TEMPLETS, INC.<br>614 Allegheny River Blvd.<br>Oakmont, Pennsylvania 15139 | x | x | Std.,Ctm.<br>Acet.,Adh. | |
| UNITED SCALE MODELS<br>P. O. Box 222<br>Concordville, Pa. 19331 | | | Std.<br>Plas.,Adh. | Std.<br>& Ctm. |
| VISUAL INDUSTRIAL PRODUCTS, INC.<br>Rt. 910<br>Indianola, Pennsylvania 15051 | | x | | Std.<br>& Ctm. |

Plas.--Plastic
Adh.--Adhesive Backing
Acet.--Acetate
Std.--Standard
Ctm.--Custom

x--Available
(1)--Special kits & other equipment available
(2)--Custom made systems

## Appendix VIII—Preparing Film Templates

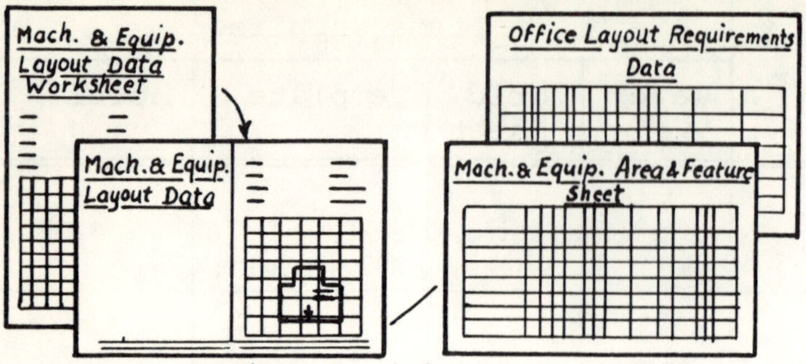


(1) Get measurements and features of machine or equipment.

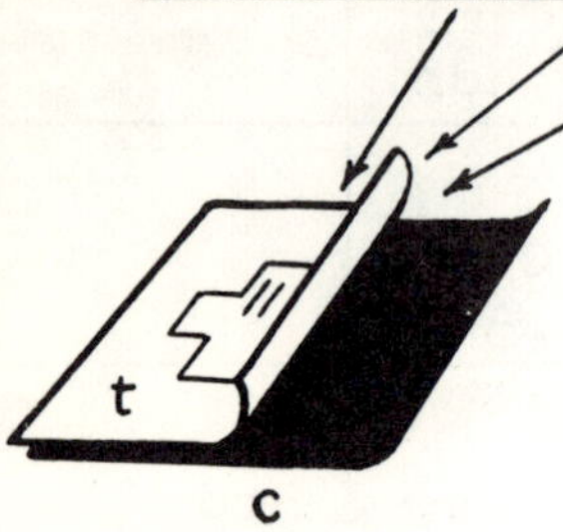


(2) Prepare plan view of each piece of equipment as a template master. Use ink, or use pencil on tracing paper (t) with orange carbon paper (c) face up. Spray back lightly with clear lacquer to keep carbon from smudging.

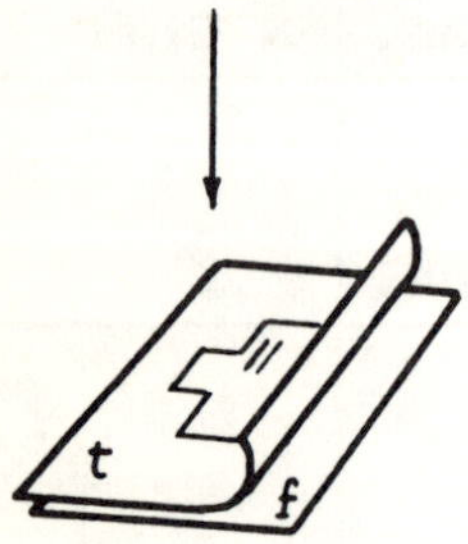


(3) Reproduce template on light-sensitive plastic film (f) by running through print machine. Light-sensitive polyester or mylar film (0.002 to 0.005 thick). For ammonia process (dry fume type), use F.Post--208PEL or 200B; Dietzgen--GF-247CE-Z; K&E--451505; Tecnifax--12P3M or 12P5M; Ozalid--102-PC3; GAF--102V. For liquid developer (wet chemical type) use Bruning--26-520.

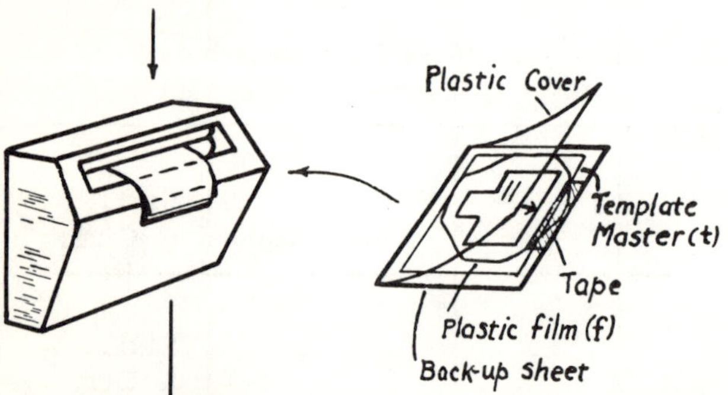


Note: When making just one or a few templates, to conserve film, cut it just large enough to cover the required outline of the template drawing. Then tack the small piece of film to a back-up sheet with any adhesive tape. A clear plastic cover will protect the master and help carry the small sheets through the print machine.

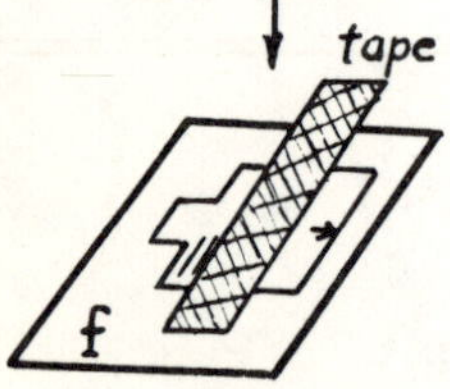


(4) Apply pressure-sensitive, double-faced, transparent tape (3M-666) to glossy surface, which should be back side of the film template.

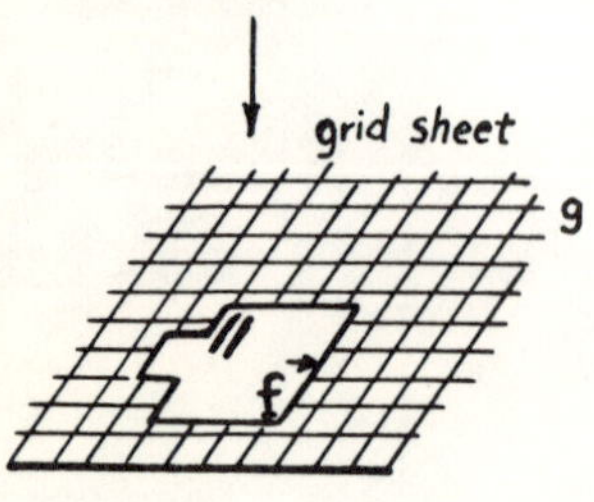


(5) Cut out template, avoiding sharp corners; remove protection "white"; and fasten to layout "grid sheet" or other plastic under-sheet (g). Or, remove part of white protection only and tack down on grid sheet, rearranging different alternatives; remove balance of white, fasten securely for making print by running grid sheet through print machine.

## Appendix IX—Drawing Number System

This self-describing numbering system is designed with both numbers and letters. Working from left to right, they progressively refine the drawing being considered.

The following diagram sets forth the sequence of symbols:

| Plant Prefix | Type of Facility | Drawing Description | Drawing Series No. | Revision Number |
|---|---|---|---|---|
| 23 | L | 11 | 12 | 2 |
| (Chicago plant) | (Lighting) | (Floor Plan of building) | (Twelfth drawing of lighting at Chicago) | (Second revision) |

*Plant Prefix:*

First digit is used to designate the region (2–Midwest, for example). The second digit is used to designate a specific site, site division, or building.

*Type of Facility:*

A. Property, Property Improvement, Fencing
B. Buildings
C. Circulation Methods. Highways, roads, streets, runways
D. Drainage Structures, Retaining Walls, Culverts, Bridges, Ditches, Channels, Dams
E. Equipment
F. Fuel, Oil, Lubrication Systems
G. Gas Lines Systems
H. Heating and Heat Distribution
J.
K. Power Distribution
L. Lighting
M. Miscellaneous
N.
P. Pits, Foundations, Test Holes
Q. Paint, Acid, Caustic Systems
R. Railways, Sidings
S. Sewer Systems
T. Telephone, Telegraph, Radio, Television
U.
V. Ventilating, Ducts, Air Handling (when not part of or combined with H)
W. Water System
X. Procedure Explanation
Y. Flow Charts
Z. Organization Charts

*Drawing Description:*

01. Location and Site Plans
02. Survey Plats
03. Topography
04. Artist Renderings or other Perspective Drawings
05. Landscape Plans
06. Aerial Photographs and Maps
07. Core and Test Drill Locations and Sections
08. Cross Section or Profiles
09.

11. Building or Floor Plans (Without Equipt. Layout)
12. Overall Layouts
13. Detail Layout Plans
14. Installation Drawings for Layouts
15. Elevations or Sections
16. Footings, Pedestals, and Wall Details
17. Building Details
18. Structural Steel Details
19.

21. Equipment- Raw or Bulk Storage
22. Equipment- Processing or Production
23. Equipment- Tools, Fixtures, Gauges
24. Equipment- Foundations

25. Equipment- Material Handling
26. Equipment- Service (Boilers, Air Compressors, Dust Collector Systems, and the like)
27. Equipment- Furniture, Benches, Office Equipment
28.
29.

Note: Expand to include items pertinent to particular company.

*Drawing Series Number:*

The drawings are numbered sequentially for every drawing series starting with one (1).

*Revisions:*

The original drawing is marked "O". Each revision is numbered in sequence—starting with one. A brief description of the nature of each revision shall be shown on the drawing.

Dashes or hyphens set off the drawing number and revision number from the description number and from each other.

Examples: 27E22-14-0 41L05-3-2

*Filing:*

Drawings or tracings of each plant will be filed separately. The filing sequence will then be type-of-facility letter, drawing description number, and then drawing series number. The latest revision of the drawing to be in front of the older revisions.

*Index:*

A numerical index with a cross index in alphabetical arrangement should be kept of all drawings.

## Appendix X—Land-To-Building Ratios

### SITE REQUIREMENTS ACCORDING TO SIZE OF STRUCTURE AND LAND-TO-BUILDING RATIO*

| SIZE OF STRUCTURE | | ACRES REQUIRED FOR SELECTED LAND-TO-BUILDING RATIOS | | | | | | | | | |
|---|---|---|---|---|---|---|---|---|---|---|---|
| Square Feet | Acreage Equivalent | 1 to 1 | 2 to 1 | 3 to 1 | 4 to 1 | 5 to 1 | 6 to 1 | 7 to 1 | 8 to 1 | 9 to 1 | 10 to 1 |
| 5,000 | .11 | .23 | .34 | .46 | .57 | .69 | .80 | .92 | 1.03 | 1.15 | 1.26 |
| 10,000 | .23 | .46 | .69 | .92 | 1.15 | 1.38 | 1.61 | 1.84 | 2.07 | 2.29 | 2.52 |
| 20,000 | .46 | .92 | 1.38 | 1.84 | 2.29 | 2.75 | 3.21 | 3.67 | 4.13 | 4.59 | 5.05 |
| 30,000 | .69 | 1.38 | 2.07 | 2.75 | 3.44 | 4.13 | 4.82 | 5.51 | 6.20 | 6.89 | 7.57 |
| 40,000 | .92 | 1.84 | 2.75 | 3.67 | 4.59 | 5.51 | 6.43 | 7.35 | 8.26 | 9.18 | 10.10 |
| 50,000 | 1.14 | 2.29 | 3.44 | 4.59 | 5.74 | 6.89 | 8.03 | 9.18 | 10.33 | 11.48 | 12.63 |
| 60,000 | 1.37 | 2.75 | 4.13 | 5.51 | 6.89 | 8.26 | 9.64 | 11.02 | 12.40 | 13.77 | 15.15 |
| 70,000 | 1.60 | 3.21 | 4.82 | 6.43 | 8.03 | 9.64 | 11.25 | 12.85 | 14.46 | 16.07 | 17.68 |
| 80,000 | 1.83 | 3.67 | 5.51 | 7.35 | 9.18 | 11.02 | 12.85 | 14.69 | 16.53 | 18.36 | 20.20 |
| 90,000 | 2.07 | 4.13 | 6.20 | 8.26 | 10.33 | 12.40 | 14.46 | 16.53 | 18.59 | 20.66 | 22.73 |
| 100,000 | 2.29 | 4.59 | 6.89 | 9.18 | 11.48 | 13.77 | 16.07 | 18.36 | 20.66 | 22.96 | 25.25 |
| 200,000 | 4.59 | 9.18 | 13.77 | 18.36 | 22.96 | 27.55 | 32.14 | 36.73 | 41.32 | 45.91 | 50.50 |
| 300,000 | 6.89 | 13.77 | 20.66 | 27.55 | 34.43 | 41.32 | 48.21 | 55.10 | 61.98 | 68.87 | 75.76 |
| 400,000 | 9.18 | 18.36 | 27.55 | 36.73 | 45.91 | 55.10 | 64.28 | 73.46 | 82.64 | 91.83 | 101.01 |
| 500,000 | 11.48 | 22.96 | 34.43 | 45.91 | 57.39 | 68.87 | 80.35 | 91.83 | 103.30 | 114.78 | 126.26 |
| 1,000,000 | 22.96 | 45.91 | 68.87 | 91.83 | 114.78 | 137.74 | 160.70 | 183.65 | 206.61 | 229.57 | 252.52 |

*Exhibit 3 in Address by T. K. Pasma, Office of Area Development U. S. Department of Commerce Before Industrial Development Conference, Bowling Green State University, March 23, 1956

## Appendix XI—Installation Symbols

Although there is no universal set of standard installation symbols used by layout planners, the symbols on the two pages of this appendix are probably the most used and accepted. Source: Time-Saver Standards, An Architectural Record Book, F. W. Dodge Corporation, New York, N. Y. Each company should agree upon and formalize in writing its own code of symbols based on its present practice, these standards, what is available in pre-printed tapes or sheet form, and the like.

# ELECTRICAL SYMBOLS FOR ARCHITECTURAL PLANS

**Prepared by American Standards Association and published by Industry Committee on Interior Wiring Design**

### GENERAL OUTLETS

| Description | Ceiling | Wall |
|---|---|---|
| Outlet | ○ | –○ |
| Capped Outlet | Ⓒ | –Ⓒ |
| Drop Cord | Ⓓ | |
| Electrical Outlet: *use when plain circle may be confused with column or other symbols* | Ⓔ | –Ⓔ |
| Fan Outlet | Ⓕ | –Ⓕ |
| Junction Box | Ⓙ | –Ⓙ |
| Lamp Holder | Ⓛ | –Ⓛ |
| Lamp Holder with Pull Switch | Ⓛ PS | –Ⓛ PS |
| Pull Switch | Ⓢ | –Ⓢ |
| Outlet for Vapor Discharge Lamp | Ⓥ | –Ⓥ |
| Exit Light Outlet | Ⓧ | –Ⓧ |
| Clock Outlet (Lighting Voltage) | ◷ | –◷ |

### CONVENIENCE OUTLETS

| Description | Symbol |
|---|---|
| Duplex Convenience Outlet | =⊖ |
| Convenience Outlet other than Duplex<br>1 = Single, 3 = Triplex, etc. | =⊖ 1, 3 |
| Weatherproof Conv. Outlet | =⊖ WP |
| Range Outlet | =⊜ R |
| Switch and Convenience Outlet | =⊖–$ |
| Radio and Convenience Outlet | =⊖ [R] |
| Special Purpose Outlet (desc. in Spec.) | ◬ |
| Floor Outlet | ⊙ |

### SWITCHES

| Description | Symbol |
|---|---|
| Single Pole Switch | $ |
| Double Pole Switch | $\$_2$ |
| Three Way Switch | $\$_3$ |
| Four Way Switch | $\$_4$ |
| Automatic Door Switch | $\$_D$ |
| Electrolier Switch | $\$_E$ |
| Key Operated Switch | $\$_K$ |
| Switch and Pilot Lamp | $\$_P$ |
| Circuit Breaker | $\$_{CB}$ |
| Weatherproof Circuit Breaker | $\$_{WCB}$ |
| Momentary Contact Switch | $\$_{MC}$ |
| Remote Control Switch | $\$_{RC}$ |
| Weatherproof Switch | $\$_{WP}$ |

### SPECIAL OUTLETS

| Description | Symbol |
|---|---|
| *Any standard symbol, with a lower case subscript added, may be used for special indications. When so used, a legend on each drawing and description in specifications are strongly recommended* | ○ a,b,c etc.<br>=⊖ a,b,c etc.<br>$ a,b,c etc. |

### PANELS, CIRCUITS, MISC.

| Description | Symbol |
|---|---|
| Lighting Panel | ▬ |
| Power Panel | ▨ |
| *Branch Circuit — Ceiling or Wall | —— |
| *Branch Circuit — Floor<br>**without other designation indicates 2-wire circuit.*<br>*For 3 wires, use*<br>*For 4 wires* | - - - -<br><br>—///—<br>—////— |
| Feeders: *use heavy lines and designate by number for quick reference* | ━━ |
| Underfloor Duct & Junction Box - Triple System: *for double or single systems use two lines or one. Symbol also adaptable to Auxiliary or Low Voltage systems* | ≡□≡ |

### PANELS, CIRCUITS, MISC.

| Description | Symbol |
|---|---|
| Generator | Ⓖ |
| Motor | Ⓜ |
| Instrument | Ⓘ |
| Transformer | Ⓣ |
| Controller | ⊠ |
| Isolating Switch | ⧄ |

### AUXILIARY or Low Voltage SYSTEMS

| Description | Symbol |
|---|---|
| Push Button | ⊡ |
| Buzzer | □/ |
| Bell | □○ |
| Annunciator | –◇ |
| Telephone | ◁ |
| Telephone Switchboard | \|◁ |
| Clock (Low Voltage) | [◷] |
| Electric Door Opener | [D] |
| Fire Alarm Bell | [F]○ |
| Fire Alarm Station | [F] |
| City Fire Alarm Station | ⧗ |
| Fire Alarm Central Station | [FA] |
| Automatic Fire Alarm Device | [FS] |
| Watchman's Station | [W] |
| Watchman's Central Station | [[W]] |
| Horn | [H] |
| Nurse's Signal Plug | [N] |
| Maid's Signal Plug | [M] |
| Radio Outlet | [R] |
| Signal Central Station | [SC] |
| Interconnection Box | ▭ |
| Battery | \|ı\|ı\|ı\|ı\| |
| Auxiliary System Circuits: *without other designation indicates 2-wire circuit. For others, use numbers, as:* — — —12 - #18 W - ¾″C. *or designate by number for quick reference* | -·-·- |
| Special Auxiliary Outlets: *subscripts refer to notes in plans, schedules or specifications* | □ a,b,c |

# STANDARD SYMBOLS FOR PLUMBING, PIPING AND VALVES

(Approved by the American Standards Association, November, 1935)

## PLUMBING

|  |  |
|---|---|
| Corner Bath | |
| Recessed Bath | |
| Roll Rim Bath | |
| Sitz Bath | SB |
| Foot Bath | FB |
| Bidet | B |
| Shower Stall | |
| Shower Head | (Plan) (Elev.) |
| Overhead Gang Shower | (Plan) (Elev.) |
| Pedestal Lavatory | PL |
| Wall Lavatory | WL |
| Corner Lavatory | LAV |
| Manicure Lavatory Medical Lavatory | ML |
| Dental Lavatory | DENTAL LAV |
| Plain Kitchen Sink | S |
| Kitchen Sink, R & L Drain Board | |
| Kitchen Sink, L H Drain Board | |
| Combination Sink & Dishwasher | |
| Combination Sink & Laundry Tray | S & T |
| Service Sink | SS |
| Wash Sink (Wall Type) | |
| Wash Sink | |
| Laundry Tray | L T |
| Water Closet (Low Tank) | |
| Water Closet (No Tank) | |
| Urinal (Pedestal Type) | |
| Urinal (Wall Type) | |
| Urinal (Corner Type) | |
| Urinal (Stall Type) | |
| Urinal (Trough Type) | TU |
| Drinking Fountain (Pedestal Type) | DF |
| Drinking Fountain (Wall Type) | DF |

## PLUMBING (continued)

|  |  |
|---|---|
| Drinking Fountain (Trough Type) | DF |
| Hot Water Tank | HW T |
| Water Heater | WH |
| Meter | M |
| Hose Rack | HR |
| Hose Bibb | HB |
| Gas Outlet | G |
| Vacuum Outlet | |
| Drain | D |
| Grease Separator | G |
| Oil Separator | O |
| Cleanout | C O |
| Garage Drain | |
| Floor Drain With Backwater Valve | |
| Roof Sump | |

## PIPING

|  |  |
|---|---|
| Piping, in general (Lettered with name of material conveyed) | |
| Non-intersecting Pipes | |
| Steam | |
| Condensate | |
| Cold Water | |
| Hot Water | |
| Air | |
| Vacuum | |
| Gas | |
| Refrigerant | |
| Oil | |

## PIPE FITTINGS

| For Welded or Soldered Fittings, use joint indication shown in Diagram A | Screwed | Bell and Spigot |
|---|---|---|
| Joint | | |
| Elbow - 90 deg | | |
| Elbow - 45 deg | | |
| Elbow - Turned Up | | |
| Elbow - Turned Down | | |

## PIPE FITTINGS (continued)

| For Welded or Soldered Fittings, use joint indication shown in Diagram A | Screwed | Bell and Spigot |
|---|---|---|
| Elbow - Long Radius | | |
| Side Outlet Elbow - Outlet Down | | |
| Side Outlet Elbow - Outlet Up | | |
| Base Elbow | | |
| Double Branch Elbow | | |
| Single Sweep Tee | | |
| Double Sweep Tee | | |
| Reducing Elbow | | |
| Tee | | |
| Tee - Outlet Up | | |
| Tee - Outlet Down | | |
| Side Outlet Tee Outlet Up | | |
| Side Outlet Tee Outlet Down | | |
| Cross | | |
| Reducer | | |
| Eccentric Reducer | | |
| Lateral | | |
| Expansion Joint Flanged | | |

## VALVES

| For Welded or Soldered Fittings, use joint indication shown in Diagram A | Screwed | Bell and Spigot |
|---|---|---|
| Gate Valve | | |
| Globe Valve | | |
| Angle Globe Valve | | |
| Angle Gate Valve | | |
| Check Valve | | |
| Angle Check Valve | | |
| Stop Cock | | |
| Safety Valve | | |
| Quick Opening Valve | | |
| Float Opening Valve | | |
| Motor Operated Gate Valve | | |

## Appendix XII—Combining Flow and Other-Than-Flow Relationships

Flow of materials is usually not the only reason for closeness ratings. In most industrial plants, relationships among activity-areas involve both flow of materials and service (or other-than-flow) requirements. These two types of relationships must be combined in a meaningful way to determine the resultant combined relationships.

Combining flow and other-than-flow relationships involves joining Box 1 and Box 2 of the SLP Pattern of Procedures. In probably 60 to 70 percent of all layout-planning projects, such combining is necessary. That is, in perhaps 15 to 20 percent of the projects, flow of materials so dominates the conditions that the planner can go directly from Box 1 to his diagram in Box 3; and in another 15 to 20 percent of all projects, there is no significant flow of materials involved so the planner can proceed from his other-than-flow relationships (Box 2) directly to his diagram.

When combining the two, use the following general procedure:

1. Develop the planned intensity of flow over each route—between each pair of activity areas. List pairs in the order of increasing identification number and keep the lower activity-number at left.
2. Calibrate or convert the flow intensities between each activity pair into the appropriate vowel-letter rating.
3. Determine the other-than-flow relationships between the same activity pairs; most frequently they are documented on a relationship chart.
4. Determine the *relative importance* of flow and other-than-flow relationships. Usually the layout planner discusses this with others immediately involved in the operations.
5. Combine the flow and other-than-flow relationship ratings into a resultant relationship.
6. Use the resultant combined relationship as a basis for making the relationship diagram (combined flow and other-than-flow relationship diagram)—Box 3.

Figure 1 shows the procedure of combining using numerical values for the vowel-letter ratings and recalibrating the resultant, combined relationships by ranking, plotting, and dividing the total combined values.

Figure 2 uses the same data to show the procedure of combining by following the matrix method of making the same combined relationship ratings.

Figure 3 provides a series of preprinted matrix tables for ready use when different ratios of relative importance of flow and other-than-flow are involved. Note that it is usually not practical to spread the relative-importance ratio between flow and other-than-flow more than 3:1 or 1:3.

Figure 4 shows a working form that is practical to use on major projects. On this work sheet, the planner can summarize the flow (in columns b through f); record the other-than-flow ratings, who says so, and why (in columns g through l and in the coded reason block at right); enter the resultant combined relationships (in subdivisions of column m); and account for the number of occurrences of each vowel-letter

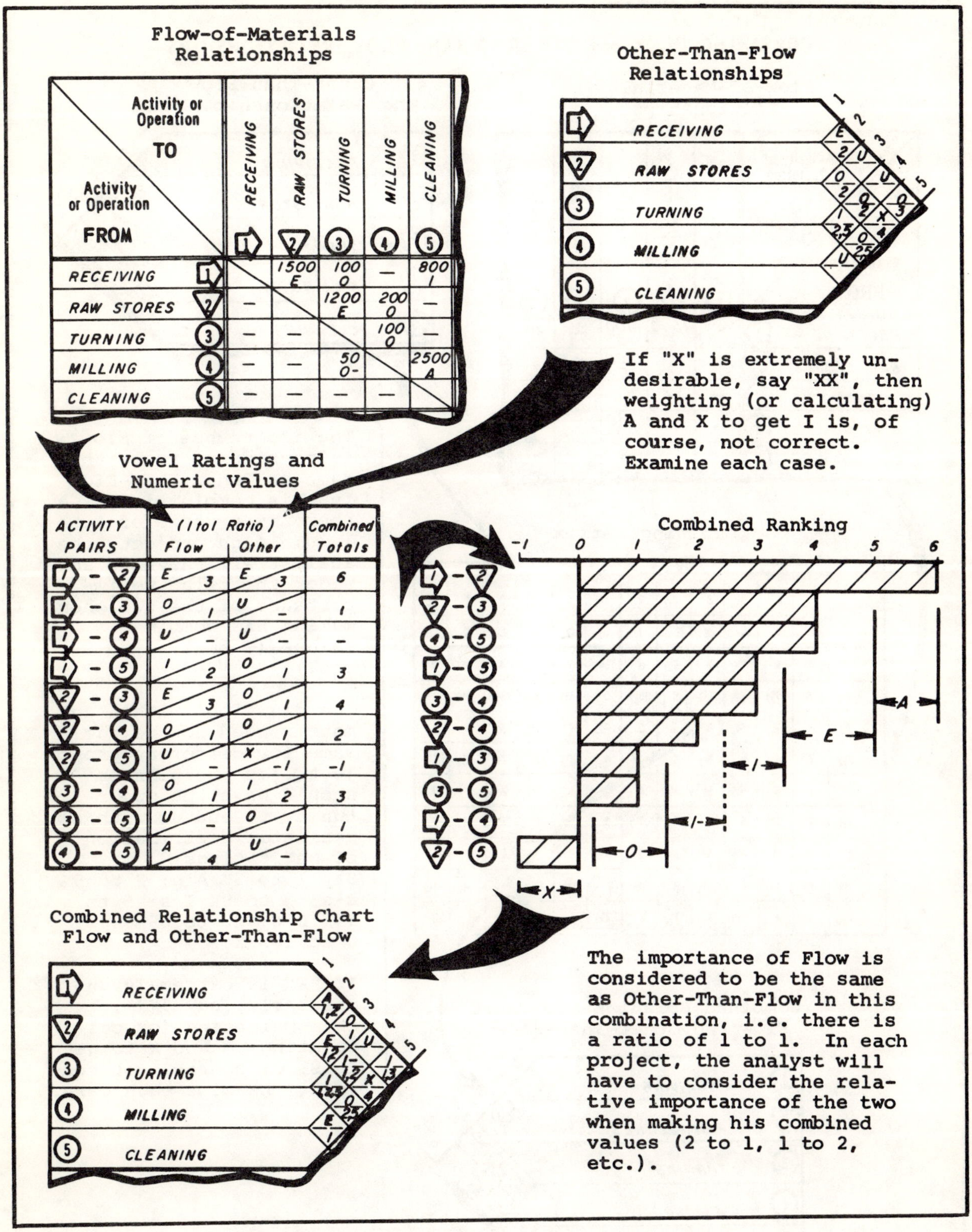


**Figure 1.** Combining by Re-Calibrating Numeric Values

## COMBINING FLOW and OTHER-THAN-FLOW RELATIONSHIPS

### Flow-of-Materials Relationships

| Activity or Operation FROM \ TO | 1 RECEIVING | 2 RAW STORES | 3 TURNING | 4 MILLING | 5 CLEANING |
|---|---|---|---|---|---|
| 1 RECEIVING | | 1500 E | 100 O | — | 800 I |
| 2 RAW STORES | — | | 1200 E | 200 O | — |
| 3 TURNING | — | — | | 100 O | — |
| 4 MILLING | — | — | 50 O- | | 2500 A |
| 5 CLEANING | — | — | — | — | |

### Other-Than-Flow Relationships

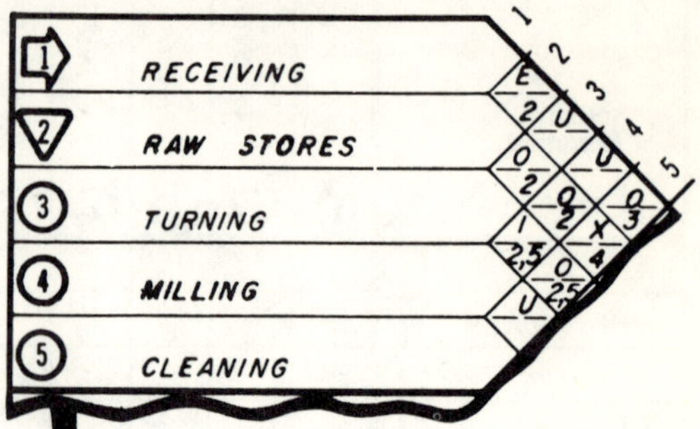


### Combining Matrix

| FLOW \ OTHER-THAN-FLOW | | 4 | 3½ | 3 | 2½ | 2 | 1½ | 1 | ½ | 0 | -1 | -2 |
|---|---|---|---|---|---|---|---|---|---|---|---|---|
| | | A | A- | E | E- | I | I- | O | O- | U | X | XX |
| 4 | A | 8 | 7½ | 7 | 6½ | 6 | 5½ | 5 | 4½ | 4 | 3 | 2 |
| 3½ | A- | 7½ | 7 | 6½ | 6 | 5½ | 5 | 4½ | 4 | 3½ | 2½ | 1½ |
| 3 | E | 7 | 6½ | 6 | 5½ | 5 | 4½ | 4 | 3½ | 3 | 2 | 1 |
| 2½ | E- | 6½ | 6 | 5½ | 5 | [illegible] | 4 | 3½ | 3 | 2½ | 1½ | ½ |
| 2 | I | 6 | 5½ | 5 | 4½ | 4 | [illegible] | 3 | 2½ | 2 | 1 | 0 |
| 1½ | I- | 5½ | 5 | 4½ | 4 | 3½ | 3 | [illegible] | 2 | 1½ | ½ | -½ |
| 1 | O | 5 | 4½ | 4 | 3½ | 3 | 2½ | 2 | [illegible] | 1 | 0 | -1 |
| ½ | O- | 4½ | 4 | 3½ | 3 | 2½ | 2 | 1½ | 1 | [illegible] | -½ | -1½ |
| 0 | U | 4 | 3½ | 3 | 2½ | 2 | 1½ | 1 | ½ | 0 | [illegible] | -2 |

### Combined Relationship Chart Flow and Other-Than-Flow

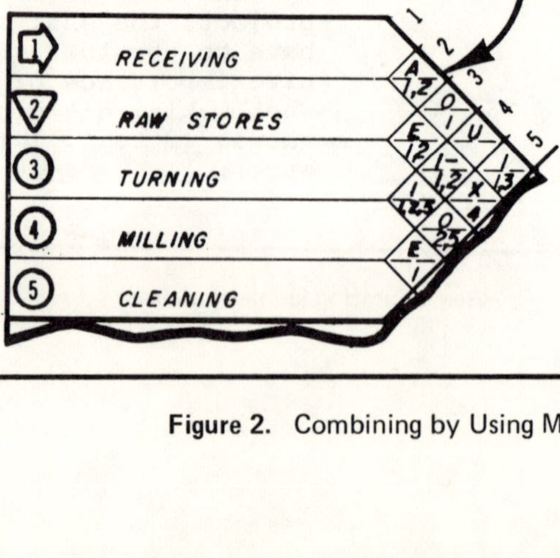


The importance of Flow is considered to be the same as Other-Than-Flow in this combination, i.e. there is a ratio of 1 to 1. In each project, the analyst will have to consider the relative importance of the two when making his combined values (2 to 1, 1 to 2, etc.).

The divisions within any matrix need not be the same for different projects, as the planner should have the opportunity to end up with a realistic distribution of net ratings: Say 1 to 3% A's; 2 to 5% E's; 3 to 8% I's; 5 to 15% O's; 20 to 85% U's; 0 to 10% X's.

If "X" is extremely undesirable, say "XX", then weighting (or calculating) A and X to get I is, of course, not correct. Examine each case.

**Figure 2.** Combining by Using Matrix of Values

## COMBINING FLOW AND OTHER-THAN-FLOW BY MATRIX METHOD

IN VARIOUS SITUATIONS, "FLOW OF MATERIALS" CARRIES DIFFERENT DEGREES OF IMPORTANCE RELATIVE TO FACTORS "OTHER THAN FLOW". ON EACH PROJECT, THE PLANNER OR ANALYST SHOULD DETERMINE THIS DEGREE OF RELATIVE IMPORTANCE, AND THEN ASSIGN A COMBINED CLOSENESS RATING FOR EACH PAIR OF ACTIVITY-AREAS.

THIS CAN BE DONE BY BUILDING A MATRIX OF FLOW AND OTHER-THAN-FLOW; ESTABLISHING A RELATIVE WEIGHT BETWEEN THE TWO; MULTIPLYING THE WEIGHT VALUE BY THE EQUIVALENT CLOSENESS-DESIRED NUMBER VALUE; AND THEN DIVIDING THE TOTALS AT CONSISTENT BREAK-OFF PLACES.

THE DIVISIONS WITHIN ANY MATRIX NEED NOT BE THE SAME FOR DIFFERENT PROJECTS, AS THE PLANNER SHOULD HAVE THE OPPORTUNITY TO END UP WITH A REALISTIC DISTRIBUTION OF NET RATINGS: SAY 1 TO 3% A's; 2 TO 5% E's; 3 TO 8% I's; 5 TO 15% O's; 20 TO 85% U's; 0 TO 10% X's.

### FLOW AND OTHER-THAN-FLOW AT RELATIVE IMPORTANCE OF 1 TO 1

| FLOW | | OTHER-THAN-FLOW | | | | | | | | | | |
|---|---|---|---|---|---|---|---|---|---|---|---|---|
| | | 4 | 3½ | 3 | 2½ | 2 | 1½ | 1 | ½ | 0 | -1 | -2 |
| | | A | A- | E | E- | I | I- | O | O- | U | X | XX |
| 4 | A | 8 | 7½ | 7 | 6½ | 6 | 5½ | 5 | 4½ | 4 | 3 | 2 |
| 3½ | A- | 7½ | 7 | 6½ | 6 | 5½ | 5 | 4½ | 4 | 3½ | 2½ | 1½ |
| 3 | E | 7 | 6½ | 6 | 5½ | 5 | 4½ | 4 | 3½ | 3 | 2 | 1 |
| 2½ | E- | 6½ | 6 | 5½ | 5 | 4½ | 4 | 3½ | 3 | 2½ | 1½ | ½ |
| 2 | I | 6 | 5½ | 5 | 4½ | 4 | 3½ | 3 | 2½ | 2 | 1 | 0 |
| 1½ | I- | 5½ | 5 | 4½ | 4 | 3½ | 3 | 2½ | 2 | 1½ | ½ | -½ |
| 1 | O | 5 | 4½ | 4 | 3½ | 3 | 2½ | 2 | 1½ | 1 | 0 | -1 |
| ½ | O- | 4½ | 4 | 3½ | 3 | 2½ | 2 | 1½ | 1 | ½ | -½ | -1½ |
| 0 | U | 4 | 3½ | 3 | 2½ | 2 | 1½ | 1 | ½ | 0 | -1 | -2 |

### FLOW CONSIDERED TWICE AS IMPORTANT AS OTHER-THAN-FLOW.

| FLOW | | OTHER-THAN-FLOW | | | | | | | | | | |
|---|---|---|---|---|---|---|---|---|---|---|---|---|
| | | 4 | 3½ | 3 | 2½ | 2 | 1½ | 1 | ½ | 0 | -1 | -2 |
| | | A | A- | E | E- | I | I- | O | O- | U | X | XX |
| 8 | A | 12 | 11½ | 11 | 10½ | 10 | 9½ | 9 | 8½ | 8 | 7 | 6 |
| 7 | A- | 11 | 10½ | 10 | 9½ | 9 | 8½ | 8 | 7½ | 7 | 6 | 5 |
| 6 | E | 10 | 9½ | 9 | 8½ | 8 | 7½ | 7 | 6½ | 6 | 5 | 4 |
| 5 | E- | 9 | 8½ | 8 | 7½ | 7 | 6½ | 6 | 5½ | 5 | 4 | 3 |
| 4 | I | 8 | 7½ | 7 | 6½ | 6 | 5½ | 5 | 4½ | 4 | 3 | 2 |
| 3 | I- | 7 | 6½ | 6 | 5½ | 5 | 4½ | 4 | 3½ | 3 | 2 | 1 |
| 2 | O | 6 | 5½ | 5 | 4½ | 4 | 3½ | 3 | 2½ | 2 | 1 | 0 |
| 1 | O- | 5 | 4½ | 4 | 3½ | 3 | 2½ | 2 | 1½ | 1 | 0 | -1 |
| 0 | U | 4 | 3½ | 3 | 2½ | 2 | 1½ | 1 | ½ | 0 | -1 | -2 |

### OTHER-THAN-FLOW TWICE AS IMPORTANT AS FLOW

| FLOW | | OTHER-THAN-FLOW | | | | | | | | | | |
|---|---|---|---|---|---|---|---|---|---|---|---|---|
| | | 8 | 7 | 6 | 5 | 4 | 3 | 2 | 1 | 9 | -2 | -4 |
| | | A | A- | E | E- | I | I- | O | O- | U | X | XX |
| 4 | A | 12 | 11 | 10 | 9 | 8 | 7 | 6 | 5 | 4 | 2 | 0 |
| 3½ | A- | 11½ | 10½ | 9½ | 8½ | 7½ | 6½ | 5½ | 4½ | 3½ | 1½ | -½ |
| 3 | E | 11 | 10 | 9 | 8 | 7 | 6 | 5 | 4 | 3 | 1 | -1 |
| 2½ | E- | 10½ | 9½ | 8½ | 7½ | 6½ | 5½ | 4½ | 3½ | 2½ | ½ | -1½ |
| 2 | I | 10 | 9 | 8 | 7 | 6 | 5 | 4 | 3 | 2 | 0 | -2 |
| 1½ | I- | 9½ | 8½ | 7½ | 6½ | 5½ | 4½ | 3½ | 2½ | 1½ | -½ | -2½ |
| 1 | O | 9 | 8 | 7 | 6 | 5 | 4 | 3 | 2 | 1 | -1 | -3 |
| ½ | O- | 8½ | 7½ | 6½ | 5½ | 4½ | 3½ | 2½ | 1½ | ½ | -1½ | -3½ |
| 0 | U | 8 | 7 | 6 | 5 | 4 | 3 | 2 | 1 | 0 | -2 | -4 |

### FLOW CONSIDERED 1½ TIMES AS IMPORTANT AS OTHER-THAN-FLOW.

(FLOW 3; OTHER THAN FLOW 2)

| FLOW | | OTHER-THAN-FLOW | | | | | | | | | | |
|---|---|---|---|---|---|---|---|---|---|---|---|---|
| | | 8 | 7 | 6 | 5 | 4 | 3 | 2 | 1 | 0 | -2 | -4 |
| | | A | A- | E | E- | I | I- | O | O- | U | X | XX |
| 12 | A | 20 | 19 | 18 | 17 | 16 | 15 | 14 | 13 | 12 | 10 | 8 |
| 10½ | A- | 18½ | 17½ | 16½ | 15½ | 14½ | 13½ | 12½ | 11½ | 10½ | 8½ | 6½ |
| 9 | E | 17 | 16 | 15 | 14 | 13 | 12 | 11 | 10 | 9 | 7 | 5 |
| 7½ | E- | 15½ | 14½ | 13½ | 12½ | 11½ | 10½ | 9½ | 8½ | 7½ | 5½ | 3½ |
| 6 | I | 14 | 13 | 12 | 11 | 10 | 9 | 8 | 7 | 6 | 4 | 2 |
| 4½ | I- | 12½ | 11½ | 10½ | 9½ | 8½ | 7½ | 6½ | 5½ | 4½ | 2½ | ½ |
| 3 | O | 11 | 10 | 9 | 8 | 7 | 6 | 5 | 4 | 3 | 1 | -1 |
| 1½ | O- | 9½ | 8½ | 7½ | 6½ | 5½ | 4½ | 3½ | 2½ | 1½ | -½ | -2½ |
| 0 | U | 8 | 7 | 6 | 5 | 4 | 3 | 2 | 1 | 0 | -2 | -4 |

### OTHER-THAN-FLOW CONSIDERED 1½ TIMES AS IMPORTANT AS FLOW.

(FLOW 2; OTHER THAN FLOW 3)

| FLOW | | OTHER-THAN-FLOW | | | | | | | | | | |
|---|---|---|---|---|---|---|---|---|---|---|---|---|
| | | 12 | 10½ | 9 | 7½ | 6 | 4½ | 3 | 1½ | 0 | -3 | -6 |
| | | A | A- | E | E- | I | I- | O | O- | U | X | XX |
| 8 | A | 20 | 18½ | 17 | 15½ | 14 | 12½ | 11 | 9½ | 8 | 5 | 2 |
| 7 | A- | 19 | 17½ | 16 | 14½ | 13 | 11½ | 10 | 8½ | 7 | 4 | 1 |
| 6 | E | 18 | 16½ | 15 | 13½ | 12 | 10½ | 9 | 7½ | 6 | 3 | 0 |
| 5 | E- | 17 | 15½ | 14 | 12½ | 11 | 9½ | 8 | 6½ | 5 | 2 | -1 |
| 4 | I | 16 | 14½ | 13 | 11½ | 10 | 8½ | 7 | 5½ | 4 | 1 | -2 |
| 3 | I- | 15 | 13½ | 12 | 10½ | 9 | 7½ | 6 | 4½ | 3 | 0 | -3 |
| 2 | O | 14 | 12½ | 11 | 9½ | 8 | 6½ | 5 | 3½ | 2 | -1 | -4 |
| 1 | O- | 13 | 11½ | 10 | 8½ | 7 | 5½ | 4 | 2½ | 1 | -2 | -5 |
| 0 | U | 12 | 10½ | 9 | 7½ | 6 | 4½ | 3 | 1½ | 0 | -3 | -6 |

**Figure 3.** Preprinted Matrix Tables for Different Ratios

**COMBINING FLOW and SERVICE RELATIONSHIPS**

Plant *Hevi-Duti Truck Co.* Project ______
By *LDT* With ______
Date *12-18* Sheet *1* of *1*

| Activity Pair (Identification) [a] | FLOW-OF-MATERIAL INTENSITY: From-To [b] | To-From [c] | Total 2-Way [d] | Order (Rank) [e] | Vowel Rating [f] | OTHER-THAN-FLOW RELATIONSHIPS: Consol. Rating [g] | Rating "Voice" [h] | Vowel Rating [j] | Reason Code [k] | Reference [l] | RESULTANT RELATIONSHIP (Not Calibrated) "√" (Incl. Ref. ltr. if needed) [m]: A | E | I | O | U | X |
|---|---|---|---|---|---|---|---|---|---|---|---|---|---|---|---|---|
| 10-1 | | | 40 | 4a | E | | | | | | | E | | | | |
| 1-1 | | | 40 | 4b | E | | | | | | | E | | | | |
| 1-2 | | | 15 | 11a | I | | | | | | | | I | | | |
| 1-3 | | | 10 | 16a | O | | | | | | | | | O | | |
| 1-4 | | | 15 | 11b | I | | | | | | | | I | | | |
| 2-3 | | | 15 | 11c | I | | | | | | | | I | | | |
| 10-2 | | | 6 | 19 | O | | | | | | | | | O | | |
| 11-2 | | | 9 | 18 | O | | | | | | | | | O | | |
| 11-3 | | | 45 | 3 | E | | | | | | | E | | | | |
| 2-5 | | | 15 | 11d | I | | | | | | | | I | | | |
| 3-6 | | | 10 | 16b | O | | | | | | | | | O | | |
| 3-7 | | | 20 | 9a | I | | | | | | | | I | | | |
| 3-8 | | | 20 | 9b | I | O | 3/8 | O/I | 3/3,5 | (a) | | | I | | | |
| 3-6 | | | 25 | 7 | I | | | | | | | | I | | | |
| 4-6 | | | 15 | 11e | I | | | | | | | | I | | | |
| 5-3 | | | 5 | 20 | O | | | | | | | | | O | | |
| 5-6 | | | 2 | 23 | O | | | | | | | | | O | | |
| 5-7 | | | 4 | 21a | O | | | | | | | | | O | | |
| 5-8 | | | 4 | 21b | O | | | | | | | | | O | | |
| 6-7 | | | 22 | 8 | I | | | | | | | | I | | | |
| 6-8 | | | 30 | 6 | I | | | | | | | | I | | | |
| 7-8 | | | 46 | 2 | E | I | 7/8 | I/I | 2/2 | | A- | | | | | |
| 8-1 | | | 100 | 1 | A | | | | | | A | | | | | |
| M-5 | | | | | | O | M/5 | O/O | 4,5/3 | | | | | O | | |
| M-6 | | | | | | E | M/6 | E/I | 3/3 | | | E | | | | |
| M-7 | | | | | | O | M/7 | O/O | 3,5/3 | | | | | O | | |
| M-8 | | | | | | O | M/8 | O/I | 3,5/3 | | | | | O | | |
| M-11 | | | | | | O | M/- | O/- | 5/- | | | | | O | | |
| 1-8 | | | | | | X | 7/8 | U/X | -/6 | | | | | | | X |
| 3-4 | | | | | | I | 3/4 | I/I | 2/2 | | | | I | | | |
| 4-5 | | | | | | O | 4/5 | O/O | 4/3 | | | | | O | | |
| - | | | | | | | | | | | | | | | | |
| - | | | | | | | | | | | | | | | | |
| - | | | | | | | | | | | | | | | | |
| - | | | | | | | | | | | 2 | 4 | 11 | 13 | | 1 |

| Vowel Rating | CLOSENESS | No. of Occur's |
|---|---|---|
| A | Absolutely Necessary | 2 |
| E | Especially Important | 4 |
| I | Important | 11 |
| O | Ordinary Close. OK | 13 |
| U | Unimportant | (6) |
| X | Not Desirable | 1 |

| Code | REASON |
|---|---|
| 1 | Flow of Materials |
| 2 | Same Supervisor |
| 3 | Freq'cy of contact |
| 4 | Share same equip. |
| 5 | Movem't of supplies |
| 6 | Dirt & contamination |
| 7 | |
| 8 | |
| 9 | |
| 10 | |
| 11 | |
| 12 | |
| 13 | |
| 14 | |
| 15 | |

Reference Notes:
a) Work Sampling study #WS-6155 indicates urgency of delivery not a problem.
b) "U" rated relationships not recorded.

NOTE:
- Col. d = b + c
- Col. e based on d; f based on e
- Col. g is independent of b thru f, but supported by j thru l
- Col. m posted to Relationship Chart



**Figure 4.** Worksheet for Combining Relationships

rating. The combined resultant relationships are then posted to a "combined" relationship chart, which, when colored, is a ready basis for preparing the relationship diagram.

It is possible to prepare the relationship diagram directly from the combining-flow-and-service-relationships worksheet (Figure 4). However, it is awkward and less reliable to do so (even though coloring the vowel-letter entries in column m) because all potential relationships are not automatically accounted for since the pairs of activities on each line are usually not listed in numerical sequence (inasmuch as flow is usually entered first and other-than-flow relationships entered later). Also the planner must repeatedly shuffle through several worksheets since the postings are spread out. It is usually best, therefore, to repost the resultant combined ratings on a combined relationship chart, before making the relationship diagram.

There is at least one other method of arriving at a resultant relationship between activities for flow and other-than-flow reasons. It involves rating for other-than-flow reasons using the flow relationships as a basis for comparison and then taking the more demanding of the flow and other-than-flow closeness ratings for each pair of activities. This procedure assumes that honoring the dominating relationship more than satisfies the requirements of the lesser one. The specific steps in this method are

1. Calibrate flow intensities between activity pairs by vowel-letter ratings.
2. Establish several typical benchmark pairs of activities for each of the vowel-letter flow ratings: A, E, I, O, and U.
3. Rate all pairs of activities for other-than-flow reasons using the flow benchmarks as a direct comparative guide.
4. Select the dominant rating between the flow and other-than-flow relationships for each pair of activities.

## Appendix XIII—Systematic Handling Analysis (SHA)

*Materials handling* is *moving*, transporting, or physically relocating *materials*, products, items, substances, or things, by *methods* that include equipment, containers, and a working system of routes, people, and procedures.

The fundamentals of materials handling—and therefore, the basis on which any handling analysis must rest—are *MATERIALS, MOVES*, and *METHODS*.

### SHA—What Is It?

*Systematic Handling Analysis* is an organized, universally applicable approach to any materials handling project. SHA consists of

1. Framework of Phases
2. Pattern of Procedures
3. Set of Conventions

### The Four Phases of SHA

As each handling project runs its course—from the initial statement of objectives to installed physical reality—it passes through four phases:

| | |
|---|---|
| Phase I | External Integration |
| Phase II | Overall Handling Plan |
| Phase III | Detailed Handling Plans |
| Phase IV | Installation |

Refer to the upper lefthand portion of the SHA Capsule Summary, Figure 1. Phase I is *external integration*. Here the movements to and from the total area, or areas, in question are appraised. First consider the movement of material outside the problem area. This way, the specific handling problem is correlated with outside situations or external conditions, over which the planner may or may not have control.

Phase II involves the *overall handling plan*, in which the method(s) of moving the materials among the major areas is established. Overall decisions must be made as to the basic routing system, the general type(s) of equipment, and the transport units or containers to be used.

Phase III is the *detailed handling plans*. It concerns the movements of materials from one specific workplace or piece of equipment to another within each major area. Decisions about detailed handling methods, such as the specific system, equipment, and containers to be used, must be made.

Phase IV is *installation*. No layout project is complete until it is installed. Here plan the necessary make-ready; procure the equipment; complete the training of indi-

# SYSTEMATIC HANDLING ANALYSIS
## (CAPSULE SUMMARY)

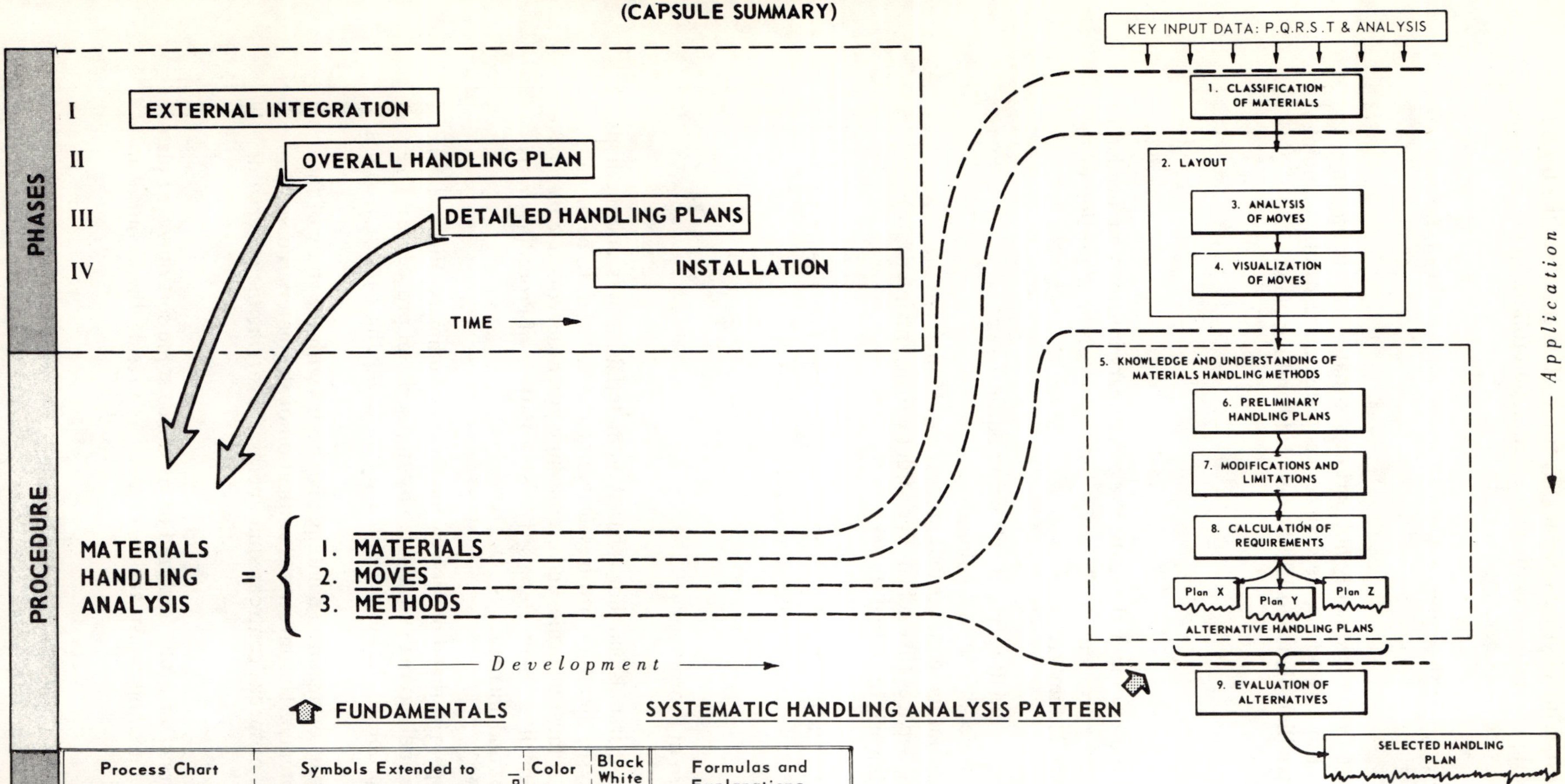


**CONVENTIONS**

| Process Chart Symbols & Action | Symbols Extended to Identify Activities & Areas | Color Ident. | Black White ** | Formulas and Explanations |
|---|---|---|---|---|
| * ○ Operation | ○ Forming or Treating Areas | Green** | [diagonal hatching] | Intensity (I) = $\frac{nP}{t}$ Designated by width of flow line |
| | ○ Assembly, incl. Sub-Assem. and Dis-Assembly | Red** | [vertical lines] | |
| * ⇨ Transportation | ⇨ Transport-related Activities/Areas | Orange Yellow** | [dense dots] | Distance (D) Designated by length of flow line |
| ◯ Handling | ◯ Handling Areas | Orange Yellow** | [dense dots] | Transport Work (TW) TW = I×D = $\frac{nP}{t}$ ×D |
| * ▽ Storage | ▽ Storage Activities/Areas | Yellow** Orange | [light dots] | Designated by area of flow line |
| * D Delay | D Set-down or Hold Areas | Yellow** Orange | [light dots] | Example: D, I, TW |
| * □ Inspection | □ Inspect, Test, Check Areas | Blue** | [horizontal lines] | |
| *A.S.M.E. Standard **I.M.M.S. Standard (Adopted as basic to SHA procedure) | Service & Support Activities/Areas | Blue** | [horizontal lines] | |
| | ⇧ Office or Building Areas, Building Features | Brown** (Gray) | [grid] | M-H-Systems: D-Direct C-Central K-Channel (Kanal) |

Use of color identification of areas is optional

| Vowel Letter | No. Value | No. of Lines* | Intensity Rating of Material Moves | Color Code | Evaluating Description | Vowel Letter |
|---|---|---|---|---|---|---|
| A | 4 | //// | Abnormally High Intensity of Moves | Red** | Almost Perfect - (Excellent) | A/4 |
| E | 3 | /// | Especially High Intensity of Moves | Orange Yellow** | Especially Good - (Very good) | E/3 |
| I | 2 | // | Important Intensity | Green** | Important Results Obtained - (Good) | I/2 |
| O | 1 | / | Ordinary Intensity | Blue** | Ordinary Results, OK - (Fair) | O/1 |
| U | 0 | | Unimportant Moves of Negligible Intensity | – | Unimportant Results - (Poor) | U/0 |
| | | | | | Not Acceptable (Not Satisfactory) | X/-1 |

* Dash (dotted) line indicates half a line, or half-way between two ratings — recorded by minus sign behind vowel letter.





Figure 1.

viduals; schedule and install the physical handling facilities. Then complete the try-out of planned handling methods, verify operating procedures, and follow up on the completed installation to be sure it is operating properly.

These four phases come in sequence, from a time standpoint. And, for best results, they should overlap.

Phase I and Phase IV are frequently not part of the materials-handling engineer's specific problem. For this reason, we shall concentrate on the strictly planning phases: II, the overall handling plan, and III, the detailed handling plans.

### Key Input Data

Let us examine the key input elements, the preliminary information, needed to analyze a materials-handling problem. As shown in the upper righthand portion of the Capsule Summary, Figure 1, they are:

P Products or materials (parts, items, commodities)
Q Quantities (sales-or-contract volume)
R Routing (operating sequence and process requirements)
S Supporting Services (like scheduling, order writing, maintenance)
T Time (timing and operating times)

These key input data are virtually identical with the input data required for SLP—or for almost any facilities planning project for that matter.

### The SHA Pattern of Procedures

As noted above, materials handling is based on materials, moves, and methods. Materials-handling analysis, therefore, involves analyzing the *materials* to be moved, analyzing the *moves* that must be made, and establishing practical economic *methods* to accomplish the movement of the materials. The SHA Pattern of Procedures rests squarely on these three fundamentals.

Refer to the upper righthand portion of the Capsule Summary, Figure 1. The Pattern of Systematic Handling Analysis is a step-by-step series of procedures. The more complicated the problem, the more useful and timesaving this pattern becomes.

The analytical part of making a handling plan begins with the study of the materials (products or items). This involves classification of the materials based on their quantities, timing, or special control requirements. *Classifying the materials* is Box 1 of the Pattern.

Before the moves can be fully analyzed or visualized some sort of *layout* plan with which the handling methods must work is needed. As a result, the layout plan—

real or on paper—is Box 2 of the Pattern of SHA. Actually, it forms a larger "box" into which the analysis and visualization of moves are fitted.

Next comes the *analysis of moves*, Box 3 in the SHA Pattern. Basically the analysis of moves involves determining the intensity and character of moves required to be made for each material on each route (origin to destination).

The next step is to *visualize the moves*, Box 4. Here the analysis is translated into a visual "picture." This is commonly done by a flow diagram or a distance-intensity plot.

Before a solution can be developed, the planner needs an *understanding of materials-handling methods*, Box 5. This is indicated by the large fenced-off box within which the methods are selected.

Box 6 is where we set *preliminary handling plans* regarding system, equipment, and transport unit (or container); and it is here all the information gathered is related and translated into physical methods. Actually, the planner develops a number of logical preliminary possibilities.

Based on various *modifications and limitations*, Box 7, preliminary plans are adjusted. Here, modify and adjust each plan, converting what is possible to what is practical.

The purpose of setting plans and adjusting or modifying them is to eliminate all ideas which are not workable. But before making any real selection of the best methods, the planner needs to calculate the number of pieces of equipment or transport units, the costs involved, and the operating time involved. This we call *calculation of requirements*, Box 8.

The last step in the SHA Pattern is to make an *evaluation of alternative* plans, Box 9. This is usually an evaluation of costs and intangibles. As a result of this evaluation, a choice of one of the alternatives is made—although frequently a combination of two or more plans results from the evaluation.

The alternative plan chosen is the *selected handling plan.*

### SHA Pattern Applies to Both Phases II and III

The SHA Pattern applies to both Phase II, Overall Handling Plan, and Phase III, Detailed Handling Plans. That is, the same pattern of steps prevails, although the degree of application will be different in the two phases. Of course, the handling methods for the detailed areas themselves will have to integrate with the overall handling methods selected in Phase II.

### SHA Conventions

Certain *conventions* are used in applying the techniques in each step of the SHA Pattern. There are various symbols, colors, letters, lines, and numbers to identify various origins and destinations, to visualize moves, to rate alternatives, and the like. These conventions are shown in the lower portion of the Capsule Summary, Figure 1.

### Illustrated Example

To help understand the SHA Pattern, an actual example is shown in Figure 2. For this project the SHA Pattern was used in planning the handling methods—although all details are not shown. The case example is intentionally oversimplified for ease of understanding.

### SHA—an Organized Approach

In this synopsis, we have examined the basic methodology of Systematic Handling Analysis: an approach to solving problems, a series of steps to follow, and a set of recording, rating, and visualizing conventions. Basically, it includes a *framework of four phases* of analysis, each of which overlaps the preceding and following phases. It includes a *pattern of step-by-step procedures* to follow in Phase II and Phase III.

The pattern recognizes and processes the key input data and integrates layout plans as part of the handling-plan analyses. And SHA has a *set of conventions* to aid and speed its application, conventions that are essentially the same as those adopted by the American Society of Mechanical Engineers in its Standard on Operation and Flow Process Charts and by the International Materials Management Society in its Standard Color Codes for Plant Layout and Materials Handling Analysis.

SHA is the development of Richard Muther and Knut Haganaes. A book entitled *Systematic Handling Analysis (SHA)* is available in English language from Cahners Books, 89 Franklin Street, Boston, Massachusetts, 02110. Editions are also available in German, Czech, and Japanese.

Figure 2

# S.H.A. in Action

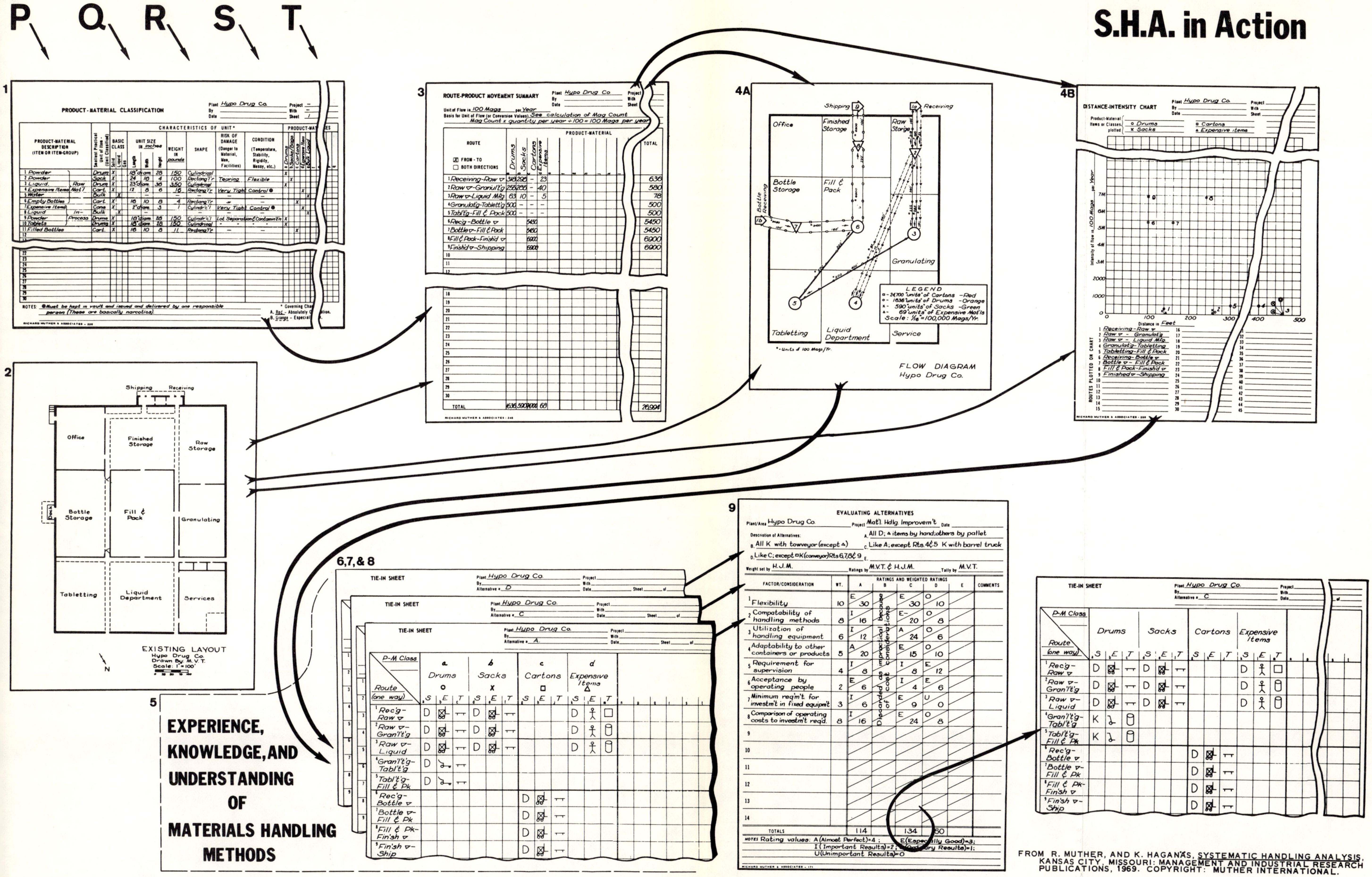


FROM R. MUTHER, AND K. HAGANÄS, SYSTEMATIC HANDLING ANALYSIS, KANSAS CITY, MISSOURI: MANAGEMENT AND INDUSTRIAL RESEARCH PUBLICATIONS, 1969. COPYRIGHT: MUTHER INTERNATIONAL.

## Appendix XIV—SLP and Layout by Computer

Basically, there are four computer-aided layout planning programs available in the United States today that print-out preliminary or adjusted layout plans. They are CRAFT (Computerized Relative Allocation of Facilities Technique), CORELAP (Computerized Relationship Layout Planning), ALDEP (Automated Layout Design Program), and RMA Comp I (Richard Muther & Associates' contribution to the field). There are other programs, both in the United States and abroad, but they are less readily available to industry.

The input requirements for all four programs are similar. All four require the fundamental inputs of relationships and space.

RMA Comp I is designed to do by computer what Systematic Layout Planning (SLP) does manually. Further manual adjustment of the output is necessary with all existing computer programs in order to develop a practical workable layout. RMA Comp I, therefore, does not make a print out with the activities shown adjoining each other.

The inputs for RMA Comp I are essentially the same as the data developed during the mental-manual analysis. As a result, relationships will have already been developed and documented on the Relationship Chart (RMA-131). Space requirements also will have been developed and documented. The Space Specifications Sheet (RMA-158) provides a good summary of the area requirement for each activity and also provides for a numerical code describing the "type" or kind of space needed. Hence, the three computer inputs—relationships, space, and type of area—can be conveniently summarized on the Relationship Chart and the Space Specifications Sheet, and submitted for key punching. See Figure 1. Note the similarity between forms RMA-131 and RMA-158 (referred to in this paragraph) and their companion, smaller forms RMA-130 and RMA-150 respectively.

Basically, RMA Comp I develops a theoretical best activity relationship diagram. It then explodes the diagram, adds the space, and prints out a space relationship diagram, without connecting relationship lines.

RMA Comp I first calculates which of the activities in the layout is the busiest or most related. The sums of each activity's closeness relationships with all other activities are compared, and the activity with the highest total closeness relationship (TCR) count is selected and located first in the layout matrix.

Each subsequent activity is placed so that all its relationships are considered before it is located. In other words, before a given activity can be located, its relationships to activities not yet placed and to those activities already placed, are conconsidered. When placing each activity, room is left for related activities to be placed later. At the same time, a check is made to see that X relationships are satisfied. In this manner, A-related activities are selected and placed, followed by E-, I-, and O-related activities. A relationship diagram is thus formed—still exclusive of space requirements. This diagram is then exploded and the required amount and type of space are assigned to each activity in the diagram. The print-out takes the form of a space relationship diagram. See Figure 2.

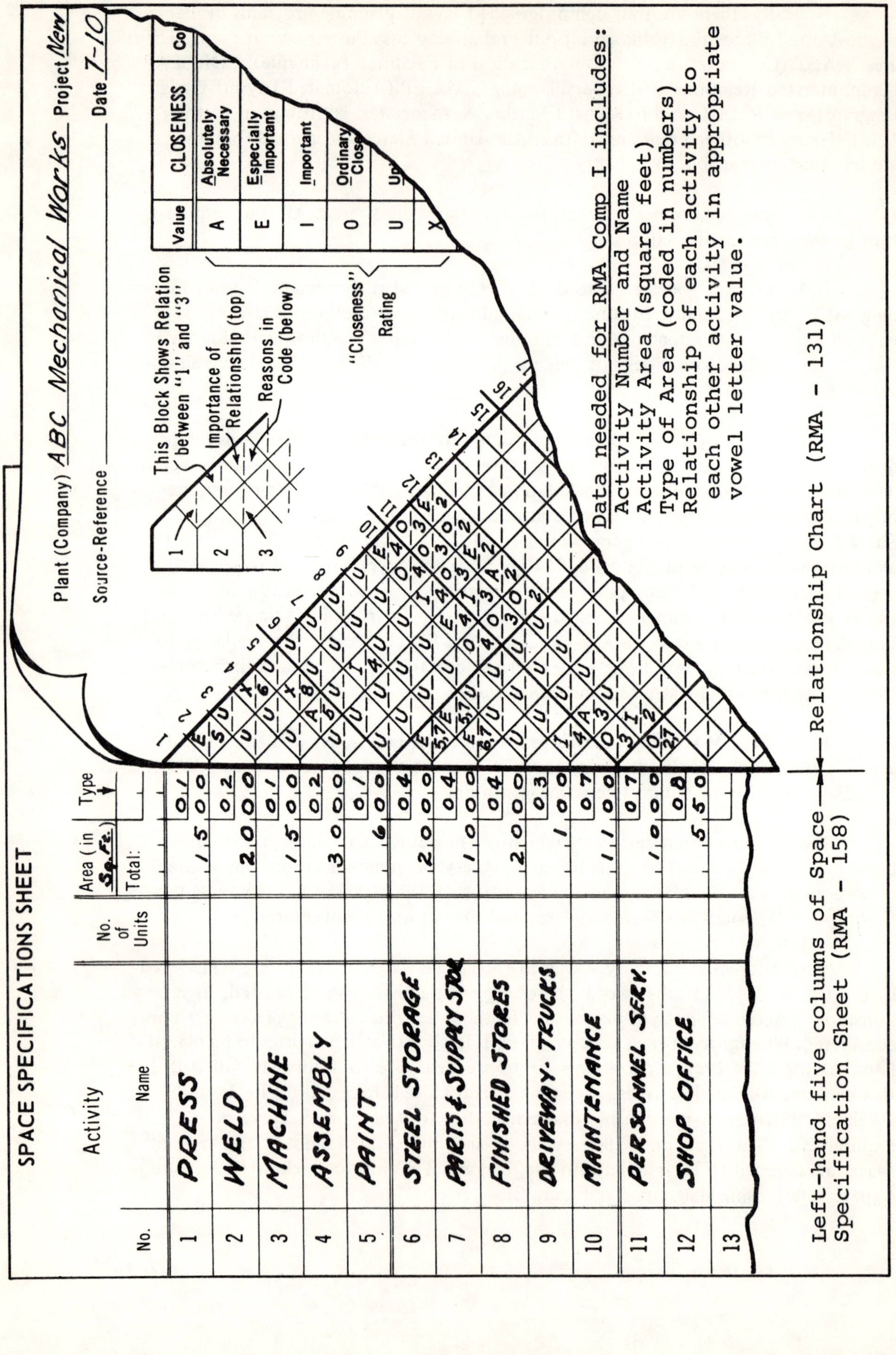


Figure 1. Input Data for RMA Comp I

ABC MECHANICAL WORKS
SPACE RELATIONSHIP DIAGRAM -- RMA Comp I

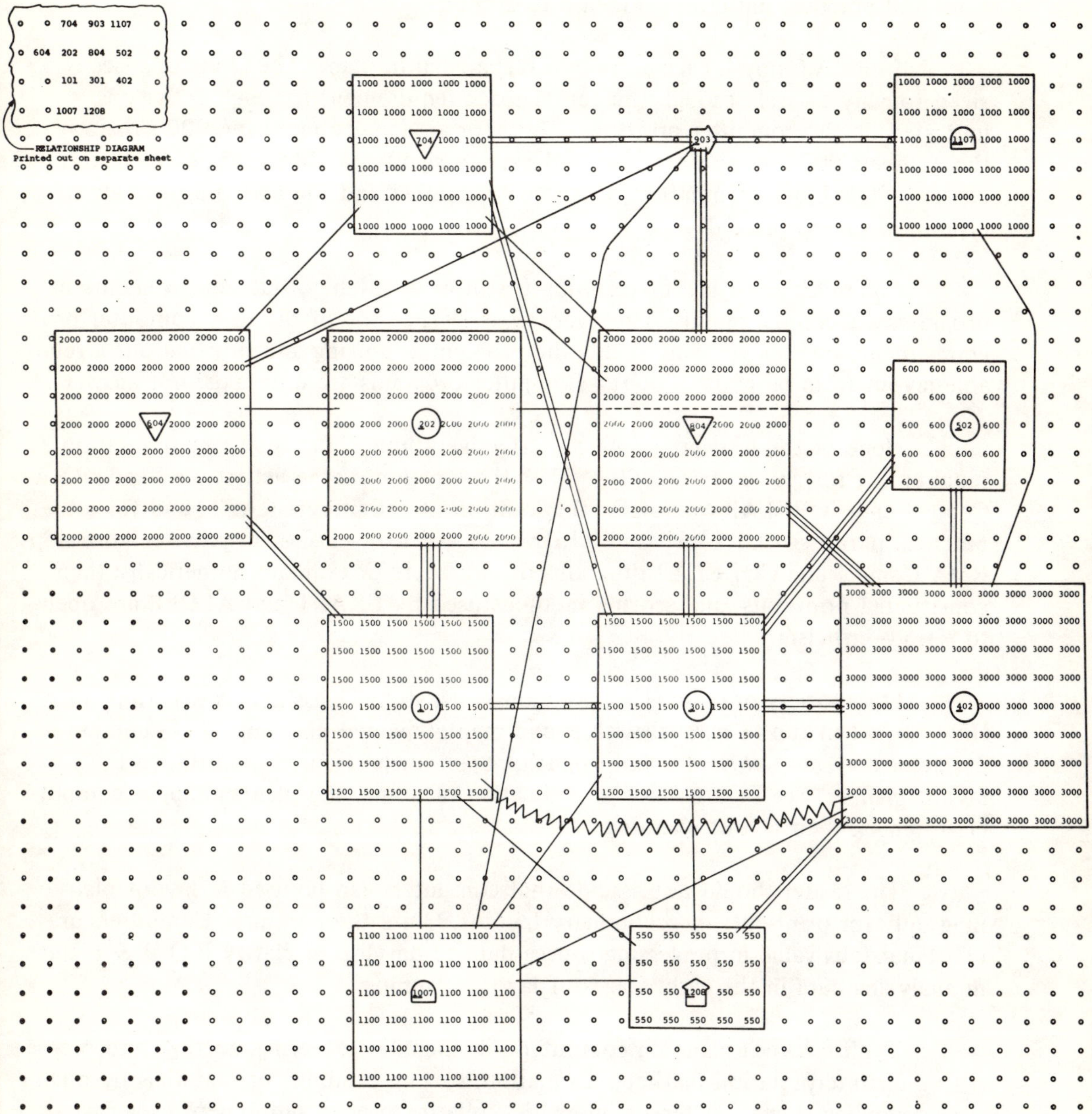


Space Relationship Diagram developed by RMA Comp I computer program from the input data shown in Figure 1. Note that lines outlining each area, have been added, as have relationship lines, type of activity symbols, and underlining of activity area identification number. A relationship diagram for the same problem (not printed out on the same sheet by the computer) is shown in the upper left-hand corner.

**Figure 2.** RMA Comp I Printout

The space-relationship diagram, equivalent to Box 6 of the SLP Pattern of Procedures, must then be manually adjusted into a layout, and modifying considerations and practical limitations taken into account.

In RMA Comp I, each activity is rectangular in shape. The block of space is sized roughly to scale though the actual space requirement for each activity is indicated on the computer print-out. Thus, for an activity requiring 900 square feet, the numbers 900 are repeated over and over to signify the block of space. In the center of this block of numbers, the activity number and the code number for the type of space are printed.

Computer-aided layout planning is still in its infancy, but improvements in programs are coming rapidly. However, as of 1972, none of the four computer programs mentioned above really meets the problem of solving and printing out a reliable layout from basic data inputs. Computer programs are only aids; not answers.

None of the four programs can effectively honor a shape or configuration requirement for any given activity, unless the location of the activity is fixed. RMA Comp I and CORELAP cannot fix the location of activities. CRAFT, and for all practical purposes CORELAP and ALDEP, do not really honor negative relationships. RMA Comp I and CORELAP programs do not score or evaluate numerically their constructed print-outs; and scoring methods used by CRAFT and ALDEP are open to serious criticism.

For a variety of reasons, all programs must still be classified as experimental. However, when run in parallel with sound mental-manual planning, these computer programs may help prevent an oversight or open unforeseen possibilities to the layout planner. See Figure 3 for an indication of how some planners apply computers to check their layout planning.

The reader should not assume that computers can be used in layout planning only for print-outs of preliminary layouts. Quite the contrary. Computers are of tremendous value in processing source data to develop or clarify P Q R S T and in analyzing data in the early stages of layout planning.

Typical layout-planning computer aids include: gathering, sorting, and preparing for interpretation market information; sorting products or materials by quantity; compiling flow-of-materials intensities; relating process equipment capacity and/or space requirements to product lines, quantity forecasts, and market areas.

## USE OF COMPUTERS IN LAYOUT PLANNING

They are an aid only!

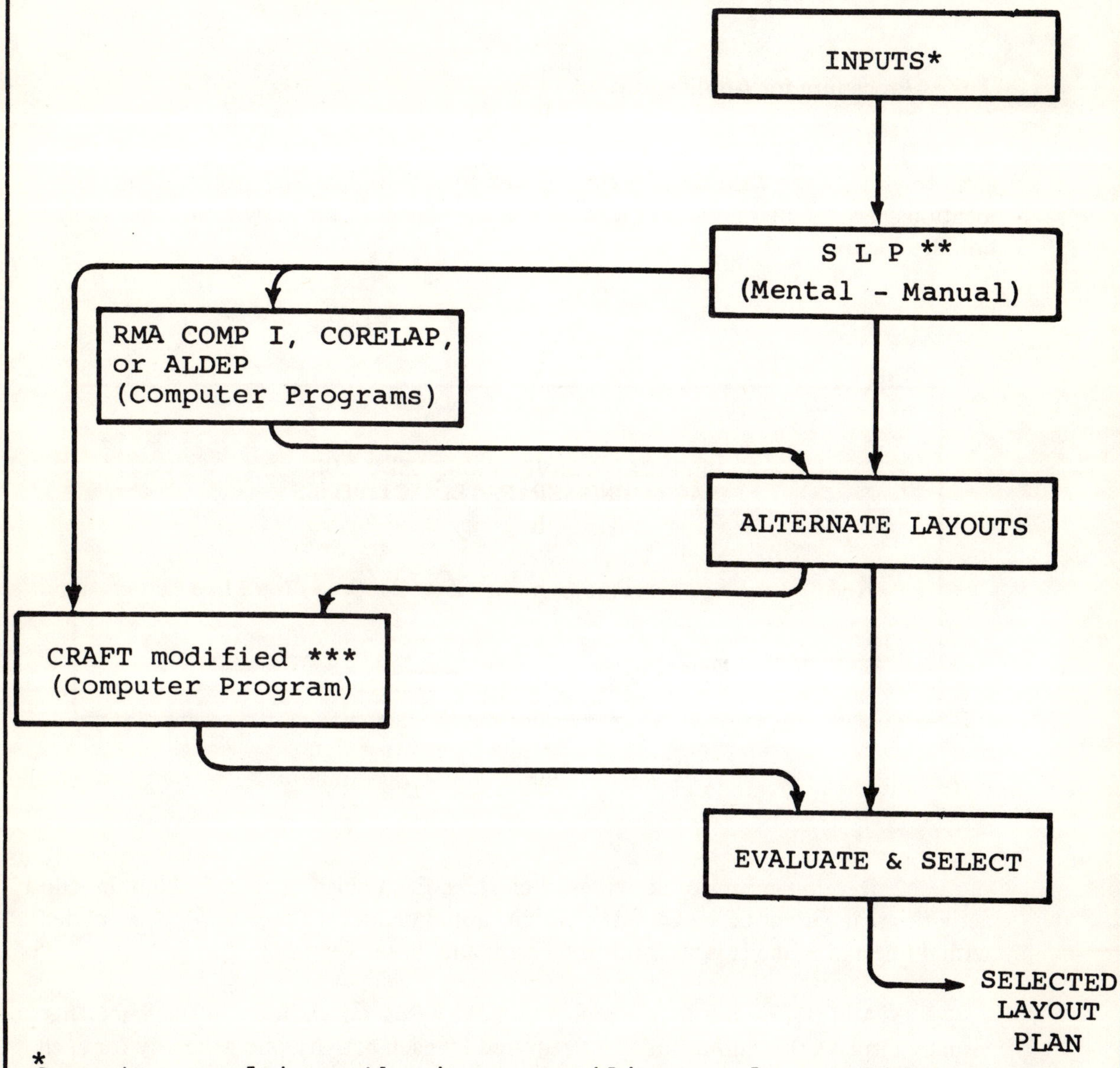


* Computer used in gathering, compiling, and consolidating data, as appropriate.

** On layout planning or site planning projects longer than six weeks (in Phase II, or one area in Phase III) --- run computer program in parallel with "straight S L P". It may offer an overlooked alternative.

*** Use modified CRAFT computer program to check the most likely alternative as insurance against having missed conceiving all logical alternatives.

Figure 3.

## Appendix XV—SLP in Multi-Story Buildings

SYSTEMATIC LAYOUT PLANNING (SLP) is an excellent technique for correctly assigning activities (work groups) to the various floors of multi-story buildings or to the various buildings in a multi-building complex.

### Procedure for Application

1. Follow SLP Pattern to Box 3. That is, make an Activity Relationship Diagram to visually show the desired closeness of all activities to each other. Mandatory locations of any activities are indicated by coding beneath the activity number or symbol. See Figure 1.

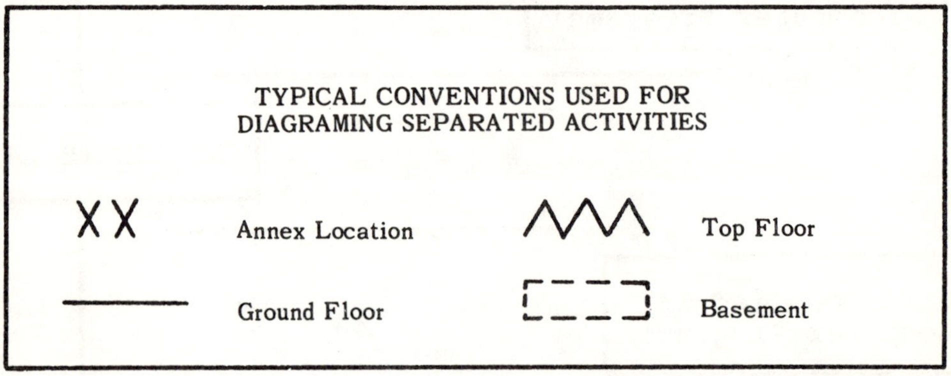


Figure 1.

2. Develop groups of activities or clusters. Each cluster generally contains those activities that cannot be separated or which should be located together—those related with 4-line and 3-line closeness ratings. For example, see Figure 2.

3. Add floor space requirements to make a Space Relationship Diagram. This visually displays the amount of space required by each activity and generally for each cluster as well. See Figure 3.

4. Make a Cluster Relationship Diagram to indicate the desired closeness among all clusters and the non-clustered individual activities. On simple projects, this step is often omitted or is used as an alternate for making a space relationship diagram. See Figure 4.

5. Assign activities to specific floors based on the above relationships. See Figure 5. The kind of space or type of activity can become a more dominant consideration than the closeness relationships when balancing activities against the available space per floor. This assignment of activities generally results in two or more alternative arrangements.

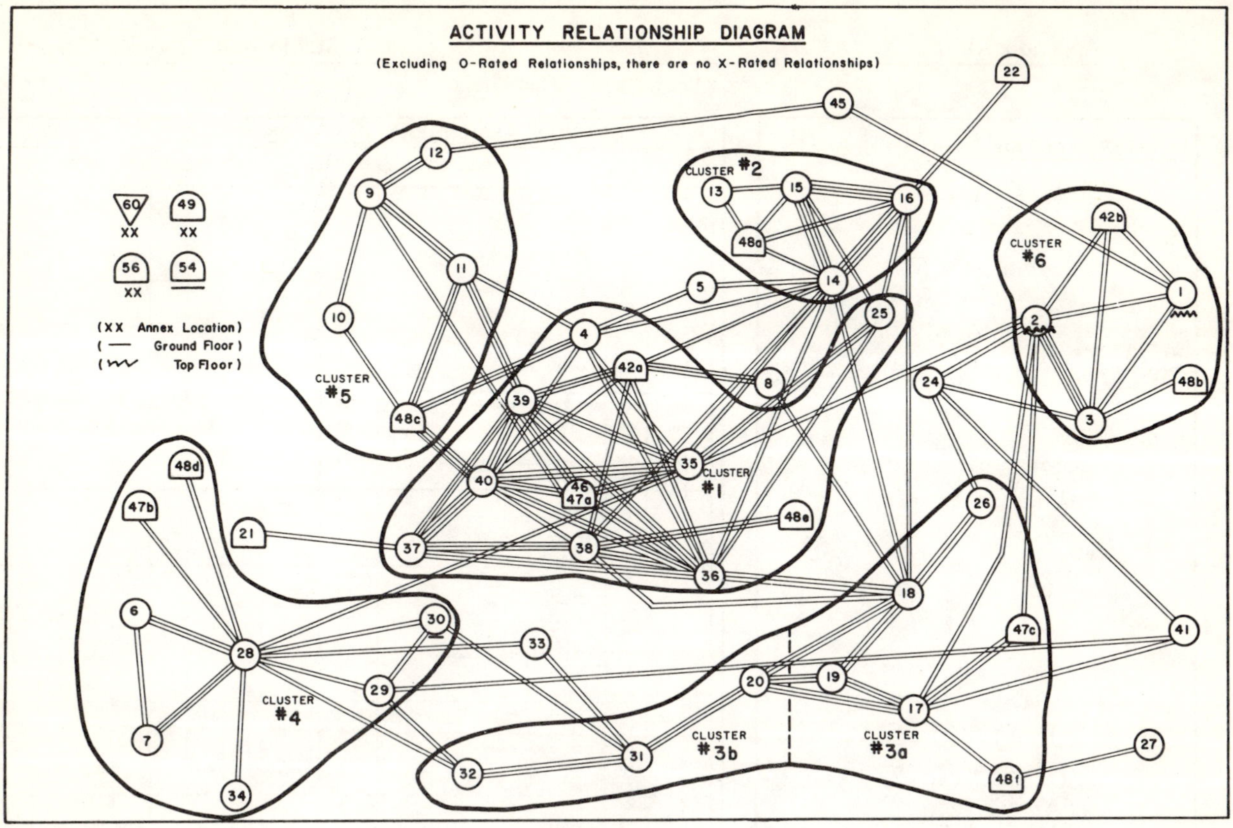


Figure 2.

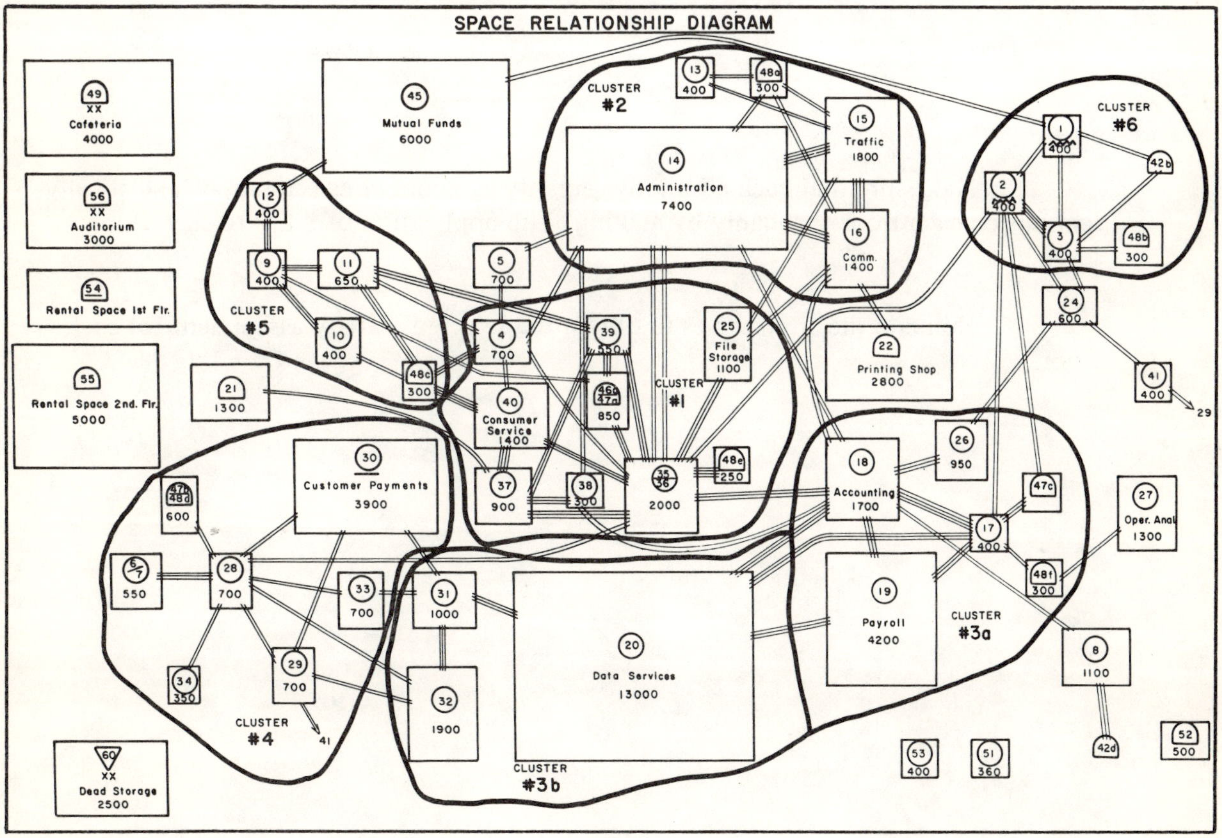


Figure 3.

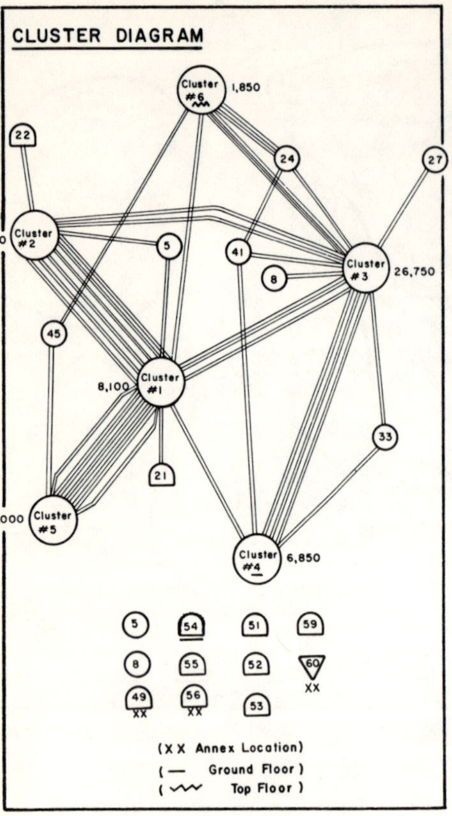


Figure 4.

**ACTIVITY ASSIGNMENT BY FLOOR** Total of 7 floors 14,000 sq. ft. per floor (net)

| Cluster No. | Activity No. | Activity Name | Sq. Ft. (x1000) | Sub-Tot. Sq. Ft. (x1000) | Annex | 1 | 2 | 3 | 4 | 5 | 6 | 7 |
|---|---|---|---|---|---|---|---|---|---|---|---|---|
| | 4 | Market Research | 0.7 | | | | | | | 0.7 | | |
| | 35 | V-P Marketing | 0.4 | | | | | | | 0.4 | | |
| | 36 | Mkt. Admin. | 1.6 | | | | | | | 1.6 | | |
| | 37 | Mkt. Services | 0.9 | | | | | | | 0.9 | | |
| | 38 | Advertising | 0.3 | | | | | | | 0.3 | | |
| | 39 | Public Relations | 0.6 | | | | | | | 0.6 | | |
| | 40 | Consumer Service | 1.4 | | | | | | | 1.4 | | |
| #1 | 46 | Large Conf. Room | 0.5 | | | | | | | 0.5 | | |
| | 47 | Medium Conf. Room | 0.4 | | | | | | | 0.4 | | |
| | 48e | Small Conf. Room | 0.3 | | | | | | | 0.3 | | |
| | 25 | File Room | 1.1 | 8.2 | | | | | | 1.1 | | |
| | 13 | V.P. Services | 0.4 | | | | | | | | 0.4 | |
| #2 | 14 | Administration | 7.4 | | | | | | | | 7.4 | |
| | 15 | Traffic | 1.8 | | | | | | | | 1.8 | |
| | 16 | Communications | 1.4 | | | | | | | | 1.4 | |
| | 48a | Small Conf. Room | 0.3 | 11.3 | | | | | | | 0.3 | |
| | 17 | V.P.-Finance | 0.4 | | | | | | 0.4 | | | |
| #3-a | 18 | Accounting | 1.7 | | | | | | 1.7 | | | |
| | 19 | Payroll | 4.2 | | | | | | 4.2 | | | |
| | 26 | Ins. & Taxes | 0.95 | 7.25 | | | | | 0.95 | | | |
| | 47 | Medium Conf. Room | 0.4 | | | | | | 0.4 | | | |
| Not in any Cluster | 5 | Statistics | 0.6 | | | | | | | 0.6 | | |
| | 8 | H and A Collections | 1.1 | | | | | | 1.1 | | | |
| | 41 | V.P. and Gen. Counsel | 0.4 | | | | | | | | | 0.4 |
| | 45 | Mutual Funds | 6.0 | | | | | | | | | 6.0 |
| | 49 | Cafeteria | 4.0 | | 4.0 | | | | | | | |
| | 51 | Credit Union | 0.4 | | | | | | | | | 0.4 |
| | 52 | First Aid | 0.5 | | | | | | 0.5 | | | |
| | 53 | Employee Store | 0.4 | | | | | | | | | 0.4 |
| | 54 | Rental Space - 1st. floor | 2.5 | | | 2.5 | | | | | | |
| | 55 | Rental Space - 2nd. floor | 5.0 | | | | 5.0 | | | | | |
| | 56 | Auditorium | 3.0 | | 3.0 | | | | | | | |
| | 59 | Vault | 0.3 | | | | | | 0.3 | | | |
| | 60 | Dead Storage | 2.5 | | 2.5 | | | | | | | |
| | | Total | 87.4 | | 9.5 | 9.7 | 12.6 | 6.1 | 12.7 | 13.5 | 12.6 | 11.0 |

Figure 5.

6. Make sure that each alternative activity assignment can in fact be realistically laid out on each floor—probably by making a sub-application of S L P to each floor level.

7. Evaluate the alternatives to obtain the selected activity assignment for each floor.

## Appendix XVI—Synopsis of Simplified SLP

Simplified Systematic Layout Planning is an abbreviated version of Systematic Layout Planning (SLP). This simplified version of SLP is best suited to office areas up to about 3,000 square feet, to shop or laboratory areas up to about 5,000 square feet, and in storage areas up to about 8,000 or 10,000 square feet. Basically, it comprises a set of six procedures or steps that encompass the three fundamentals of any layout planning project.

1. *Relationships* between various functions, activities, or work stations
2. *Space* in a certain amount, kind, and/or shape for each activity
3. *Adjustment* of the relationships and spaces into a layout plan(s)

The six steps form a logical sequence which takes the planner from raw data to a workable layout plan. Figure 1 illustrates each of the six steps, and each is described below in more detail. Note that a different symbol is used to help identify each step, though the symbol has no significance itself in the planning.

### Step 1. Chart the Relationships

To determine which activities (or departments or work groups) should be located next to others, the relative closeness required must be measured, and recorded in a simple way that shows the relative closeness desired between each pair of activities. In addition, we should have some way of recording *why* that particular closeness rating is assigned.

The Relationship Chart (RMA-130) is a device for recording, in an organized way, the results of these decisions. A vowel-letter closeness-desired rating (A, E, I, O, and U in descending order of value) is used; X is used to show a not desirable relationship. The closeness-rating letter is posted in the upper half of the diamond-shaped block that shows the relationship between a pair of activities. Additionally, the "reason" is coded by numbers, entering the number in the lower half of the block. The reason coded to each number is explained in a separate section of the form. In this way, the individual relationship between every pair of activities is rated, substantiated, and recorded.

### Step 2. Establish Space Requirements

In the next step, establishing space requirements, the same activities as in Step 1 are listed. Then establish and record the space required to support each activity. These amounts of space can be established in several ways. One is to determine, for each individual activity, the amount of space now used; then, apply a factor for the effectiveness of utilization of existing space and another factor for any projected change in the level of operations to be supported. Another is to sketch out to scale, or otherwise tally, the space for each piece of equipment (allowing for working area, access aisles, maintenance room and the like); calculate the area required to house each work area; and total the space required for the entire operation. For rest rooms, private offices, and parking areas, standard space requirements may be used.

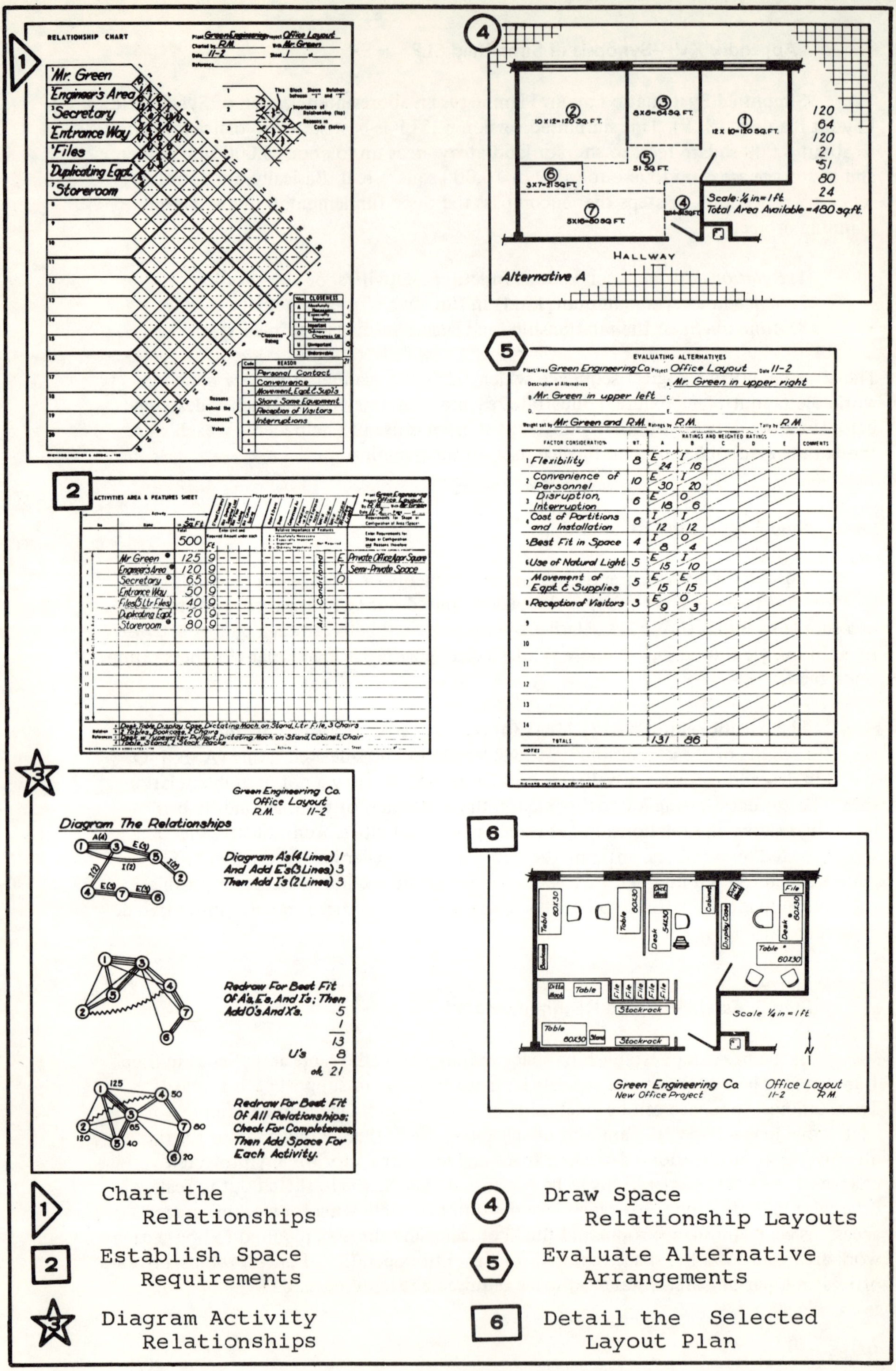


**Figure 1.** The Six Steps of Simplified S L P

The center portion of the Activities Area and Features Sheet (RMA-150) provides a way of entering pertinent data about physical features that are required. And, on the righthand side of the form, there is space for noting any specific shape or configuration—and the reasons for it.

### Step 3. Diagram Activity Relationships

So far, we have merely tabulated accumulated data. Now convert those data into a more usable form by preparing an activity relationship diagram. Circles, numbered for identification, are used to represent the activities. Each pair of activities is connected by parallel lines corresponding to the ratings on the relationship chart prepared in Step 1. The highest rated relationships (A) are represented by four connecting lines; the E's and I's with three and two lines, respectively. The diagram is then redrawn for the best fit of relationships, and the O's and X's added. X's are shown as zig-zag or wiggly lines. The diagram may be rearranged a second or third time for best fit of all relationships.

The goal is to place activities with the highest closeness ratings nearest each other, and lower ratings relatively farther away. When the best arrangement is worked out, the required floor space is marked on the diagram beside each activity circle. This completed diagram represents the theoretical ideal arrangement of activities without consideration of the actual space involved for each activity. The relationship diagram is the most important step in the entire procedure. If it is not properly done, the planner cannot be certain of getting a layout with the best overall arrangement.

### Step 4. Draw Space Relationship Layouts

The first two fundamental problems, relationships and space, having been solved, the diagram can be adjusted into a layout.

Draw the preliminary layout to scale on a sheet of graph paper or grid-line tracing paper laid over a floor plan or building print. The space required for each activity is blocked in, the relative position of each activity being maintained according to the relationships diagramed in Step 3. Ordinarily, the layout will need some adjustments to allow for such factors as availability of utilities, personnel convenience, procedures and controls, building features, equipment configurations, access ways, and the like. Such considerations influence the layout, whether the problem is a rearrangement of existing facilities or a completely new area.

Several workable layout plans usually result from this planning, and normally the two, three, or four best alternatives are drawn up.

### Step 5. Evaluate Alternative Arrangements

An evaluation procedure is now used to assess objectively the relative value of the alternative layouts. First identify each alternative plan on the form used—RMA-171 is designed specifically for this purpose. Then list all the objectives to be achieved and factors influencing the choice. Assign relative weights to each factor—starting with 10 for the most important. Then rate the effectiveness of each arrangement for each factor, again using A, E, I, O, and U, in descending order of results provided by each plan.

After all ratings are thus made and recorded, convert the letter ratings to corresponding numerical values: A = 4, E = 3, I = 2, O = 1, and U = 0. The numerical values are then multiplied, each by its factor's weight value, and the weighted, rated values down totaled for each alternative arrangement. The highest total value should indicate the most suitable layout.

### Step 6. Detail the Selected Layout Plan

The final step is to redraw the layout to scale, identify the areas, show major features and equipment, draw in the details of each item of equipment and/or machinery, and show individual utilities as required. Reappraise the fit of these details; adjust as necessary; provide appropriate identification data and dimensions; and reproduce the required number of copies. After final approval, the planner should be able to turn the plans over to those persons who will actually install the equipment.

Simplified SLP should not be used for large or complex problems. Specifically, it is not suited to layouts where a high or dominant flow of materials is involved. In such cases, the full Systematic Layout Planning procedure is preferable. But the simplified method is ideal for small office, laboratory, or service areas. It is often used to train service department managers so they can plan their own Phase III detail layouts, and so save the planners time for the more complex areas. Banks, small businesses, and all manner of service and office operations have put Simplified SLP to good practical use. Indeed, just as Work Simplification takes the basic elements of cost reduction right to the foreman and supervisors, so Simplified S L P takes the basic elements of layout planning right to the non-technically trained supervisor or business manager.

A more complete explanation of Simplified SLP is available in the booklet, *Simplified Systematic Layout Planning*, by Richard Muther and John Wheeler. English-language copies of this booklet can be obtained from Management and Industrial Research Publications, P. O. Box 7133, Kansas City, Missouri 64113.

## Working Forms

Following this page are fresh copies of many of the forms illustrated throughout the text. They may be removed and put to use when planning your next specific layout project.

You may reproduce copies of these forms for your own use, provided you recognize their original source and hold the use of the forms within the copyright restrictions covering this book. Or, additional forms are available from:

Richard Muther & Associates, Inc.
6155 Oak Street
Kansas City, Missouri 64113

Below is a ruler with die-cut symbols. This should be used when charting and diagraming flow and activity relationships. Also on the rule, for easy reference, are listed the chief rating and recording conventions used in Systematic Layout Planning (S L P).

See reverse side for list of Working Forms

**List of Working Forms**

Each of the forms included in this section is explained in the text in the figure number referenced, and is listed in its order of appearance in the text. However, for ease in finding the working forms in this section, they are arranged numerically by form number.

| Fig. No. | Form No. | Form Title |
|---|---|---|
| 4–12 | 136 | From–To Chart |
| 5-2 | 130 | Relationship Chart (8½ x 11) |
| 5-7 | 131 | Relationship Chart (11 x 17) |
| 7-1 | 154 | Machinery & Equipment Layout Data |
| 7-2 | 164 | Machinery & Equipment Layout Data Worksheet |
| 7-3 | 153 | Machinery & Equipment Area & Features Sheet |
| 7–5 | 151 | Office Layout Requirements Data |
| 7–6 | 155 | Space Requirements–Converting |
| 7–9 | 150 | Activities Area & Features Sheet |
| 9–1 | 901 | Universal Problem-Solving Procedure |
| 9–5 | 240 | Route Chart |
| 9–6 | 140 | Flow-In Flow-Out Chart |
| 9–8 | 202 | M. H. Equipment Evaluation |
| 10–7 | 171 | Evaluating Alternatives |
| 11–2 | 227 | Flow Process Chart |
| 11–5 | 305 | Review and Approval Record |
| 14–1 | 302 | Installation Cost Summary |
| 14–2 | 310 | Installation Coordination Worksheet |
| 14–4 | 315 | Installation Instructions Summary |
| 14–6 | 320 | Installation Record |
| 15–6 | 705 | Project Schedule Sheet |
| 15–7 | 710 | Coordination and Progress Summary |
| Appendix XII | 142 | Combining Flow & Service Relationships |
| Appendix XIV | 158 | Space Specifications Sheet |

# RELATIONSHIP CHART

Plant (Company)________________ Project________________

Charted by________________ With________________

Date________________ Sheet________ of________

Reference________________________________

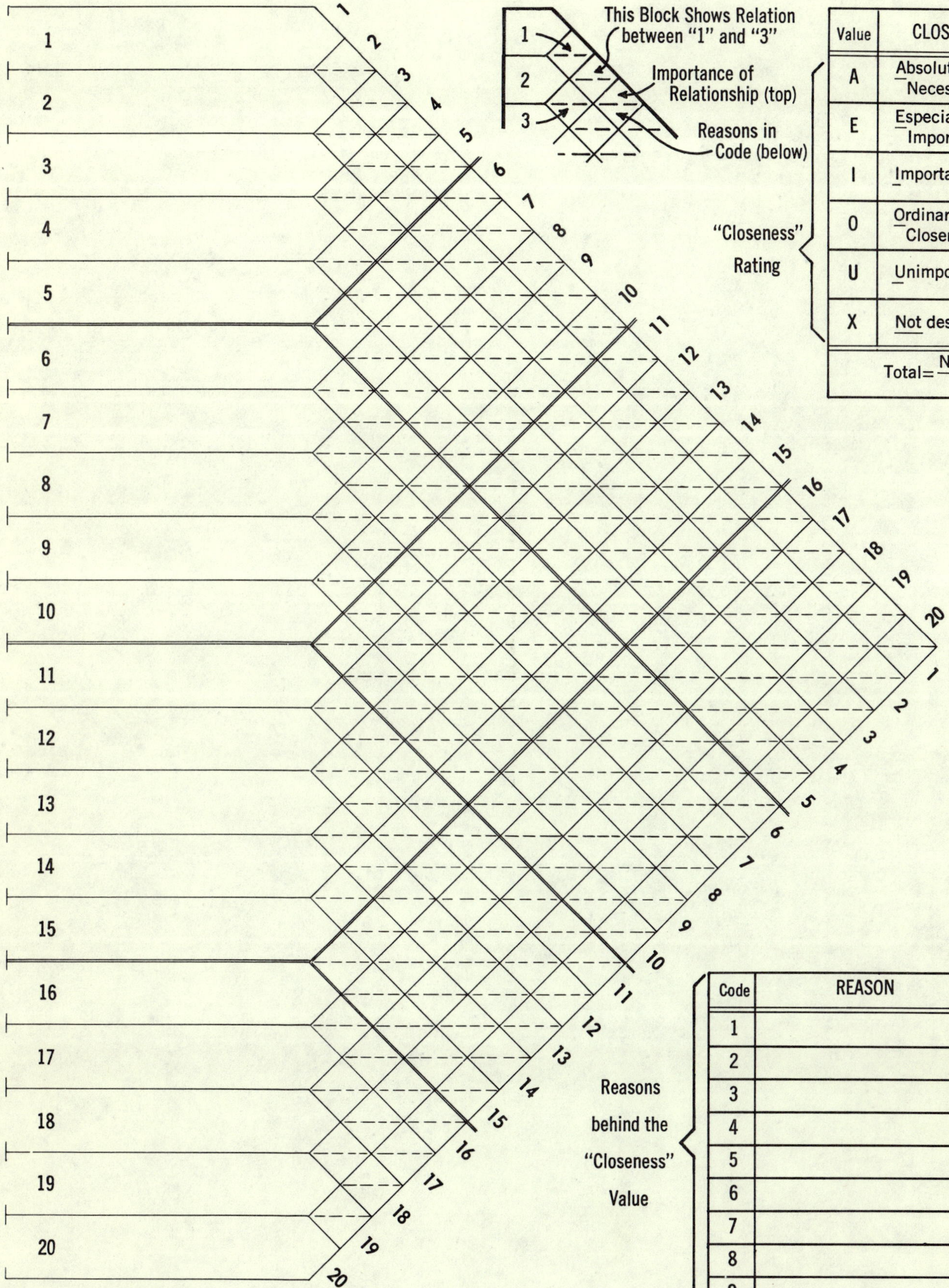


"Closeness" Rating

| Value | CLOSENESS | No. of Ratings |
|---|---|---|
| A | Absolutely Necessary | |
| E | Especially Important | |
| I | Important | |
| O | Ordinary Closeness OK | |
| U | Unimportant | |
| X | Not desirable | |
| | Total= $\frac{N \times (N-1)}{2}$ = | |

Reasons behind the "Closeness" Value

| Code | REASON |
|---|---|
| 1 | |
| 2 | |
| 3 | |
| 4 | |
| 5 | |
| 6 | |
| 7 | |
| 8 | |
| 9 | |

## RELATIONSHIP CHART

Plant (Company) ______________________ Project ______________________

Source-Reference ______________________ Date __________ Charted by ______________________

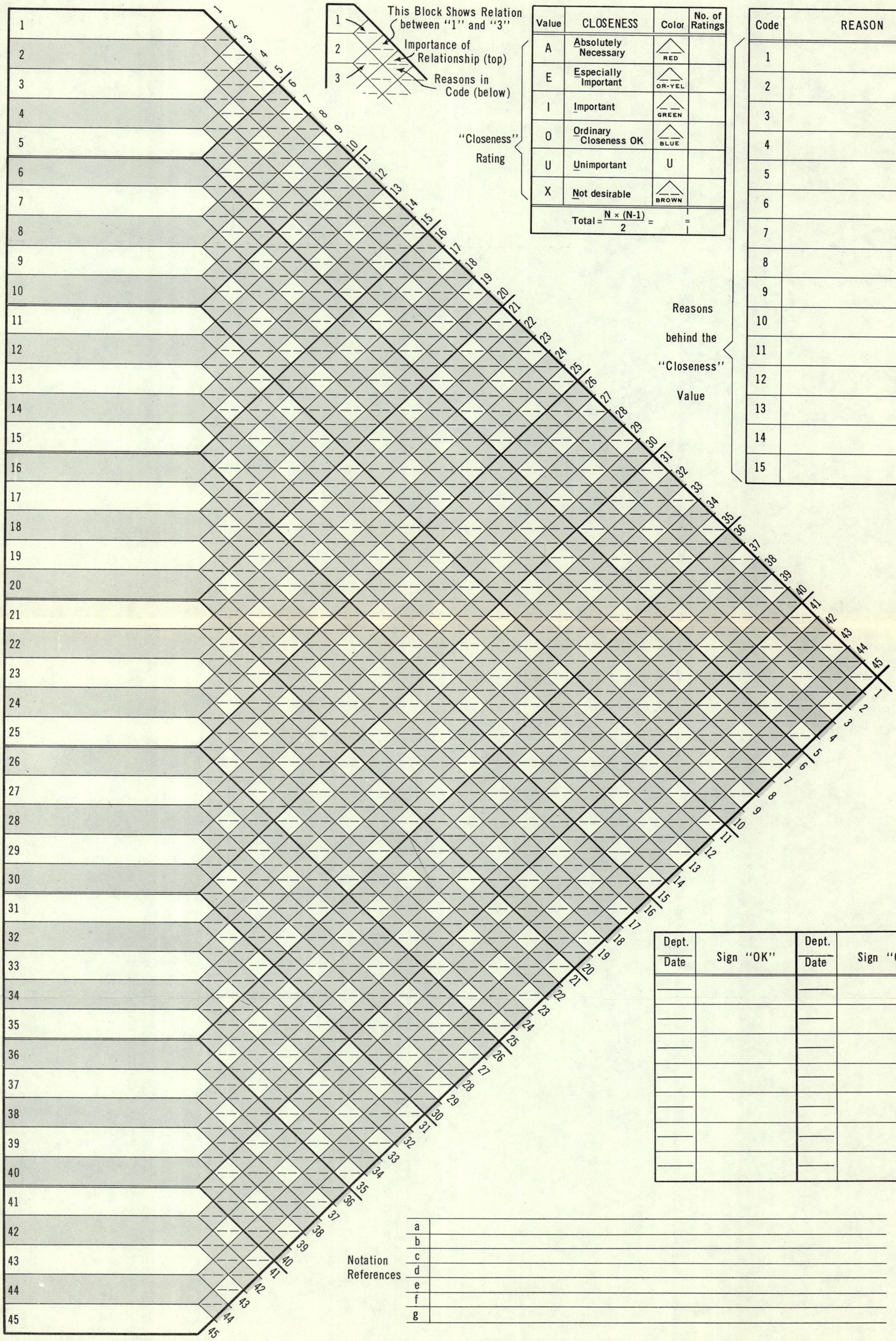




"Closeness" Rating

| Value | CLOSENESS | Color | No. of Ratings |
|---|---|---|---|
| A | Absolutely Necessary | RED | |
| E | Especially Important | OR-YEL | |
| I | Important | GREEN | |
| O | Ordinary Closeness OK | BLUE | |
| U | Unimportant | U | |
| X | Not desirable | BROWN | |
| | Total = $\frac{N \times (N-1)}{2}$ = | | = |

Reasons behind the "Closeness" Value

| Code | REASON |
|---|---|
| 1 | |
| 2 | |
| 3 | |
| 4 | |
| 5 | |
| 6 | |
| 7 | |
| 8 | |
| 9 | |
| 10 | |
| 11 | |
| 12 | |
| 13 | |
| 14 | |
| 15 | |

| Dept. / Date | Sign "OK" | Dept. / Date | Sign "OK" |
|---|---|---|---|
| | | | |
| | | | |
| | | | |
| | | | |
| | | | |
| | | | |
| | | | |

Notation References

| | |
|---|---|
| a | |
| b | |
| c | |
| d | |
| e | |
| f | |
| g | |

## FROM – TO CHART

Plant ______ Project ______

By ______ With ______

Item(s) Charted ______ Basis of Values ______ Date ______ Page ______ of ______

| Activity or Operation TO / Activity or Operation FROM | 1 | 2 | 3 | 4 | 5 | 6 | 7 | 8 | 9 | 10 | 11 | 12 | 13 | 14 | 15 | 16 | 17 | 18 | 19 | 20 | TOTALS |
|---|---|---|---|---|---|---|---|---|---|---|---|---|---|---|---|---|---|---|---|---|---|
| 1 | | | | | | | | | | | | | | | | | | | | | |
| 2 | | | | | | | | | | | | | | | | | | | | | |
| 3 | | | | | | | | | | | | | | | | | | | | | |
| 4 | | | | | | | | | | | | | | | | | | | | | |
| 5 | | | | | | | | | | | | | | | | | | | | | |
| 6 | | | | | | | | | | | | | | | | | | | | | |
| 7 | | | | | | | | | | | | | | | | | | | | | |
| 8 | | | | | | | | | | | | | | | | | | | | | |
| 9 | | | | | | | | | | | | | | | | | | | | | |
| 10 | | | | | | | | | | | | | | | | | | | | | |
| 11 | | | | | | | | | | | | | | | | | | | | | |
| 12 | | | | | | | | | | | | | | | | | | | | | |
| 13 | | | | | | | | | | | | | | | | | | | | | |
| 14 | | | | | | | | | | | | | | | | | | | | | |
| 15 | | | | | | | | | | | | | | | | | | | | | |
| 16 | | | | | | | | | | | | | | | | | | | | | |
| 17 | | | | | | | | | | | | | | | | | | | | | |
| 18 | | | | | | | | | | | | | | | | | | | | | |
| 19 | | | | | | | | | | | | | | | | | | | | | |
| 20 | | | | | | | | | | | | | | | | | | | | | |
| TOTALS | | | | | | | | | | | | | | | | | | | | | |

NOTES: ______

# FLOW-IN FLOW-OUT CHART

Area: ______ Plant ______ Project ______
By ______ With ______
Date ______ Sheet ______ of ______

| | FLOW IN → | | | | OPERATION OR AREA | FLOW OUT → | | | | |
|---|---|---|---|---|---|---|---|---|---|---|
| PRODUCT-MATERIAL DESCRIPTION (ITEM OR ITEM GROUP) | QUANTITY PER____ | | | FROM | | TO | QUANTITY PER____ | | | PRODUCT-MATERIAL DESCRIPTION (ITEM OR ITEM GROUP) |
| | UNIT | AV. | MAX. | | | | UNIT | AV. | MAX. | |
| 1 | | | | | | | | | | |
| 2 | | | | | | | | | | |
| 3 | | | | | | | | | | |
| 4 | | | | | | | | | | |
| 5 | | | | | | | | | | |
| 6 | | | | | | | | | | |
| 7 | | | | | | | | | | |
| 8 | | | | | | | | | | |
| 9 | | | | | | | | | | |
| 10 | | | | | | | | | | |
| 11 | | | | | | | | | | |
| 12 | | | | | | | | | | |
| 13 | | | | | | | | | | |
| 14 | | | | | | | | | | |
| 15 | | | | | | | | | | |
| 16 | | | | | | | | | | |
| 17 | | | | | | | | | | |
| 18 | | | | | | | | | | |
| 19 | | | | | | | | | | |
| 20 | | | | | | | | | | |
| 21 | | | | | | | | | | |
| 22 | | | | | | | | | | |
| 23 | | | | | | | | | | |
| 24 | | | | | | | | | | |
| 25 | | | | | | | | | | |
| 26 | | | | | | | | | | |
| 27 | | | | | | | | | | |
| 28 | | | | | | | | | | |
| 29 | | | | | | | | | | |
| 30 | | | | | | | | | | |

NOTES: ______

# COMBINING FLOW and SERVICE RELATIONSHIPS

Plant ______________ Project ______________
By ______________ With ______________
Date ______________ Sheet ______ of ______

| Activity Pair (Identi-fication) a | ← FLOW-OF-MATERIAL INTENSITY → From-To b | To-From c | Total 2-Way d | Order (Rank) e | Vowel Rating f | ← OTHER-THAN-FLOW RELATIONSHIPS → Consol. Rating g | Rating "Voice" h | Vowel Rating j | Reason Code k | Refer-ence l | RESULTANT RELATIONSHIP (Enter letter of combined rating in appropriate sub-column.) m A | E | I | O | U | X |
|---|---|---|---|---|---|---|---|---|---|---|---|---|---|---|---|---|
| – | | | | | | | | | | | | | | | | |
| – | | | | | | | | | | | | | | | | |
| – | | | | | | | | | | | | | | | | |
| – | | | | | | | | | | | | | | | | |
| – | | | | | | | | | | | | | | | | |
| – | | | | | | | | | | | | | | | | |
| – | | | | | | | | | | | | | | | | |
| – | | | | | | | | | | | | | | | | |
| – | | | | | | | | | | | | | | | | |
| – | | | | | | | | | | | | | | | | |
| – | | | | | | | | | | | | | | | | |
| – | | | | | | | | | | | | | | | | |
| – | | | | | | | | | | | | | | | | |
| – | | | | | | | | | | | | | | | | |
| – | | | | | | | | | | | | | | | | |
| – | | | | | | | | | | | | | | | | |
| – | | | | | | | | | | | | | | | | |
| – | | | | | | | | | | | | | | | | |
| – | | | | | | | | | | | | | | | | |
| – | | | | | | | | | | | | | | | | |
| – | | | | | | | | | | | | | | | | |
| – | | | | | | | | | | | | | | | | |
| – | | | | | | | | | | | | | | | | |
| – | | | | | | | | | | | | | | | | |
| – | | | | | | | | | | | | | | | | |
| – | | | | | | | | | | | | | | | | |
| – | | | | | | | | | | | | | | | | |
| – | | | | | | | | | | | | | | | | |
| – | | | | | | | | | | | | | | | | |
| – | | | | | | | | | | | | | | | | |
| – | | | | | | | | | | | | | | | | |
| – | | | | | | | | | | | | | | | | |
| – | | | | | | | | | | | | | | | | |
| – | | | | | | | | | | | | | | | | |
| – | | | | | | | | | | | | | | | | |
| – | | | | | | | | | | | | | | | | |
| – | | | | | | | | | | | | | | | | |
| – | | | | | | | | | | | | | | | | |
| – | | | | | | | | | | | | | | | | |
| – | | | | | | | | | | | | | | | | |
| – | | | | | | | | | | | | | | | | |
| – | | | | | | | | | | | | | | | | |
| – | | | | | | | | | | | | | | | | |
| – | | | | | | | | | | | | | | | | |
| – | | | | | | | | | | | | | | | | |

| Vowel Rating | CLOSENESS | No. of Occur's |
|---|---|---|
| A | Absolutely Necessary | |
| E | Especially Important | |
| I | Important | |
| O | Ordinary Close. OK | |
| U | Unimportant | |
| X | Not Desirable | |

| Code | REASON |
|---|---|
| 1 | *Flow of Materials* |
| 2 | |
| 3 | |
| 4 | |
| 5 | |
| 6 | |
| 7 | |
| 8 | |
| 9 | |
| 10 | |
| 11 | |
| 12 | |
| 13 | |
| 14 | |
| 15 | |

Reference Notes: ______________

NOTE:
- Col. d = b + c
- Col. e based on d; f based on e
- Col. g is independent of b thru f, but supported by j thru l
- Col. m posted to Relationship Chart

# ACTIVITIES AREA & FEATURES SHEET

Plant ______
Project ______
By ______ With ______
Date ______ Page ___ of ___

| Activity: No. | Activity: Name | Activity: Area in ___ Total: | Physical Features Required: O'Head Clearance | Max. Overhead Supported Load | Max. Floor Loading | Min. Column Spacing | Water & Drains | Steam | Compressed Air | Foundations –or Pits | Fire or Explosion Hazard | Special Ventilation | Special Electrification | Requirements for Shape or Configuration of Area (Space) |
|---|---|---|---|---|---|---|---|---|---|---|---|---|---|---|
| | | | Enter Unit and Required Amount under each | | | | Relative Importance of Features<br>A – Absolutely Necessary<br>E – Especially Important<br>I – Important<br>O – Ordinary Importance<br>— - Not Required | | | | | | | Enter Requirements for Shape or Configuration and Reasons therefore |
| 1. | | | | | | | | | | | | | | |
| 2. | | | | | | | | | | | | | | |
| 3. | | | | | | | | | | | | | | |
| 4. | | | | | | | | | | | | | | |
| 5. | | | | | | | | | | | | | | |
| 6. | | | | | | | | | | | | | | |
| 7. | | | | | | | | | | | | | | |
| 8. | | | | | | | | | | | | | | |
| 9. | | | | | | | | | | | | | | |
| 10. | | | | | | | | | | | | | | |
| 11. | | | | | | | | | | | | | | |
| 12. | | | | | | | | | | | | | | |
| 13. | | | | | | | | | | | | | | |
| 14. | | | | | | | | | | | | | | |
| 15. | | | | | | | | | | | | | | |

Sub-Activities or Areas

Notation References
a ______
b ______
c ______
d ______



No. ______ Activity ______ Sheet ______ of ______

## OFFICE LAYOUT REQUIREMENTS DATA

☐ Present ☐ Projected

Company________ Bldg.________ Floor________ Page**________

Sq. ft. now Assigned**________ Dept.________ Sect'n________ Sheet____ of____

| Room or Area Number | Identification: Personal Name or Work Group. (List each individual -- arranged by work group—or list name of work group or function. If individual's name is used, check appropriate columns; if working group, enter numbers.) | Identification: Job title or Description | Space Req'd: Male | Space Req'd: Female | Space Req'd: Type of Space # | Space Req'd: Square Feet* (Net Area) | Equipment Involved (Fill in Quant'y, Size & Description as req'd.) — Regular Equipment: Standard Desk | Typist Desk | Bench or Table | Size | Chair or Stool | Letter File | Legal File | | Telephone | Special Light | Dict./Trans. Eqpt. | Tel./Dict. on Stand | Elec. Mach on Desk | Elec. Mach. on Stand | Other Equipment: No. | Description | Left to Right | Front to Back | Height | Special Utilities or Requirements (Describe) |
|---|---|---|---|---|---|---|---|---|---|---|---|---|---|---|---|---|---|---|---|---|---|---|---|---|---|---|
| | Totals | | | | | | | | | | | | | | | | | | | | | | | | | |

#Type of Space: P-Private E-Semi-Private Enclosure O-Open or Semi-Open S-Service or Special Areas (For special areas furnish sketch and dimensions)
G-General areas not specifically assigned. (Shared areas not considered service or spcl.)

*Net Area: Excludes stairways, restrooms, central corridors, etc. (except where specifically listed), but includes working and access area or apportioned share thereof.

**Filled in by layout engineer (or office layout analyst)

Reference Notes:________

Data Recorded by________ Date________

# MACHINERY & EQUIPMENT AREA & FEATURES SHEET

Company/Plant ______________ Bldg/Dept/Area ______________

Project ______________
By ______________ With ______________
Date ______________ Sheet ______ of ______

| Identification Data | | Space | | | | | | | | | Physical Features Required | | | | | | | | | | | | | | | | |
|---|---|---|---|---|---|---|---|---|---|---|---|---|---|---|---|---|---|---|---|---|---|---|---|---|---|---|---|
| Machine or Equipment Identification Number | Name and/or Description | Left – Right | Front – Back | Height | Area for Machinery or Eqpt. in ____ | Operator(s) Work & Maintenance Area | Material Set-down Area | Total Area each Machine or piece of Eqpt. | No. of Mchs/Eqpt. | Total Net Area* (in ____ ) | 110–A.C. | 220–A.C. | Other Power | Ampere Rating | | Water | Steam | Drains | Compressed Air | Other Piping | Foundation/Pit | Exhaust/Hood | Dust Collector | | | | Comments - - - - - Shape/Configuration Special Requirements |
| | | | | | | | | | | | | | | | | | | | | | | | | | | | |
| | | | | | | | | | | | | | | | | | | | | | | | | | | | |
| | | | | | | | | | | | | | | | | | | | | | | | | | | | |
| | | | | | | | | | | | | | | | | | | | | | | | | | | | |
| | | | | | | | | | | | | | | | | | | | | | | | | | | | |
| | | | | | | | | | | | | | | | | | | | | | | | | | | | |
| | | | | | | | | | | | | | | | | | | | | | | | | | | | |
| | | | | | | | | | | | | | | | | | | | | | | | | | | | |
| | | | | | | | | | | | | | | | | | | | | | | | | | | | |
| | | | | | | | | | | | | | | | | | | | | | | | | | | | |

* Required space for main or delivery aisles and service areas not included.

Total Net Area Required* (in________ ) ________

Aisles ............... ________
Services ........... ________
Other ................. ________
Total Area Rq'd. ________

Reference Notes:
a. ______________
b. ______________
c. ______________
d. ______________
e. ______________
f. ______________

# MACHINERY & EQUIPMENT LAYOUT DATA

Plant ______ Project ______
Prepared by ______ With ______ Date ______

Water ______ Steam ______ Drains ______
Comp. Air ______ Gas ______ ______
Foundation ______ Pit ______ Level/Lag ______
Exhaust ______ ______ Sp'l. Elect. ______
Max. Height ______
Weight (less attach.) ______

| Electrical | H.P. | Volts | Cycle | Phase | Amps. | | |
|---|---|---|---|---|---|---|---|
| Drive Motor | | | | | | | |
| Auxiliary Motor | | | | | | | |
| Auxiliary Motor | | | | | | | |

Name/Type ______ File ______

Manufacturer ______ Size/Model ______
Speed/Capacity ______ Signif. Ident. ______

Left-Right ______
Front-Back ______
Net Floor Area ______
Worker/Maint. Area ______
Material Set-down ______
Area for Aisles ______
Service/Other ______
Gross Area (plan for) ______

| Co. Mach/Eqpt. Ident. Numbers Covered by this Sheet | | |
|---|---|---|
| | | |
| | | |
| | | |
| | | |
| | | |

ELEVATION SKETCH OR PHOTO    SOURCE ______ DATE ______

PLAN VIEW    SCALE ______ INCH = 1 Ft.

Reference Notations/changes ______

## SPACE REQUIREMENTS – – CONVERTING

Plant ________ Project ________

By ________ With ________

Basis (year, period, quant'y) of Cols. e, f, g ________ Cols. h, j, k ________ Date ________ Sheet ____ of ____

| a<br>Activity – –<br>Area or Dept. | b<br>Area Now<br>Occupied<br>( ) | c<br>+ or –<br>Adjstmt.<br>( ) | d<br>Should<br>Have Now<br>( ) | e<br>Increase<br>Decrease<br>( ) | f<br>Req'd Area<br>Determined<br>( ) | g<br>Plan-For<br>Area<br>( ) | h<br>Increase<br>Decrease<br>( ) | j<br>Req'd Area<br>Determined<br>( ) | k<br>Plan-For<br>Area<br>( ) |
|---|---|---|---|---|---|---|---|---|---|
| Unit → | | | | | | | | | |
| 1. | | | | | | | | | |
| 2. | | | | | | | | | |
| 3. | | | | | | | | | |
| 4. | | | | | | | | | |
| 5. | | | | | | | | | |
| 6. | | | | | | | | | |
| 7. | | | | | | | | | |
| 8. | | | | | | | | | |
| 9. | | | | | | | | | |
| 10. | | | | | | | | | |
| 11. | | | | | | | | | |
| 12. | | | | | | | | | |
| 13. | | | | | | | | | |
| 14. | | | | | | | | | |
| 15. | | | | | | | | | |
| 16. | | | | | | | | | |
| 17. | | | | | | | | | |
| 18. | | | | | | | | | |
| TOTALS | | | | | | | | | |

NOTES:

SPACE SPECIFICATIONS SHEET

Plant ____________
Project ____________
By ____________ With ____________
Date ____________

| No. | Activity Name | No. of Units | Area (in ____) Total: | O'Head Clearance | Max. Overhead Supported Load | Max. Floor Loading | Min. Column Spacing | Water & Drains | Steam | Compressed Air | Foundations –or Pits | Fire or Explosion Hazard | Special Ventilation | Special Electrification | Requirements for Shape or Configuration of Area (Space) and Reasons therefore |
|---|---|---|---|---|---|---|---|---|---|---|---|---|---|---|---|
| | | | | Enter Unit and Amount under each | | | | Enter Relative Importance { A-ABSOL. NECESS. E-ESPEC. IMPORT. I-IMPORTANT O-ORDIN. IMPORT. —-NOT REQUIRED | | | | | | | |
| 1 | | | | | | | | | | | | | | | |
| 2 | | | | | | | | | | | | | | | |
| 3 | | | | | | | | | | | | | | | |
| 4 | | | | | | | | | | | | | | | |
| 5 | | | | | | | | | | | | | | | |
| 6 | | | | | | | | | | | | | | | |
| 7 | | | | | | | | | | | | | | | |
| 8 | | | | | | | | | | | | | | | |
| 9 | | | | | | | | | | | | | | | |
| 10 | | | | | | | | | | | | | | | |
| 11 | | | | | | | | | | | | | | | |
| 12 | | | | | | | | | | | | | | | |
| 13 | | | | | | | | | | | | | | | |
| 14 | | | | | | | | | | | | | | | |
| 15 | | | | | | | | | | | | | | | |
| 16 | | | | | | | | | | | | | | | |
| 17 | | | | | | | | | | | | | | | |
| 18 | | | | | | | | | | | | | | | |
| 19 | | | | | | | | | | | | | | | |
| 20 | | | | | | | | | | | | | | | |
| 21 | | | | | | | | | | | | | | | |
| 22 | | | | | | | | | | | | | | | |
| 23 | | | | | | | | | | | | | | | |
| 24 | | | | | | | | | | | | | | | |
| 25 | | | | | | | | | | | | | | | |
| 26 | | | | | | | | | | | | | | | |
| 27 | | | | | | | | | | | | | | | |
| 28 | | | | | | | | | | | | | | | |
| 29 | | | | | | | | | | | | | | | |
| 30 | | | | | | | | | | | | | | | |
| 31 | | | | | | | | | | | | | | | |
| 32 | | | | | | | | | | | | | | | |
| 33 | | | | | | | | | | | | | | | |
| 34 | | | | | | | | | | | | | | | |
| 35 | | | | | | | | | | | | | | | |
| 36 | | | | | | | | | | | | | | | |
| 37 | | | | | | | | | | | | | | | |
| 38 | | | | | | | | | | | | | | | |
| 39 | | | | | | | | | | | | | | | |
| 40 | | | | | | | | | | | | | | | |
| 41 | | | | | | | | | | | | | | | |
| 42 | | | | | | | | | | | | | | | |
| 43 | | | | | | | | | | | | | | | |
| 44 | | | | | | | | | | | | | | | |
| 45 | | | | | | | | | | | | | | | |

# MACHINERY & EQUIPMENT LAYOUT DATA WORK SHEET

File ________

Plant ________ Bldg. ________ Dept. ________
Area ________ Nearest Column Location ________ as of ________
Compiled by ________ With ________ Date ________

Company Identification No. ________
Size/Model/Style ________
Speed ________ Capacity ________
Sgnif. Identfn. (Serial No., Code) ________

Utilities and Service Reqmts:
Water ________ Steam ________
Drains ________
Comprsd. Air ________ Gas ________
Other Piping ________
Foundation ________ Pit ________
Level ________ Lag ________
Exhaust ________
Elect. Outlet ________ Direct Conctn. ________
Spcl. Electr. ________

| Motors | HP | Volt | Cycle | Phase | Amp | |
|---|---|---|---|---|---|---|
| Drive | | | | | | |
| Auxiliary | | | | | | |
| Auxiliary | | | | | | |
| | | | | | | |

Name/type ________

Manufacturer ________

Dimensions and Area Reqmts:
Height at max. point ________
Weight (less attachments) ________

Left-Right (base) ________
Left-Right (max) ________
Front-Back (base) ________
Front-Back (max) ________
Net Floor Area Req'd. ________

Worker/Maint. Area ________
Material Setdown Area ________
Area for Aisles ________
Support/Other Areas ________
Gross area allocated ________
(on current ☐ basis
projected ☐ basis) ________

Elevation Sketch (Scale ________ )

Template Sketch (Scale ________ )

¼"

File ________

Reference or Explanatory Notes: ________

# EVALUATING ALTERNATIVES

Plant/Area ________________ Project ________________ Date ________________

Description of Alternatives: A. ________________

B. ________________ C. ________________

D. ________________ E. ________________

Weight set by ________________ Ratings by ________________ Tally by ________________

| FACTOR/CONSIDERATION | WT. | RATINGS AND WEIGHTED RATINGS A | B | C | D | E | COMMENTS |
|---|---|---|---|---|---|---|---|
| 1 | | | | | | | |
| 2 | | | | | | | |
| 3 | | | | | | | |
| 4 | | | | | | | |
| 5 | | | | | | | |
| 6 | | | | | | | |
| 7 | | | | | | | |
| 8 | | | | | | | |
| 9 | | | | | | | |
| 10 | | | | | | | |
| 11 | | | | | | | |
| 12 | | | | | | | |
| 13 | | | | | | | |
| 14 | | | | | | | |
| TOTALS | | | | | | | |

***NOTES***

## M. H. EQUIPMENT EVALUATION

Plant ____________ Project ____________
By ____________ With ____________
Date ____________ Sheet ______ of ______
Handling Situation ____________

Alternative A: ____________
Alternative B: ____________
Alternative C: ____________
Alternative D: ____________
Alternative E: ____________

| OBJECTIVE – WHAT IS WANTED FROM ANY HANDLING EQUIPMENT? Does it -- | Enter the requirements (what is required of this specific equipment?)–That is, the importance to the project of each Objective and Sub-Objective, using numerical values. Rate each alternative by rating letter; convert, extend, total. | REQMTS. Importance To the Project O | S-O | ALTERNATIVE A | B | C | D | E |
|---|---|---|---|---|---|---|---|---|
| MOVEMENT --move materials 1. | a. freely? | | | | | | | |
| | b. to the right place? | | | | | | | |
| | c. | | | | | | | |
| DIRECTNESS --move materials 2. | a. without transfers? | | | | | | | |
| | b. directly to point of use? | | | | | | | |
| | c. without unnecessary delays? | | | | | | | |
| | d. | | | | | | | |
| CONVENIENCE --allow minimum 3. | a. loading time? | | | | | | | |
| | b. unloading time? | | | | | | | |
| | c. rehandling time? | | | | | | | |
| | d. | | | | | | | |
| SAFETY AND QUALITY 4. | a. keep materials safe from breakage or damage? | | | | | | | |
| | b. keep material free from contamination, deterioration? | | | | | | | |
| | c. avoid hazard to workers or facilities? | | | | | | | |
| | d. | | | | | | | |
| SPACE --accomplish more 5. | a. without consuming much floor-space? | | | | | | | |
| | b. without obstructing workers, machines, storage services? | | | | | | | |
| | c. | | | | | | | |
| COMBINED ACTIVITY --allow equipt. to be used as 6. | a. a worktable or work-holding device? | | | | | | | |
| | b. a storage device? | | | | | | | |
| | c. an inspection or checking device? | | | | | | | |
| | d. a pace setter? | | | | | | | |
| | e. | | | | | | | |
| FLEXIBILITY 7. | a. handle several different materials, products, containers? | | | | | | | |
| | b. have adaptability to change, removal, or relocation? | | | | | | | |
| | c. have adaptability for increased volume, weight, size? | | | | | | | |
| | d. | | | | | | | |
| PRODUCTIVE AID --keep materials 8. | a. in sequence or easily rearrangeable? | | | | | | | |
| | b. from getting lost? | | | | | | | |
| | c. timed, scheduled, or synchronized? | | | | | | | |
| | d. free of workers' watching or attention? | | | | | | | |
| | e. easy to count, check, oversee? | | | | | | | |
| | f. readily available to operators? | | | | | | | |
| | g. | | | | | | | |
| OPERATING ECONOMY --operate with optimum 9. | a. operating man-hours? | | | | | | | |
| | b. maintenance and repair cost? | | | | | | | |
| | c. power or fuel cost? | | | | | | | |
| | d. | | | | | | | |
| DEPRECIATION INVESTMENT 10. | a. carry optimum rate of depreciation? | | | | | | | |
| | b. require reasonable outlay of funds? | | | | | | | |
| | c. | | | | | | | |
| OTHERS 11. | | | | | | | | |
| | | | | | | | | |
| | | | | | | | | |
| | TOTAL | | | | | | | |

NOTES/ADDITIONAL SOLUTION SUGGESTED: ____________

# FLOW PROCESS CHART

Process charted: ____________________

☐ Man or ☐ Material
Starting point ____________________
Ending point ____________________

| Conversions for Charted Unit to End Unit | | |
|---|---|---|
| Charted Unit | Size/Weight | Quantity/End Unit |
| | | |
| | | |
| | | |
| | | |
| | | |
| | | |
| | | |

Plant ____________ Project ____________
By ____________ With ____________
Date ____________ Sheet ______ of ______

Quantity of End Unit (unit of end item) per (time) ____________

☐ Present ☐ Proposed (Alternative # ____________)
Description of Alternative: ____________

| CHARTED UNIT (Unit of Product or Material Charted) | UNITS PER LOAD | Operation / Handling / Transport / Inspection / Delay / Storage | DESCRIPTION OF ACTION | Weight or Size of Load in ___ | Number of trips per ___ | Distance in ___ | Time in ___ per ___ | Cost in ___ per ___ | NOTES Verify: Product-Quantity-Route-Support-Time Analyse: Why-What-Where-When-Who-How Eliminate-Combine-Rearrange |
|---|---|---|---|---|---|---|---|---|---|
| 1 | | ○ ⬡ ⇨ □ D ▽ | | | | | | | |
| 2 | | ○ ⬡ ⇨ □ D ▽ | | | | | | | |
| 3 | | ○ ⬡ ⇨ □ D ▽ | | | | | | | |
| 4 | | ○ ⬡ ⇨ □ D ▽ | | | | | | | |
| 5 | | ○ ⬡ ⇨ □ D ▽ | | | | | | | |
| 6 | | ○ ⬡ ⇨ □ D ▽ | | | | | | | |
| 7 | | ○ ⬡ ⇨ □ D ▽ | | | | | | | |
| 8 | | ○ ⬡ ⇨ □ D ▽ | | | | | | | |
| 9 | | ○ ⬡ ⇨ □ D ▽ | | | | | | | |
| 10 | | ○ ⬡ ⇨ □ D ▽ | | | | | | | |
| 11 | | ○ ⬡ ⇨ □ D ▽ | | | | | | | |
| 12 | | ○ ⬡ ⇨ □ D ▽ | | | | | | | |
| 13 | | ○ ⬡ ⇨ □ D ▽ | | | | | | | |
| 14 | | ○ ⬡ ⇨ □ D ▽ | | | | | | | |
| 15 | | ○ ⬡ ⇨ □ D ▽ | | | | | | | |
| 16 | | ○ ⬡ ⇨ □ D ▽ | | | | | | | |
| 17 | | ○ ⬡ ⇨ □ D ▽ | | | | | | | |
| 18 | | ○ ⬡ ⇨ □ D ▽ | | | | | | | |
| 19 | | ○ ⬡ ⇨ □ D ▽ | | | | | | | |
| 20 | | ○ ⬡ ⇨ □ D ▽ | | | | | | | |
| 21 | | ○ ⬡ ⇨ □ D ▽ | | | | | | | |
| 22 | | ○ ⬡ ⇨ □ D ▽ | | | | | | | |
| 23 | | ○ ⬡ ⇨ □ D ▽ | | | | | | | |
| 24 | | ○ ⬡ ⇨ □ D ▽ | | | | | | | |
| 25 | | ○ ⬡ ⇨ □ D ▽ | | | | | | | |
| 26 | | ○ ⬡ ⇨ □ D ▽ | | | | | | | |
| 27 | | ○ ⬡ ⇨ □ D ▽ | | | | | | | |
| 28 | | ○ ⬡ ⇨ □ D ▽ | | | | | | | |
| 29 | | ○ ⬡ ⇨ □ D ▽ | | | | | | | |
| 30 | | ○ ⬡ ⇨ □ D ▽ | | | | | | | |
| | Totals | | | | Totals | | | | |

# ROUTE CHART

From ______________ To ______________

Distance ______________

Plant ______________ Project ______________

By ______________ With ______________

Date ______________ Sheet ______ of ______

| | PRODUCT-MATERIAL DESCRIPTION (ITEM OR ITEM GROUP) | P-M CLASS | QUANTITY PER ______ | | | | NOTES |
|---|---|---|---|---|---|---|---|
| | | | UNIT | AV. | MIN. | MAX. | |
| 1 | | | | | | | |
| 2 | | | | | | | |
| 3 | | | | | | | |
| 4 | | | | | | | |
| 5 | | | | | | | |
| 6 | | | | | | | |
| 7 | | | | | | | |
| 8 | | | | | | | |
| 9 | | | | | | | |
| 10 | | | | | | | |
| 11 | | | | | | | |
| 12 | | | | | | | |
| 13 | | | | | | | |
| 14 | | | | | | | |
| 15 | | | | | | | |
| 16 | | | | | | | |
| 17 | | | | | | | |
| 18 | | | | | | | |
| 19 | | | | | | | |
| 20 | | | | | | | |
| 21 | | | | | | | |
| 22 | | | | | | | |
| 23 | | | | | | | |
| 24 | | | | | | | |
| 25 | | | | | | | |

| TOTAL MOVEMENT | PRODUCT-MATERIAL CLASS DESCRIPTION | P-M CLASS | INTENSITY OF FLOW | | | INTENSITY x DISTANCE | | | COMMENTS OR SUGGESTIONS |
|---|---|---|---|---|---|---|---|---|---|
| | | | UNIT | AV. | PLAN | UNIT | AV. | PLAN | |
| | | | | | | | | | |
| | | | | | | | | | |
| | | | | | | | | | |
| | | | | | | | | | |
| | | | | | | | | | |
| | | | | | | | | | |
| | | | | | | | | | |
| | | | | | | | | | |

## INSTALLATION COST SUMMARY

Description ____________________ Plant ____________ Project ____________

Estimated by ____________ Extension by ____________

Date ____________ Sheet ______ of ______

| NATURE OF WORK | Estimated Mat'l. | Estimated Hours | Rate/ Hour | Labor Cost | Total Cost | Outside Bid |
|---|---|---|---|---|---|---|
| 1. Clear and prepare new area - - - including marking aisles, columns, locations. | | | | | | |
| 2. Building repairs, alterations, or construction | | | | | | |
| 3. Paint new areas, before and after move. | | | | | | |
| 4. Clean up and repair machines and equipment now in storage. | | | | | | |
| 5. | | | | | | |
| 6. Disconnect utilities - - - electric, water, air, gas, etc. | | | | | | |
| 7. Disconnect auxiliaries - - - vents, drains, ducts, conveyors, other handling equipment, etc. | | | | | | |
| 8. Move out operating machines and equipment. | | | | | | |
| 9. Move out all service and miscellaneous equipment. | | | | | | |
| 10. Move out materials, work-in-process, stores, tools, supplies. | | | | | | |
| 11. Remove operating machines and equipment not to be relocated. | | | | | | |
| 12. Remove service and miscellaneous equipment not to be relocated. | | | | | | |
| 13. Prepare non-movers for storage, sale, or other disposition. | | | | | | |
| 14. Fill-in pits, close holes, and clean up old area. | | | | | | |
| 15. | | | | | | |
| 16. Install pits, foundations, openings, special enclosures. | | | | | | |
| 17. Install conveyors, hoists, other handling equipment, racks, shelving, storage equipment. | | | | | | |
| 18. Install electric power equipment, leads, outlets, lighting fixtures, etc. | | | | | | |
| 19. Install utility lines and outlets - - - water, air, gas - - drains and sewers. | | | | | | |
| 20. Install heating, ventilating, air conditioning, ducts, fans, filters, dust collectors, etc. | | | | | | |
| 21. Move in operating machines - - - spot, level, lag, mount. | | | | | | |
| 22. Move in all service and miscellaneous equipment (not already installed). | | | | | | |
| 23. Move in all materials, work-in-process, stores, tools, supplies. | | | | | | |
| 24. Hook-up or connect, straighten-up, try out prior to operating. | | | | | | |
| 25. | | | | | | |
| Total | | | | | | |

NOTES: ____________________

## REVIEW AND APPROVAL RECORD

Plant ______________ Project ______________
By ______________ With ______________
Date ______________ Sheet ______ of ______

Project Identification Name and Description:

______________________________________________

______________________________________________

______________________________________________

According to good practice and established procedure all persons having responsibility for subsequent performance of any installation should review and approve the plans beforehand.

I, the undersigned, have carefully reviewed the plans for the above mentioned installation. It is complete and workable as far as I can tell and I will fully support having this project installed as planned and function as designed.

| Title or Department / Name | Approval Signature | Date | Drawing/Print Number and Revision or Model Identification |
|---|---|---|---|
| 1. | | | |
| 2. | | | |
| 3. | | | |
| 4. | | | |
| 5. | | | |
| 6. | | | |
| 7. | | | |
| 8. | | | |
| 9. | | | |
| 10. | | | |
| 11. | | | |
| 12. | | | |
| 13. | | | |
| 14. | | | |
| 15. | | | |
| 16. | | | |
| 17. | | | |
| 18. | | | |

# INSTALLATION COORDINATION WORKSHEET

Plant ______ Project ______
By ______ With ______
Date ______ Sheet ______ of ______

| | | WHAT | WHEN | WHO | STATUS | | | | |
|---|---|---|---|---|---|---|---|---|---|
| MAKE READY | PLAN | 1. Start planning the installation | | | | | | | |
| | | 2. Establish sequence of moves | | | | | | | |
| | | 3. Inventory everything to move | | | | | | | |
| | | 4. Get disposition of non-movers | | | | | | | |
| | | 5. Schedule moves in detail | | | | | | | |
| | | 6. Assign move numbers; check vs. inventory & equipment (tag) number | | | | | | | |
| | PROVIDE | 1. Decide who will make moves | | | | | | | |
| | | 2. Secure bids as necessary | | | | | | | |
| | | 3. Determine and reserve moving equipment required | | | | | | | |
| | | 4. Set up communications for both ends of move | | | | | | | |
| | | 5. Appoint key men for each area | | | | | | | |
| | | 6. Get work order (s) for move (s) | | | | | | | |
| | PREPARE | 1. Prepare new locations - - physical area, conditions, auxiliaries | | | | | | | |
| | | 2. Broadcast plans | | | | | | | |
| | | 3. Brief personnel specifically involved | | | | | | | |
| | | 4. Mark everything to move: identification, move no., destination | | | | | | | |
| | | 5. Disconnect or ready equipment | | | | | | | |
| | | 6. Check out equipment release to movers | | | | | | | |
| DO | MOVE | 1. Move equipment intact to reduce reassembly time | | | | | | | |
| | | 2. Move close to spot to reduce line-up and hook-up time | | | | | | | |
| | | 3. Post move performance as accomplished | | | | | | | |
| | | 4. Keep moving crew informed, coordinated | | | | | | | |
| | | 5. Be on hand - - layout interpretation | | | | | | | |
| | | 6. Be on hand - - auxiliaries interpretation | | | | | | | |
| PUT AWAY | HOOK-UP | 1. Spot equipment; check location | | | | | | | |
| | | 2. Temporary hook-ups where needed | | | | | | | |
| | | 3. Check and release for permanent connections | | | | | | | |
| | | 4. Inspect the installation & release for tryout | | | | | | | |
| | | 5. Maintenance tryout | | | | | | | |
| | | 6. Release to operating group; secure acceptance | | | | | | | |
| | CLEAN-UP | 1. Survey-inspect old and new areas | | | | | | | |
| | | 2. Schedule & assign clean-up - - old and new areas | | | | | | | |
| | | 3. Verify layout as installed | | | | | | | |
| | | 4. Verify auxiliary service as installed | | | | | | | |
| | | 5. Verify or adjust layout & service-specification records | | | | | | | |
| | | 6. Recap installation costs and performance | | | | | | | |

RICHARD MUTHER & ASSOCIATES - 310

# INSTALLATION INSTRUCTIONS SUMMARY

Description ______________________________________
_______________________________________________

Plant ____________ Project ____________
By ____________ With ____________
Date ____________ Sheet ______ of ______

| ITEM No. | DESCRIPTION / MACH./EQPT. No. | MOVE From / To | UTILITIES Air | Water | Gas | Drain | Vent | MILLWRIGHTS– MECHANICAL | ELECTRICAL Volts | Phase | KVA | Amps. | H.P. | ELECTRICIANS | OTHER WORK–(Code who and describe) C-Carpenters, M-Methods, S-Safety Eng'r., F-Fire Marshall, P-Plant Engrg., *-Out. Contrtr. WHO | WHAT |
|---|---|---|---|---|---|---|---|---|---|---|---|---|---|---|---|---|
| | | — | | | | | | | | | | | | | | |
| | | — | | | | | | | | | | | | | | |
| | | — | | | | | | | | | | | | | | |
| | | — | | | | | | | | | | | | | | |
| | | — | | | | | | | | | | | | | | |
| | | — | | | | | | | | | | | | | | |
| | | — | | | | | | | | | | | | | | |
| | | — | | | | | | | | | | | | | | |
| | | — | | | | | | | | | | | | | | |
| | | — | | | | | | | | | | | | | | |
| | | — | | | | | | | | | | | | | | |
| | | — | | | | | | | | | | | | | | |
| | | — | | | | | | | | | | | | | | |
| | | — | | | | | | | | | | | | | | |
| | | — | | | | | | | | | | | | | | |
| | | — | | | | | | | | | | | | | | |
| | | — | | | | | | | | | | | | | | |
| | | — | | | | | | | | | | | | | | |



To accompany Drwg. Nos. ____________ Work Order ____________

## INSTALLATION RECORD

Plant ______ Project ______

Bldg. ______ Area ______ Dept. ______

Nearest Column Location ______

Sheet issued by* ______

Date issued ______ Date returned ______

Mach. or Eqpt. No. ______

Mach./Eqpt. Type ______

Mfgr. ______

Size/Model ______

Signif. Identfn. ______

---

MECHANICAL, MAINTENANCE, MILLWRIGHTS

Level ______ Sign ______ Date ______

Lag ______ Sign ______ Date ______

Mount ______ Sign ______ Date ______

______ Sign ______ Date ______

Complete (sign) ______ Date ______

Comments ______

---

ELECTRICIANS

Power ______ Sign ______ Date ______

Control ______ Sign ______ Date ______

Lights ______ Sign ______ Date ______

______ Sign ______ Date ______

Complete (sign) ______ Date ______

Comments ______

---

PIPING & SHEET METAL

Water ______ Sign ______ Date ______

Steam ______ Sign ______ Date ______

Drains ______ Sign ______ Date ______

Comprsd. Air ______ Sign ______ Date ______

Gas ______ Sign ______ Date ______

Coolant, Lube ______ Sign ______ Date ______

Hoods, Ducts, Fans ______ Sign ______ Date ______

______ Sign ______ Date ______

Complete (sign) ______ Date ______

Comments ______

---

PAINT and CLEAN-UP

Paint ______ Sign ______ Date ______

Clean-Up ______ Sign ______ Date ______

______ Sign ______ Date ______

Complete (sign) ______ Date ______

Comments ______

---

SAFETY

Guards ______ Sign ______ Date ______

Fire Regs. ______ Sign ______ Date ______

Fumes, Acids ______ Sign ______ Date ______

______ Sign ______ Date ______

Complete (sign) ______ Date ______

Comments ______

---

OTHER INSTALLATION WORK SIGN OFF

______ Sign ______ Date ______

______ Sign ______ Date ______

______ Sign ______ Date ______

______ Sign ______ Date ______

Complete (sign) ______ Date ______

Comments ______

---

FINAL ACCEPTANCE (by installation coordinator or operating supervision)

Name ______ Title ______ Sign ______ Date ______

Comment ______

Name ______ Title ______ Sign ______ Date ______

Comment ______

(After final acceptance, return this sheet to person originally issuing it. See above.)

NOTES, Special Instructions, Reasons for Delay ______

*Return this sheet to person originally issuing it, after final acceptance.

# PROJECT SCHEDULE SHEET

Covering (Plant) ______________________ As of ______________

Distribution:

Originating Dept. ______________________

Prepared by ______________________

Date ______________ Sheet ____ of ______

| Proj. No. and/or Description<br>Work to do; Action to take | | | | | | | | | | | | | | | | | | | | | | | | | Further Schedule |
|---|---|---|---|---|---|---|---|---|---|---|---|---|---|---|---|---|---|---|---|---|---|---|---|---|---|
| 1. | | | | | | | | | | | | | | | | | | | | | | | | | |
| 2. | | | | | | | | | | | | | | | | | | | | | | | | | |
| 3. | | | | | | | | | | | | | | | | | | | | | | | | | |
| 4. | | | | | | | | | | | | | | | | | | | | | | | | | |
| 5. | | | | | | | | | | | | | | | | | | | | | | | | | |
| 6. | | | | | | | | | | | | | | | | | | | | | | | | | |
| 7. | | | | | | | | | | | | | | | | | | | | | | | | | |
| 8. | | | | | | | | | | | | | | | | | | | | | | | | | |
| 9. | | | | | | | | | | | | | | | | | | | | | | | | | |
| 10. | | | | | | | | | | | | | | | | | | | | | | | | | |
| 11. | | | | | | | | | | | | | | | | | | | | | | | | | |
| 12. | | | | | | | | | | | | | | | | | | | | | | | | | |
| 13. | | | | | | | | | | | | | | | | | | | | | | | | | |
| 14. | | | | | | | | | | | | | | | | | | | | | | | | | |
| 15. | | | | | | | | | | | | | | | | | | | | | | | | | |
| 16. | | | | | | | | | | | | | | | | | | | | | | | | | |
| 17. | | | | | | | | | | | | | | | | | | | | | | | | | |
| 18. | | | | | | | | | | | | | | | | | | | | | | | | | |

Gantt Chart Code: ⌐ Date work scheduled to start ¬ Date work scheduled to finish ⌐¬ Total time scheduled for work — Amount of work done

**(Each vertical period represents one unit of time. Use Gantt Chart Code or enter numbers and/or other indicators.)**

RICHARD MUTHER & ASSOCIATES – 705

## COORDINATION AND PROGRESS SUMMARY

Covering (Plant) ________________ As of ________________

Distribution:

Originating Dept. ________________

Prepared by ________________

Date ________________ Sheet ______ of ______

| Project No. and/or Description / Work to be done; action to be taken | Dept. Acting | Code | Status Remarks |
|---|---|---|---|
| 1. | | | |
| 2. | | | |
| 3. | | | |
| 4. | | | |
| 5. | | | |
| 6. | | | |
| 7. | | | |
| 8. | | | |
| 9. | | | |
| 10. | | | |
| 11. | | | |
| 12. | | | |
| 13. | | | |
| 14. | | | |
| 15. | | | |
| 16. | | | |
| 17. | | | |
| 18. | | | |

Status Code; 0, not yet begun; 1, Preliminaries underway; 2, Prelim. complete & work in process; 3, well along; 4, almost done; Compl., complete.

RICHARD MUTHER & ASSOCIATES – 710

# UNIVERSAL PROBLEM-SOLVING PROCEDURE

Problem ______________________________ Plant ______________

______________________________ Date Started ______________

Area/Dept. ______________________________ Analyst ______________

1. STATE THE PROBLEM

2. GET THE FACTS

3. RESTATE THE PROBLEM

4. ANALYZE AND DECIDE

5. TAKE ACTION - - What, Who, When

6. FOLLOW-UP